Ice and Snow Properties and Their Applications

Ice and Snow Properties and Their Applications

Guest Editors

Fang Li
Zhijun Li
Pentti Kujala
Weiping Li
Shifeng Ding

MDPI

Basel • Beijing • Wuhan • Barcelona • Belgrade • Novi Sad • Cluj • Manchester

Guest Editors

Fang Li
School of Ocean and Civil Engineering
Shanghai Jiao Tong University
Shanghai
China

Zhijun Li
State Key Laboratory of Coastal and Offshore Engineering
Dalian University of Technology
Dalian
China

Pentti Kujala
Estonian Maritime Academy
Tallinn University of Technology
Tallinn
Estonia

Weiping Li
Center for Soil and Water Environment Research in the Cold and Arid Regions of the Yellow River Basin
Inner Mongolia University of Science & Technology
Baotou
China

Shifeng Ding
School of Ocean and Civil Engineering
Shanghai Jiao Tong University
Shanghai
China

Editorial Office
MDPI AG
Grosspeteranlage 5
4052 Basel, Switzerland

This is a reprint of the Special Issue, published open access by the journal *Water* (ISSN 2073-4441), freely accessible at: https://www.mdpi.com/journal/water/special_issues/NZ1N9BSRS3.

For citation purposes, cite each article independently as indicated on the article page online and as indicated below:

Lastname, A.A.; Lastname, B.B. Article Title. *Journal Name* **Year**, *Volume Number*, Page Range.

ISBN 978-3-7258-3814-1 (Hbk)
ISBN 978-3-7258-3813-4 (PDF)
https://doi.org/10.3390/books978-3-7258-3813-4

Contents

water

MDPI

Editorial

Ice and Snow Properties and Their Applications

Fang Li [1,*], Zhijun Li [2], Pentti Kujala [1,3], Weiping Li [4] and Shifeng Ding [1]

1 School of Ocean and Civil Engineering, Shanghai Jiao Tong University, Shanghai 200240, China; pentti.kujala@taltech.ee (P.K.); shifengding@sjtu.edu.cn (S.D.)
2 State Key Laboratory of Coastal and Offshore Engineering, Dalian University of Technology, Dalian 116024, China; lizhijun@dlut.edu.cn
3 Estonian Maritime Academy, Tallinn University of Technology, 19086 Tallinn, Estonia
4 Center for Soil and Water Environment Research in the Cold and Arid Regions of the Yellow River Basin, Inner Mongolia University of Science & Technology, Baotou 014010, China; sjlwp@163.com
* Correspondence: fang.li@sjtu.edu.cn

1. Introduction

Ice and snow are essential components of the Earth's cryosphere, contributing significantly to the global climate system and human civilization [1]. While seawater constitutes 96.5% of the Earth's hydrosphere, freshwater accounts for just 3.5%, with the majority of this freshwater stored as ice or snow in polar regions, predominantly in glaciers, ice caps, and snow cover. Throughout history, ice and snow have shaped human development, influencing migration, agriculture, transportation, and even the rise and fall of civilizations [2]. The melting of glaciers and the retreat of polar ice caps are becoming increasingly prominent, which highlights the ongoing importance of understanding the properties and behaviors of ice and snow, particularly in light of the challenges posed by global warming.

The transformation of ice and snow due to climate change is having a profound impact on water resources, ecosystems, and infrastructure [3]. Rising temperatures are causing glaciers to diminish, sea ice to shrink, and snow cover to recede. These shifts are not only altering landscapes but also influencing critical functions such as water storage, hydrological cycles, and the stability of ecosystems. In the polar regions, for instance, the dynamics of ice and snow play a crucial role in freshwater storage and ecosystem health [4]. The retreat of glaciers and the shrinking of ice sheets are contributing to rising sea levels, while changes in sea ice affect marine life and species migration [5]. Furthermore, the reduction in snow cover alters seasonal freshwater flow, impacting hydrology in affected areas. These changes present challenges in managing water resources, preserving ecosystems, and maintaining infrastructure in cold regions, making the study of ice and snow properties ever more important.

In addition to their environmental significance, ice and snow play a vital role in engineering, especially in regions where human activities are becoming more widespread. The physical, mechanical properties of ice are key to designing structures such as roads, buildings, and power lines in ice-prone areas [6]. For example, in Arctic and Antarctic regions, the design of infrastructure such as ice roads, ports, and even airports require an in-depth understanding of ice behavior under different temperatures and mechanical stress. As climate change affects ice stability, engineering solutions must adapt to the changing conditions. Ice-related challenges, such as the increasing risk of ice-jam flooding, require innovative strategies in water management and flood prevention. Moreover, engineers are exploring new opportunities for renewable energy, such as offshore wind and solar power in ice-covered areas, further underscoring the importance of understanding ice dynamics in engineering applications.

Received: 25 February 2025
Accepted: 10 March 2025
Published: 25 March 2025

Citation: Li, F.; Li, Z.; Kujala, P.; Li, W.; Ding, S. Ice and Snow Properties and Their Applications. *Water* **2025**, *17*, 954. https://doi.org/10.3390/w17070954

Ecologically, the melting of ice and snow is reshaping ecosystems, particularly in polar and subpolar regions [7]. Ice serves as a platform for marine life, from algae growth to fish breeding, while also regulating species migration. The loss of ice cover threatens these ecosystems by disrupting the food chain and altering habitats. As sea ice declines, ecosystems are becoming increasingly vulnerable, affecting biodiversity and food security. Snowmelt also influences freshwater ecosystems, as it determines the timing and volume of water flow into rivers and lakes [8]. The ecological consequences of these changes extend beyond the immediate loss of habitat; they also affect migration patterns, breeding cycles, and overall ecosystem health. The research presented in this Special Issue delves into how changes in snow and ice properties are affecting ecosystems and explores ways to mitigate these impacts through the better understanding and management of frozen landscapes.

This Special Issue, "Ice and Snow Properties and Their Applications", aims to advance knowledge in the areas of hydrology, ecology, and engineering by focusing on the changing physical, thermal, and mechanical properties of ice and snow. With climate change rapidly altering the cryosphere, it is crucial to employ interdisciplinary approaches to study these changes. The research collated in this Special Issue utilizes a range of methods, including remote sensing, numerical modeling, and experimental studies, to improve our understanding of ice and snow behavior. These studies will help provide the data necessary for developing strategies to mitigate the risks associated with ice dynamics, such as flooding and ice-related disasters, and to support the sustainable development of infrastructure and ecosystems in cold regions. By bringing together research from diverse fields, this Special Issue fosters collaboration that will help inform future scientific advancements and practical solutions in the face of a rapidly changing cryosphere.

2. List and Summaries of the Contributions

This Special Issue received 14 manuscript submissions, all of which were subject to the rigorous review process of Water. In total, 12 papers were finally accepted for publication and included in this Special Issue. The contributions are listed in the List of Contributions section.

As shown in Table 1, the published papers cover broad topics such as ecology in cold regions, environmental science concerning meteorology and hydrology, ice engineering (including multiphase components), grain structure, mechanical property, and resistance to ships.

Table 1. Analysis of the contributions published in this Special Issue.

Number of Contribution	Research Area	Focus	Research Methods	Potential Applications
1	Ecology	Impact of freeze–thaw processes on spring algal blooms	Literature survey	Algal bloom prevention and lake management
2	Environmental science	Meteorological changes and water resources	Data analysis and modeling	Water resource management and climate change research
3	Ice engineering	Ship resistance in rafted ice regions	Numerical modeling	Ice resistance prediction for ship design
4	Environmental science	Sea surface temperature prediction	Data analysis and modeling	Improved SST forecasting and resource efficiency

Table 1. *Cont.*

Number of Contribution	Research Area	Focus	Research Methods	Potential Applications
5	Environmental science	Lake ice formation and breakup prediction	Remote sensing and statistical modeling	Lake ice prediction and climate impact assessment
6	Ice engineering	Ice formation and mechanical properties of columnar ice	Experiment	Ice property research and polar engineering
7	Ice engineering	Accurate segmentation of ice multiphase components	Experiment	Ice engineering and disaster prevention
8	Ice engineering	Flexural strength and fracture toughness of snow ice	Experiment	Ice engineering
9	Environmental science	Solar radiation transfer through ice and lake water temperature changes	Experiment	Climate modeling, environmental monitoring
10	Environmental science	Thickness of lake ice	Data analysis and modeling	Temporary ice runway
11	Environmental science	Snow and ice surface albedo scheme for lake	Data analysis and modeling	Climate impact assessment, environmental monitoring
12	Ice engineering	Simulation of an ice-class propeller in ice blockage	Numerical modeling	Anti-cavitation design and excitation force suppression of propellers

3. An Overview of Published Articles

As reviewed above, the published articles are mostly derived from the field of environmental science and ice engineering. An overview of these articles is provided here.

Lakes, as critical freshwater resources, are influenced by both natural processes and anthropogenic factors. At mid-to-high latitudes, the freeze—thaw cycles of lakes play a unique role in nutrient migration, water temperature changes, and algal physiology, which differ significantly from processes in low-latitude lakes [9]. The phenomenon of spring algal blooms has become more frequent and intense in these regions, demanding an understanding of its driving factors for effective prevention and management strategies [10]. Zhao et al. (Contribution 1) conducted a literature survey of publications from 2007 to 2023, identifying research trends and hotspots in the study of freeze–thaw processes and their impact on algal blooms. They identified nutrient dynamics, water temperature changes, and algal physiology during freeze–thaw periods as key factors influencing bloom formation. The study highlights phosphorus transformation during frozen periods as a critical driver and emphasizes the dual pressure of climate change and human activity in increasing bloom frequency and intensity. An integrated framework for understanding and managing algal blooms was introduced, combining principle analysis, modeling, and basin-scale management strategies. This research provides valuable insights for mitigating the ecological and water security challenges posed by algal blooms in sensitive lake regions.

The study of surface temperature changes, particularly in natural environments such as lakes, rivers, and coastal regions, is crucial for understanding climate dynamics and their impacts on ecosystems and water resources [11,12]. Traditional modeling approaches, such as physical or statistical models, often simplify real-world conditions to idealized representations, which can lead to inaccuracies when applied to complex, heterogeneous environments. However, with the rapid advancement of imaging technologies and remote sensing techniques, it is now possible to capture more detailed and accurate surface temperature variations under natural conditions. AI and ML techniques allow researchers to process vast amounts of data from various sensors and sources, uncovering intricate relationships between surface temperature and the multiple environmental factors that influence it, such as atmospheric conditions, solar radiation, and water interactions. Three papers were published on the topic of measuring sea/lake surface temperature changes using machine learning and experimental methods. Yue et al. (Contribution 2) examined the impact of climate change on water resources in the Heilongjiang (Amur) River Basin, which spans four countries and serves as an important international boundary river. Using daily temperature and precipitation data from 282 meteorological stations over a period from 1980 to 2022, their study analyzes spatial and temporal trends in temperature and precipitation changes. The results show a significant increasing trend in both temperature and precipitation within the basin. Spatially, the annual warming rate increases from the southeastern coastal regions to the northwestern plateau, while precipitation increases more significantly in the central and southern plains. Temperature and precipitation change points were identified in 2001 and 2012, respectively. The study further employs the long short-term memory (LSTM) model to predict precipitation, showing high accuracy with improved performance compared to traditional models. Jiao et al. (Contribution 4) addressed the challenge of large errors in SST predictions along the coast, focusing on improving forecast accuracy using deep learning techniques. Specifically, the study develops an optimal SST prediction model based on LSTM, using Xiaomaidao Station as a case study, and then extends it to 14 coastal stations along the Bohai Sea and Yellow Sea. The results demonstrate that the LSTM-based SST model significantly reduces forecast errors, with a 78% reduction in the mean absolute error for 1–3 day forecasts at Xiaomaidao Station and a 61% average reduction for other stations. The model not only improves forecast accuracy but also enhances computational efficiency, saving resources while increasing the reliability of short-term SST predictions. Niu et al. (Contribution 9) investigated the warming mechanisms of lake water under the ice during the frozen period in the Tibetan Plateau (TP), focusing on Qinghai Lake, the largest lake in China. This study conducted a field experiment to examine thermal conditions and radiation transfer across air–ice–water interfaces. Using high-resolution remote sensing technologies, the study identifies three stages of lake surface conditions: snow stage, sand stage, and bare ice stage. During the snow and sand stages, reduced solar radiation penetration leads to lower water temperatures. However, during the bare ice stage, increased solar radiation penetration significantly warms the water beneath the ice. The study also highlights how surface coverings (snow, sand, and ice) influence ice and water temperatures, with the bare ice stage exhibiting the greatest diurnal temperature variations. The findings enhance understanding of solar radiation transfer and temperature changes in ice-covered lakes and provide key parameters for improving models of lake dynamics in high-altitude regions. Wang et al. (Contribution 11) investigated the feasibility and safety of ice runway construction on Huhenuoer Lake, located in Chen Barag Banner, northeastern China. The study focused on the ice formation period from 2023 to 2024, utilizing field measurements and modeling approaches. Ice thickness data, collected through drilling, revealed that thickness exceeded 100 cm by 29 February 2024, with a record high of 139 cm recorded at site #2 on 14 March 2024. The Stefan equation was employed to model ice growth processes, yielding a fitted Stefan coefficient of 2.202, while a safety-adjusted coefficient of 1.870 was recommended for runway construction. Spatial analysis indicated that the northern part of the lake is most suitable for runway construction. By integrating the Stefan model with fitting techniques, the study established relationships between ice thickness, cumulative snowfall, and negative accumulated temperature. Using the P-III method, the 50-year return period values for maximum negative accumulated temperature and cumulative snowfall were determined as 2092.46 °C·d and 58.4 mm, respectively. These values were applied to predict ice thickness patterns for varying return periods. The study concludes with practical recommendations for ice runway construction on Huhenuoer Lake, introducing ice field research and growth modeling to support operational planning and safety. This work provides technical insights for the development of ice runways in similar environments. Cao et al. (Contribution 12) conducted surface albedo measurements of snow and ice on Lake Ulansu in the Central Asian arid climate zone during the winter of 2016–2017. The study categorized observations into three stages based on ice growth and surface conditions: bare ice, snow cover, and melting. During the bare ice stage, the mean surface albedo was 0.35, with a decreasing trend attributed to wind-blown sediment accumulation (range: 0.99–1.87 g/m^2). Snowfall events during the snow cover stage significantly increased albedo to 0.91, while the melting stage saw albedo decrease at a decay rate of 0.20–0.30 per day. Four existing albedo schemes were evaluated but deemed unsuitable for Lake Ulansu. A new surface albedo scheme was proposed by integrating existing models with measured data, incorporating the effect of sediment content on bare ice albedo for the first time. This scheme demonstrated a modeling efficiency of 0.933 over the three-month period, with validation against observations from other winters achieving an efficiency of 0.940. The closer the value is to 1, the higher the model's predictive accuracy and reliability. The proposed scheme shows potential applicability to other lakes in the Central Asian arid climate zone, characterized by low precipitation, frequent sandstorms, and intense solar radiation. This work provides a robust framework for improving albedo modeling in similar environments.

The seasonal dynamics of lake ice in cold-region ecosystems plays a crucial role in regulating ecosystem functions [13,14]. It affects various physical, chemical, and biological processes within lakes by altering temperature and light conditions, dissolved gas levels, and biological productivity. These changes, in turn, influence both the health of aquatic ecosystems and the livelihoods of human communities dependent on these water bodies. Kirchner et al. (Contribution 5) introduced a novel approach for predicting lake ice formation and breakup in Southwest Alaska, a region vital for both biodiversity and local communities. Due to the limited availability of consistent data for large lakes in this sparsely populated area, the study utilized optical remote sensing data from MODIS (2002–2016) to establish a phenology record for key lakes. Additionally, the researchers developed a survival model using temperature and solar radiation-based predictors to model ice formation and breakup from 1982 to 2022, including years when lakes did not freeze. The model was validated using observational data from two lakes and temperature profiles from three others. The results indicate that cumulative freeze-degree days and thaw-degree days were the strongest predictors of ice formation and breakup. The study also found that lake volume influenced ice phenology, with smaller lakes exhibiting longer and more consistent ice-cover durations. The research provides valuable insights into the future behavior of lake ice in the Boreal region, highlighting the potential for shorter ice seasons in smaller lakes and increased variability in larger ones. This study presents an innovative methodology for lake ice prediction in data-scarce regions and contributes to understanding the future of lake ice dynamics under climate change.

Ice plays a critical role in the design and construction of infrastructure in cold regions, such as ice roads, buildings, and artificial ice rinks [15,16]. Understanding the physical characteristics and mechanical properties of ice, including its formation, structure, and response to stress, is essential for improving engineering applications and mitigating ice-related hazards. Recent advancements in experimental methods and technologies, such as wind tunnels, X-ray computed tomography (CT), and three-point bending tests, have facilitated more accurate and detailed analyses of ice properties, contributing to a deeper understanding of its behavior under various environmental conditions [17]. Three papers were published on the topic of ice formation and mechanical properties using experimental ways. Zhang et al. (Contribution 6) presented the design and use of a small open-circuit wind tunnel to simulate and analyze the formation of columnar ice in laboratory conditions. The study focuses on the effects of environmental temperature and wind speed on the ice formation process. It was found that increasing wind speed led to a decrease in grain size of the columnar ice. Key findings include the validation of wind tunnel contraction sections, real-time temperature monitoring during ice formation, and a positive correlation between wind speed and grain size. The method provides a controlled environment to study the mechanical properties of polar columnar ice and offers a foundation for future research on ice behavior under windy polar conditions. This technique also facilitates the study of ice's mechanical properties in polar environments, offering valuable insights for ice engineering and structural designs in cold climates. Hu et al. (Contribution 7) explored the use of X-ray computed tomography (CT) to analyze the multiphase components of natural ice, which include gas, ice, unfrozen water, sediment, and brine. The study applies a watershed algorithm for the multi-threshold segmentation of CT images to improve the accuracy of the segmentation process and create high-precision three-dimensional models of ice. By analyzing Yellow River ice, Wuliangsuhai lake ice, and Arctic sea ice, the study demonstrates that the combined use of CT and the watershed algorithm can efficiently and non-destructively segment ice into its multiphase components. The results provide a detailed microscopic understanding of the ice's composition, with implications for ice engineering, ice remote sensing, and disaster prevention in ice-related infrastructure. The study contributes to the field by offering an advanced methodology for analyzing ice structure and composition at a microscopic level, enhancing the accuracy of ice models for scientific and engineering applications. Han et al. (Contribution 8) investigates the mechanical properties of granular snow ice, focusing on its flexural strength and fracture toughness under varying strain rates and temperatures. Through a series of three-point bending tests, the study finds that flexural strength increases at low strain rates but decreases at higher strain rates. The study also observes that temperature significantly influences the flexural strength and brittleness of granular snow ice. At colder temperatures, the ice becomes more brittle, and the strain rate at which maximum strength occurs decreases. Additionally, the study explores fracture toughness, noting that it decreases as strain rate increases and that fracture patterns remain consistent across various temperatures and strain rates, with cracks predominantly developing along prefabricated lines. These findings provide crucial insights into the mechanical behavior of granular snow ice, which is important for designing and maintaining structures in cold regions, such as ice rinks and cold-climate construction projects. The results contribute to the understanding of the tough–brittle transition in ice and its mechanical response to environmental conditions.

In ship–ice interaction studies, most existing research has primarily focused on the resistance faced by ships navigating through level ice conditions [18–20], with less attention given to the more complex conditions, such as rafted ice. Rafted ice is common in polar regions or areas with high ice concentrations, where vessels may encounter higher resistance than under typical level ice conditions. Accurately predicting ship resistance in these challenging conditions is essential for optimizing ship design, operational strategies, and ensuring the safety and efficiency of maritime activities in icy waters. Huang et al. (Contribution 3) developed a numerical model designed to predict ship resistance in areas with rafted ice, addressing a significant gap in previous research. The study used preset grid cells to simulate rafted ice conditions and validates the model's accuracy and reliability through comparisons with test results. How factors such as ice thickness, ship speed, and the bending and crushing strengths of ice affect the ice resistance encountered by ships under both level and rafted ice conditions was investigated. The results show that while ship resistance is generally higher in rafted ice than in level ice, the patterns of resistance differ between the two conditions. Specifically, ships navigating through rafted ice face more concentrated ice resistance compared to the more distributed resistance experienced under level ice conditions. Huang et al. (Contribution 10) investigated the hydrodynamics and cavitation behavior of ice-class propellers operating in ice-covered environments. The study focused on the non-uniform inflow conditions caused by ice blocks sliding along the ship hull in front of the propeller blades, which lead to increased excitation forces and significant cavitation. Using a hybrid Reynolds-averaged Navier–Stokes/large eddy simulation (RANS/LES) method combined with the Schnerr–Sauer cavitation model, the researchers analyzed hydrodynamic performance, excitation forces, cavitation evolution, and flow field characteristics under ice blockage conditions. The numerical method demonstrated a high accuracy, with hydrodynamic errors controlled within 3.0%. The results revealed that at low cavitation numbers, cavitation remains severe even with reduced blockage distance, and hydrodynamic coefficients do not increase significantly. When the blockage distance is 0.15 times the propeller diameter, the cavitation area covers 20% of the propeller blades. As the advance coefficient increases, the total cavitation area decreases, but the cavitation area behind the ice blockage persists, leading to a rise in excitation force. Ice blockage also induces backflow in the wake, with the most significant backflow occurring at the tip of the blade behind the ice. Higher advance coefficients amplify the high-pressure area on the pressure side and increase pressure differences, causing a sharp rise in excitation forces. This study provides a theoretical foundation for the anti-cavitation design and excitation force suppression of propellers operating in ice-covered regions, offering valuable insights for improving propeller performance and durability in such environments.

Author Contributions: Writing—original draft preparation, F.L.; writing—review and editing, Z.L., W.L., S.D. and P.K. All authors have read and agreed to the published version of the manuscript.

Funding: This organization of this Special Issue was supported by different projects from National Natural Science Foundation of China (grant numbers 52301331, U23A2012, 42207088, 51979024).

Conflicts of Interest: The authors declare no conflicts of interest.

List of Contributions

1. Zhao, Z.; Liu, X.; Wu, Y.; Zhang, G.; Dai, C.; Qiao, G.; Ma, Y. A Review on the Driving Mechanism of the Spring Algal Bloom in Lakes Using Freezing and Thawing Processes. *Water* **2024**, *16*, 257. https://doi.org/10.3390/w16020257.
2. Yue, Q.; Yu, G.; Miao, Y.; Zhou, Y. Analysis of Meteorological Element Variation Characteristics in the Heilongjiang (Amur) River Basin. *Water* 2024, 16, 521. https://doi.org/10.3390/w16040521.
3. Huang, J.; Diao, F.; Ding, S.; Han, S.; Kujala, P.; Zhou, L. A Study on the Ice Resistance Characteristics of Ships in Rafted Ice Based on the Circumferential Crack Method. *Water* **2024**, *16*, 854. https://doi.org/10.3390/w16060854.
4. Jiao, Y.; Li, G.; Zhao, P.; Chen, X.; Cao, Y.; Liu, G.; Wu, L.; Xu, X.; Fu, D.; Xin, R.; et al. Construction of Sea Surface Temperature Forecasting Model for Bohai Sea and Yellow Sea

Coastal Stations Based on Long Short-Time Memory Neural Network. *Water* **2024**, *16*, 2307. https://doi.org/10.3390/w16162307.

5. Kirchner, P.B.; Hannam, M.P. Volume-Mediated Lake-Ice Phenology in Southwest Alaska Revealed through Remote Sensing and Survival Analysis. *Water* **2024**, *16*, 2309. https://doi.org/10.3390/w16162309.
6. Zhang, Y.; Qian, Z.; Huang, W. Method for Producing Columnar Ice in Laboratory and Its Application. *Water* **2024**, *16*, 2558. https://doi.org/10.3390/w16182558.
7. Hu, S.; Wang, Q.; Li, C.; Li, Z. Threshold Ranges of Multiphase Components from Natural Ice CT Images Based on Watershed Algorithm. *Water* **2024**, *16*, 3330. https://doi.org/10.3390/w16223330.
8. Han, H.; Li, W.; Li, Y.; Liu, Z.; Liu, X. An Experimental Investigation of the Flexural Strength and Fracture Toughness of Granular Snow Ice Under a Three-Point Bending Test. *Water* **2024**, *16*, 3358. https://doi.org/10.3390/w16233358.
9. Niu, R.; Wen, L.; Wang, C.; Tang, H.; Leppäranta, M. Air–Ice–Water Temperature and Radiation Transfer via Different Surface Coverings in Ice-Covered Qinghai Lake of the Tibetan Plateau. *Water* **2025**, *17*, 142. https://doi.org/10.3390/w17020142.
10. Huang, Q.; Zheng, S.; Li, H.; He, X.; Li, X. Investigation on the Excitation Force and Cavitation Evolution of an Ice-Class Propeller in Ice Blockage. *Water* **2025**, *17*, 295. https://doi.org/10.3390/w17030295.
11. Wang, Y.; Zhao, Q.; Zhang, B.; Wang, Q.; Lu, P.; Wang, Q.; Bao, X.; He, J. An Investigation of the Thickness of Huhenuoer Lake Ice and Its Potential as a Temporary Ice Runway. *Water* **2025**, *17*, 400. https://doi.org/10.3390/w17030400.
12. Cao, X.; Yu, M.; Huo, P.; Lu, P.; Cheng, B.; Gao, W.; Shi, X.; Wang, L. Optimization of a Snow and Ice Surface Albedo Scheme for Lake Ulansu in the Central Asian Arid Climate Zone. *Water* **2025**, *17*, 523. https://doi.org/10.3390/w17040523.

References

1. Parkinson, C.L. Earth's Cryosphere: Current State and Recent Changes. *Annu. Rev. Environ. Resour.* **2006**, *31*, 33–60. [CrossRef]
2. Fagan, B. *The Great Warming: Climate Change and the Rise and Fall of Civilizations*; Bloomsbury Publishing USA: New York, NY, USA, 2008.
3. Voigt, T.; Füssel, H.M.; Gärtner-Roer, I.; Huggel, C.; Marty, C.; Zemp, M. *Impacts of Climate Change on Snow, Ice, and Permafrost in Europe: Observed Trends, Future Projections, and Socio-Economic Relevance*; European Topic Centre on Air and Climate Change: Bilthoven, The Netherlands, 2010.
4. Callaghan, T.V.; Johansson, M. Snow, Ice, and the Biosphere. In *Snow Ice-Relat. Hazards Risks Disasters*; Shroder, J.F., Haeberli, W., Whiteman, C., Eds.; Elsevier: Amsterdam, The Netherlands, 2015; pp. 139–165. [CrossRef]
5. Alley, R.B.; Clark, P.U.; Huybrechts, P.; Joughin, I. Ice-Sheet and Sea-Level Changes. *Science* **2005**, *310*, 456–460. [CrossRef] [PubMed]
6. Petrovik, J.J. Review Mechanical properties of ice and snow. *J. Mater. Sci.* **2003**, *38*, 1–6. [CrossRef]
7. Smetacek, V.; Nicol, S. Polar Ocean Ecosystems in a Changing World. *Nature* **2005**, *437*, 362–368. [CrossRef] [PubMed]
8. Slemmons, K.E.; Saros, J.E.; Simon, K. The influence of glacial meltwater on alpine aquatic ecosystems: A review. *Environ. Sci. Process. Impacts* **2013**, *15*, 1794–1806. [CrossRef] [PubMed]
9. Han, W.; Huang, C.; Duan, H.; Gu, J.; Hou, J. Lake Phenology of Freeze-Thaw Cycles Using Random Forest: A Case Study of Qinghai Lake. *Remote Sens.* **2020**, *12*, 4098. [CrossRef]
10. Platt, T.; Fuentes-Yaco, C.; Frank, K.T. Marine Ecology: Spring Algal Bloom and Larval Fish Survival. *Nature* **2003**, *423*, 398–399. [CrossRef] [PubMed]
11. Hao, X.; Xia, J.; Wang, R. Influence of Climate Change on Surface Water Environment. *Hydrology* **2010**, *30*, 67–72.
12. Hurrell, J.W.; Hack, J.J.; Shea, D.; Caron, J.M.; Rosinski, J. A New Sea Surface Temperature and Sea Ice Boundary Dataset for the Community Atmosphere Model. *J. Clim.* **2008**, *21*, 5145–5153. [CrossRef]
13. Magnuson, J.J.; Robertson, D.M.; Benson, B.J.; Wynne, R.H.; Livingstone, D.M.; Arai, T.; Assel, R.A.; Barry, R.G.; Card, V.; Kuusisto, E.; et al. Historical Trends in Lake and River Ice Cover in the Northern Hemisphere. *Science* **2000**, *289*, 1743–1746. [CrossRef] [PubMed]
14. Newton, A.M.; Mullan, D.J. Climate Change and Northern Hemisphere Lake and River Ice Phenology from 1931–2005. *Cryosphere* **2021**, *15*, 2211–2234. [CrossRef]

15. Cheng, P.; Li, J.; Li, X. Analysis of Carrying Capacity of Ice Cover Based on Winkler Model. *Yangtze River* **2017**, *48*, 64–68. (In Chinese) [CrossRef]
16. Makkonen, L.; Tikanmäki, M.; Sainio, P. Friction in sliding heavy objects on ice. *J. Glaciol.* **2016**, *62*, 1186. [CrossRef]
17. Hammonds, K.; Baker, I. Quantifying damage in polycrystalline ice via X-ray computed micro-tomography. *Acta Mater.* **2017**, *127*, 463–470. [CrossRef]
18. Su, B. *Numerical Predictions of Global and Local Ice Loads on Ships*; Norwegian University of Science and Technology: Trondheim, Norway, 2011.
19. Li, F.; Goerlandt, F.; Kujala, P. Numerical simulation of ship performance in level ice: A framework and a model. *Appl. Ocean Res.* **2020**, *102*, 102288. [CrossRef]
20. Gu, Y.; Zhou, L.; Ding, S.; Tan, X.; Gao, J.; Zhang, M. Numerical simulation of ship maneuverability in level ice considering ice crushing failure. *Ocean Eng.* **2022**, *251*, 111110. [CrossRef]

water

MDPI

Review

A Review on the Driving Mechanism of the Spring Algal Bloom in Lakes Using Freezing and Thawing Processes

Ziyue Zhao [1,2], Xuemei Liu [1,*], Yanfeng Wu [1], Guangxin Zhang [1], Changlei Dai [2], Guoli Qiao [3] and Yinghui Ma [3]

1 Northeast Institute of Geography and Agroecology, Chinese Academy of Sciences, Changchun 130102, China; zhaoziyue@iga.ac.cn (Z.Z.)
2 Institute of Water Conservancy and Electric Power, Heilongjiang University, Harbin 150080, China
3 Baicheng Forestry Research Institute, Baicheng 137000, China
* Correspondence: liuxuemei@iga.ac.cn; Tel.: +86-0431-85542311; Fax: +86-0431-85542298

Abstract: Spring algal blooms in mid–high-latitude lakes are facing serious challenges such as earlier outbreaks, longer duration, and increasing frequency under the dual pressure of climate warming and human activities, which threaten the health of freshwater ecosystems and water security. At present, the freeze-thaw processes is the key to distinguishing spring algal blooms in mid- to high-latitude lakes from low-latitude lakes. Based on the visualization and an analysis of the literature in the WOS database during 2007–2023, we clarified the driving mechanism of the freeze-thaw process (freeze-thaw, freeze-up, and thawing) on spring algal bloom in lakes by describing the evolution of the freeze-thaw processes on the nutrient migration and transformation, water temperature, lake transparency and dissolved oxygen, and physiological characteristics of algae between shallow lakes and deep lakes. We found that the complex phosphorus transformation process during the frozen period can better explain the spring-algal-bloom phenomenon compared to nitrogen. The dominant species of lake algae also undergo transformation during the freeze-thaw process. On this basis, the response mechanism of spring algal blooms in lakes to future climate change has been sorted out. The general framework of "principles analysis, model construction, simulation and prediction, assessment and management" and the prevention strategy for dealing with spring algal bloom in lakes have been proposed, for which we would like to provide scientific support and reference for the comprehensive prevention and control of spring algal bloom in lakes under the freezing and thawing processes.

Keywords: freezing and thawing processes; spring algal bloom; climate change; driving mechanism; prevention and control strategies

Citation: Zhao, Z.; Liu, X.; Wu, Y.; Zhang, G.; Dai, C.; Qiao, G.; Ma, Y. A Review on the Driving Mechanism of the Spring Algal Bloom in Lakes Using Freezing and Thawing Processes. *Water* **2024**, *16*, 257. https://doi.org/10.3390/w16020257

Academic Editor: Giuseppe Castaldelli

Received: 22 December 2023
Revised: 6 January 2024
Accepted: 8 January 2024
Published: 11 January 2024

1. Introduction

Lakes are important carriers of surface water resources, playing a role in protecting biodiversity, maintaining ecological balance within the watershed, and supplying fresh water [1–4]. The migration pathways and rates of nitrogen and phosphorus nutrients to lakes have exhibited diversity and variability under the dual pressure of global warming and human activities [2,5]. The algal blooms in mid- to high-latitude lakes are facing challenges such as earlier outbreak times, longer duration, and increased frequency of occurrence [3,6]. Previous studies have found that the presence of freeze-thaw processes is the key to promoting the mechanism of algal blooms in mid- to high-latitude lakes, which is different from that in low-latitude lakes [4–6]. Hence, how to reveal the impact mechanism of freeze-thaw processes on the occurrence and development of spring algal blooms is crucial for water-environment management.

Compared to low-latitude lakes, the growth and melting of lakes' ice would change the living environment of algae in mid–high-latitude lakes [5,6]. The water temperature structure, hydrodynamic conditions, sunshine conditions, and eutrophication degree will change with the freeze-thaw process. These factors will promote changes in the growth mechanism of the plankton and microbial communities in the lake [6,7]. Among them, the freeze-thaw process affects the migration and transformation of nutrients in the lake, and the competition for algae growth is more complex and variable. Taking Lake Washington in the United States as a typical seasonally covered shallow lake, the concentration of nutrients is actually the highest in winter. This situation encourages algae to receive a lot of light and nutrients after the lake's ice melts in the spring [7–9]. Subglacial water not only increases nutrients through concentration, it also increases the conductivity of the water by a factor of 1.7–2.7 compared to summer. This phenomenon increases the risk of spring algal blooms. [10,11]. However, there was insufficient research on the competitive living environment and self characteristics of algae growth during different periods such as freeze-thaw, freeze-up, and thawing [11]. Therefore, it is urgent to propose future mechanisms and prevention and control strategies for spring algal blooms by reviewing existing research on the impact of freeze-thaw processes on spring algal blooms. In the current study, the survival of plankton and nitrogen and phosphorus substances in lakes during the ice-covered period has been richly researched, but the changes in these substances during the whole process from freeze-thaw to thawing of lakes have not been sufficiently researched [5,8,9]. Meanwhile, a set of effective preventive measures for spring algal bloom in lakes has not been proposed. This paper proposed some theories and strategies to address these shortcomings.

In order to sort out the driving mechanism of the spring algal bloom in lakes with freezing and thawing processes, the literature during 2007–2023 in the Web of Science (WoS) database were summarized and analyzed. The objectives of this study are (i) to summarize the hotspots and difficulties of the freeze-thaw process on the driving mechanism of spring algal blooms through visual analysis of the number of publications and hot vocabulary; (ii) to sort out the response mechanism of spring algal blooms in lakes to the freeze-thaw process' effects using the migration and transformation of nutrient, the transparency and dissolved oxygen, and the succession and renewal of the algal community structure; (iii) to provide the strategies for prevention and control of spring algal bloom in lakes.

2. Data and Methods

CiteSpace was used to analyze the historical literature from 2007 to 2023 on the effect of freeze-thaw processes on spring algal bloom in lakes. Research hotspots and future research trends will be further revealed in our study. The advantage of CiteSpace is it uses mathematical and statistical methods to conduct in-depth literature mining, through visualizing the structural and hot-topic relationships between massive amounts of data, clarifying the development process of the field [12–16]. This study mainly utilized time-slicing technology to construct a time-varying model of time series, integrating a single network into an overview network. In addition, this study achieved the visualization effect of the literature through dynamic time-series mapping, including keyword recognition, extraction of research hotspots, and correlation between publishing units.

2.1. Data Sources

The literature was obtained from the Web of Science's (WoS's) core-collection database in the Clarivate Analytics website and was citation indexed from SCI-E (Science Citation Index Extend). The search date was 28 November 2023, and the search time was 2007–2023. The search keywords were "eutrophication", "eutrophication Lakes", "algal bloom", "water bloom", "phytoplankton", "spring algal blooms", "spring bloom", "lake bloom" and "freeze", "freeze thawing", "seasonal freeze-thaw" and "freezing and thawing". In order to accurately discover the relevant literature in this field, the retrieved studies were further screened and eliminated in terms of title, abstract and keywords.

2.2. Methods of Analysis

The study information such as year of publication, number of documents, keywords and other information were screened and extracted. The trend and keyword maps of the effect of freeze-thaw processes on spring algal blooms in lakes in various countries over the years have been drawn using a bibliometric functional analysis method. The hotspot and tendencies were analyzed using visualization software based on the year of publication, the number of papers, and keywords.

3. Hotspots Revisited

3.1. Statistical Analysis of the Volume of Publications

The number of publications on the impact of freeze-thaw processes on spring algal blooms in lakes has been increasing since 2007 based on WoS database. Especially after 2014, the number of publications has sharply increased. An analysis was conducted based on 519 papers from 2007 to 2023 on the mechanisms of the influence of freeze-thaw processes on spring algal blooms in lakes. The top 5 countries (The United States of America (USA), China, Canada, the United Kingdom (UK), Germany) with the highest number of publications and the greatest extent of impacts from spring algal blooms were extracted for statistical analysis (Figure 1) [7,12,15]. At the same time, these countries are located at high latitudes, where lake freeze-thaw phenomena and spring algal blooms are common. The results showed that the average annual publication volume in the United States was the highest in the world regarding the impact of freeze-thaw processes on spring algal blooms. The average annual publication volume of China had the highest growth rate compared to other countries. The tendency of publication numbers in Germany and the United Kingdom were relatively stable.

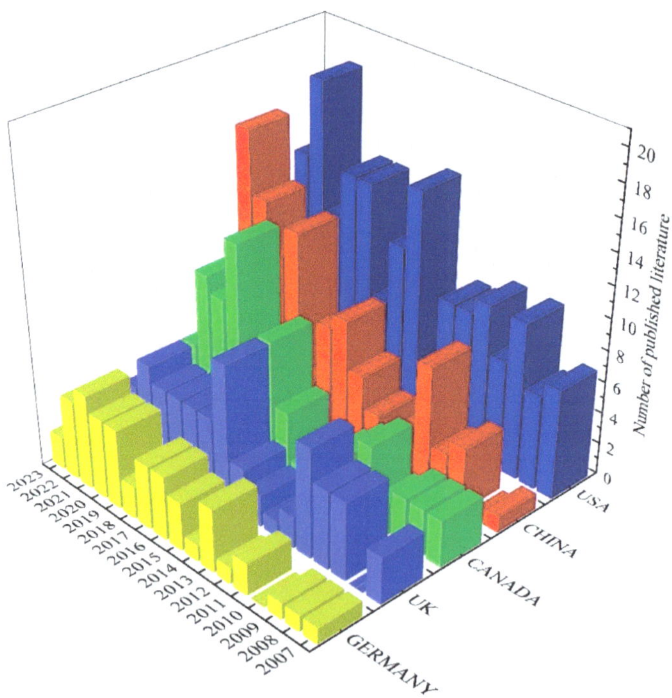

Figure 1. Statistical chart of annual publication volume in the top five countries with the highest publication volume.

3.2. Statistical Analysis of Keywords

The study on the impact of freeze-thaw processes on spring algal bloom in lakes has broad and focal differences based on keywords. There were 298 keywords that appeared during 2007–2023 in the WoS database. Among these keywords, there were nine keywords that appear at least 15 times. The top five keywords with the highest frequency of occurrence were phytoplankton (77 times), climate change (38 times), nitrogen (27 times), water (19 times), and sea ice (19 times) (Table 1). The top five keywords for centrality were phytoplankton, climate change, water, nitrogen, and temperature. The research hotspots also began to shift from the earlier hydrodynamic characteristics, and there were changes in nitrogen and phosphorus levels and phytoplankton growth mechanisms. The research hotspots gradually migrated to mechanisms affecting water-quality changes, aquatic-plant tolerance, and the phytoplankton-growth processes. Based on the keyword co-occurrence map, studies on the effects of freeze-thaw processes on spring algal blooms mainly focus on two aspects: exploring the hydrodynamic characteristics under climate change and studying the growth mechanism of phytoplankton. (Figure 2). In summary, international research on the effects of freeze-thaw processes on spring algal blooms has focused on the physicochemical properties of ice-water bodies (nutrients, light, dissolved oxygen (DO), etc.). Research is also directed at the effects of their changes for phytoplankton growth mechanisms.

Table 1. Mapping of keyword frequency, centrality, and high-frequency burst analysis in the WoS database (top9).

No.	Order of Occurrence		Order of Centrality		Sudden-Appearance Analysis				
	Keywords	Frequency	Keywords	Centrality	Keywords	Strength	Begin	End	2007–2023
1	phytoplankton	77	phytoplankton	0.67	dynamics	4.62	2009	2010	
2	climate change	38	climate change	0.23	southern ocean	2.44	2009	2017	
3	nitrogen	27	water	0.18	phytoplankton	3.15	2011	2014	
4	water	19	nitrogen	0.17	eutrophication	3.99	2018	2020	
5	sea ice	19	temperature	0.15	water quality	3.27	2019	2021	
6	eutrophication	19	harmful algal blooms	0.11	cold hardiness	3.01	2019	2023	
7	temperature	18	sea ice	0.1	loesses	2.44	2019	2021	
8	dynamics	17	ocean	0.1	damage	2.41	2019	2021	
9	growth	16	dynamics	0.07	organic matter	2.39	2020	2023	

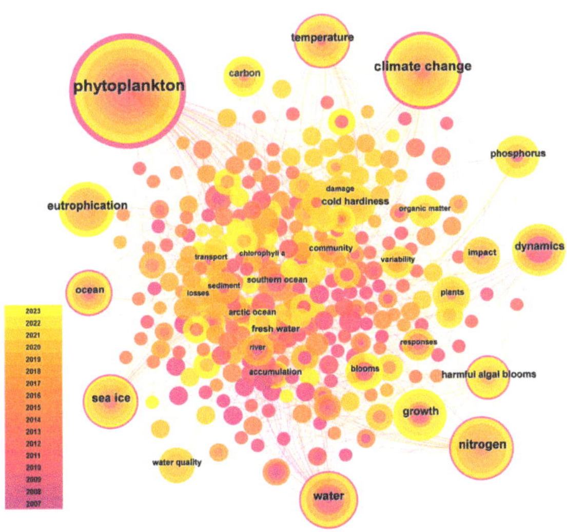

Figure 2. Keyword-co-occurrence-analysis mapping.

4. Results of Driving Mechanism Analysis

4.1. Effects of Freeze-Thaw Processes on Nutrient Migration and Transformation

Nitrogen and phosphorus were the major drivers of phytoplankton growth, competition, and succession, and directly affect primary productivity in lakes [17]. Excess nutrients could contribute to lake algal blooms [18]. Lakes' spring algal bloom has been expanding to the middle and high latitudes, and the scale, frequency, and intensity of its occurrence are all increasing under the dual pressures of climate change and human activities [19]. The algal blooms in lakes at mid to high latitudes arounds the world are also showing an increasing trend. Most of the studies focused on the mechanism of algal blooms in low-latitude lakes, with a lack of studies on the driving mechanism of algal blooms in mid–high-latitude lakes, especially those with seasonal ice and freeze-thaw phenomena. Therefore, it was important to analyze the mechanism of nutrient transport and transformation during freeze-thaw processes on spring algal bloom in lakes [20]. The freeze-thaw processes of lakes include freeze-thaw, freeze-up, and thawing [21]. During this period, the physical (water temperature, solar radiation, gas release, etc.), chemical (dissolved oxygen, CO_2 concentration, etc.), and hydrology factors (hydrodynamic conditions, water velocities, water circulation, etc.) of the lakes would change significantly, which will directly or indirectly drive the migration and transformation of nutrients [22]. The freeze-thaw effect on the migration and transformation of nutrients in lakes will affect the stability and development of the entire lake ecosystem [23]. During the freezing period of the Ulansuhai Lake in Inner Mongolia, due to the thickness of the snow cover and the shallow depth of the lake, the organisms at the bottom of the water body are able to carry out photosynthesis to promote the migration and transformation of nutrients; as represented by Woods Lake, the nutrient concentration replaces the temperature of the water body as an important controlling factor affecting the stability of the lake ecosystem (Table 2) [24,25]. The growth and melting of ice sheets altered the growth of phytoplankton by affecting physical and biogeochemical processes in the water beneath the ice [26]. The study of Norfolk Lake in the UK and Rappbode Reservoir in Germany found that the increase of nutrient concentration caused by ice sheet freezing led to the decrease of plant abundance and biomass in the water [27,28]. Therefore, the effects of the freeze-thaw processes on the nutrient-transport mechanism, transparency, and dissolved oxygen, and the physiological and ecological characteristics of algae in the water column should be considered [29].

Table 2. Eutrophication in selected high-latitude lakes.

No.	Country	Lake	Algae Bloom State	References
1	USA	Great-salt-sea	Escalation	[30,31]
2	China	Hulun	Sharply escalate	[32,33]
3	Canada	Winnipeg	Escalation	[34,35]
4	UK	Lough Neagh	In grave difficulty	[36,37]

4.1.1. Effects of Freeze-Thaw Processes on Nutrient Transport in Lakes

Nutrients were mainly distributed on the surface and bottom sediment layers of the water during the non-freezing period of lakes, which was an obvious vertical stratification phenomenon [38]. The dissolved oxygen concentration at the bottom of the water column was relatively low, while the content of organic matter and particulate matter was higher than that at the top of the water column [39]. The formation of ice sheets promoted the transportation of nutrients in ice concentration into the water, resulting in higher nutrient concentrations in the water than during the non-freezing period [40]. Of particular note is the formation of thermocline in deep-water lakes located in cold or temperate regions during freezing and thawing. It is difficult to exchange material between the upper mixed layer (epilimnion) and the lower stagnant layer (hypolimnion) within the lake. Large quantities of particulate organic matter and nutrients are difficult to resuspend into the upper layers of the water column through re-suspension after settling to the bottom of the lake. On the other hand, in shallow lakes, wind, waves, and turbulence can reach the bottom of the lake directly before the ice cap forms. There is an impact on organic matter and nutrients deposited on the lake bottom. These substances can enter the overlying water column through re-suspension, creating a nutrient cycle on the sediment–water inner surface (Figure 3) [41].

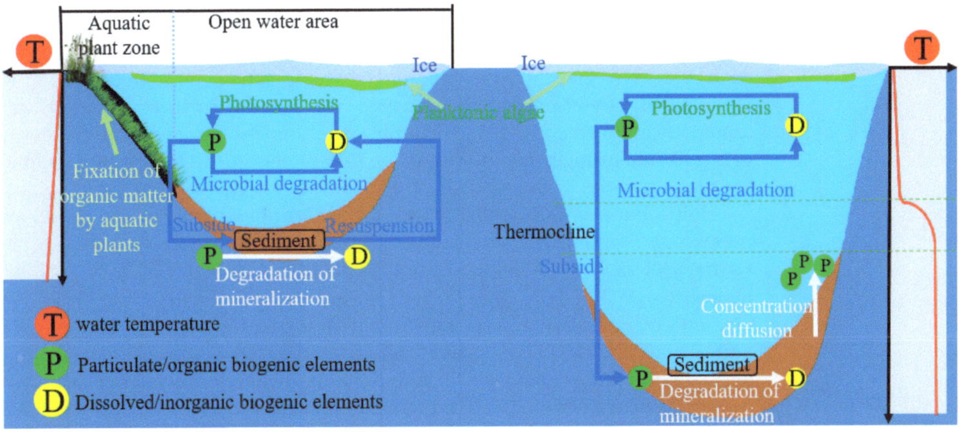

Figure 3. Patterns of nutrient cycling in shallow (left) and deep-water (right) lakes during the freezing period.

When the lake was in the frozen period, the presence of ice sheets and snow promoted significant differences in the physical and chemical environment compared to other periods [42]. The water flow rate was slow, while the nutrient concentration varied greatly in multiple media. Nutrient concentration showed a "C-shaped" distribution in sub glacial water bodies [43]. The concentration of ammonia nitrogen and nitrate nitrogen in water was higher than those in sediment, while the tendency of available phosphorus was opposite [42]. The distribution characteristics of nitrogen and phosphorus at the sediment water interface were relatively different [44]. The concentration of ammonia nitrogen and nitrate nitrogen decreased with the increase of sedimentation depth. However, the concentration of effective phosphorus showed a tendency to increase and then decrease with the increase of sedimentation depth [45]. The presence of lake ice promotes the slowed rheological behavior of ice water, changing disturbance between sediment, and promoting a different distribution of nutrients between ice, water, and sediment compared to other periods [46]. There was a critical value for external factors such as flow velocity and disturbance in water bodies under ice caps, and both above and below this threshold will have different effects on nitrogen and phosphorus releasing [47].

During the thawing period, nutrients in snow quickly entered the water body, which resulted in a sudden increasing of nutrient concentration in the water body [17]. The increasing of water temperature accelerated the metabolism of algae [21]. During the thawing period, the water flow rate increased, and the nutrient cycling rate and biogeochemical reaction rate both accelerated [48]. In addition, studies had shown that the comprehensive eutrophication index of water during the freezing and thawing periods was higher than that during non-freezing period [40]. These variations will lead to the spring algal blooms [49].

4.1.2. Effects of Freeze-Thaw Processes on Nutrient Transformation

Nitrogen and phosphorus, as important components of biogeochemical cycles in lakes, are the material basis for the growth and reproduction of phytoplankton and microorganisms [50]. At present, studies showed that a variety of factors such as temperature, dissolved oxygen content, acidity, alkalinity, and solar radiation were subject to change. Regrettably, there were fewer studies on the polymorphic transformation of nitrogen and phosphorus in lakes with the freezing and thawing process. The response of each substance was difficult to quantify, which greatly restricted an in-depth understanding of the mechanism of spring algal bloom in mid–high latitude lakes [51].

The main chemical reaction mechanisms of nitrogenous nutrients in lakes included anaerobic denitrification, anaerobic ammonium oxidation, aerobic denitrification, and anaerobic methane oxidation [52]. The transformation of nitrogen forms in water mainly includes processes of ammonification, nitrification, and denitrification [53]. The existence of freeze-thaw processes in lakes led to lower water temperature and dissolved oxygen concentration in the lake [5]. Anaerobic and low-temperature environments led to a decrease in microbial activity, promoting a decrease in the rates of nitrification, denitrification, and ammonification reactions [54]. Anaerobic environment also promoted further reduction of nitrate into nitrogen and nitrous oxide [55]. However, the contribution of freezing and thawing processes to ammonification and nitrification can not be specifically quantified.

Phosphorus is an essential macronutrient for phytoplankton growth, which plays a more important role in phytoplankton succession than nitrogen in lakes. The occurrence forms of phosphorus included orthophosphates ($H_2PO_4^-$, HPO_4^{2-}, PO_4^+), polymerized phosphates ($P_2O_7^{4-}$, $P_3O_{10}^{5-}$), and organophosphates (phosphatidylinositol) [56,57]. The rate phosphorus migration and its transformation in lakes was higher than that of nitrogen and silicon [58]. The phosphorus content decreased below the critical value required for algae growth (2 μg/L) due to the long-term low-temperature and hypoxic environment during the frozen period [57]. The process of converting organic phosphorus into inorganic phosphorus and orthophosphate into adenosine triphosphate (ATP) was greatly inhibited due to the decrease in microbial activity during the frozen period [58]. Phosphate can form insoluble precipitates with metal cations Fe_3^+, Al_3^+, Ca_2^+, Mg_2^+, and the reaction speed was also affected by the temperature and oxygen content (Figure 4). Phosphate exhibited vertical stratification due to the decomposition of dead vegetation residues by microorganisms during the frozen period [59]. The anaerobic environment promoted the transformation of insoluble $Fe(ON)_3$ into soluble $Fe(ON)_2$, providing a material-source basis for the improvement of primary productivity in spring [60]. Some anaerobic microorganisms in the frozen sediments accelerated the conversion of organic to inorganic phosphorus in the sediments. This phenomenon also increases phosphorus levels in the overlying water column [61]. Hence, the complex phosphorus transformation process during the frozen period can better explain the spring-algal-bloom phenomenon compared to nitrogen.

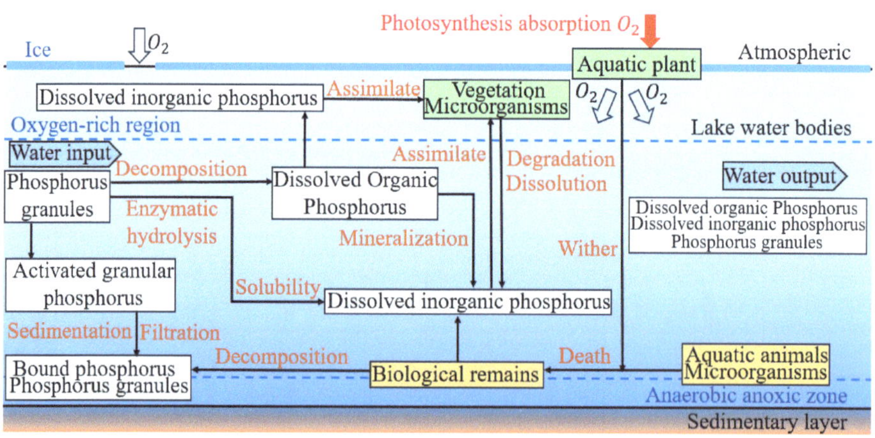

Figure 4. Mechanisms of phosphorus-containing nutrient transformation in lakes during freezing and thawing processes.

4.2. Effect of Freeze-Thaw Processes on Transparency and Dissolved Oxygen

The presence of snow and ice layers weakened the intensity of solar radiation entering sub-glacial water bodies, promoting a decrease in sub-glacial light intensity to inhibit algal photosynthesis [62,63]. The lake ice also temporarily buffered atmospheric sedimentation and reduced wind disturbance, suppressing the resuspension of sediment [64].

Ice caps are influenced by a number of factors during their formation. The freezing temperature of the ice, the rate of ice growth, and the salinity of the water together determine the density, crystal structure, and internal microstructure of the ice. They lead to a decrease in the transparency of the ice, making less light receivable under the ice [65]. The weakening of photosynthesis led to a further decrease in dissolved-oxygen content [66]. In addition, the nutrient concentration in the ice and the freezing separation coefficient also increased with the decreasing of freezing temperature, promoting the release of nutrients from the ice into the water body [67]. Ice caps redistributed nutrients between water bodies and ice layers through freezing, salt discharge, and melting dilution [68]. The research found that the formation of ice sheets redistributed nutrients among ice, water, and sediment [69]. The ice sheet weakened the disturbance of wind, blocks the exchange of substances between the atmosphere and water, and reduced the re-suspension of particles caused by wind [68,69]. The pollutants in the overlying water became more uniform during the frozen period, which promoted the increasing of transparency [69–71]. Hence, transparency of different lakes reacts differently during freeze-thaw processes, while it all led to a decreasing of dissolved oxygen due to the ice sheets.

4.3. Effect of Freeze-Thaw Processes on Algae Physiology

Phytoplankton, as the basis of material circulation and energy flow in lake ecosystems, plays an important role in maintaining the balance of the entire ecosystem [72]. Freezing and thawing had direct or indirect effects on the physiological and ecological characteristics of planktonic algae from physical (temperature, light intensity, dissolved sample content), chemical (nutrient salt concentration, metabolic rate) and hydrological (hydrodynamic conditions, water circulation) [71].

According to the previous research, the freeze-thaw process greatly limited the activity of underwater organisms [73]. However, scholars found that the diversity of benthic phytoplankton species during the frozen period was still relatively high in recent years [74–76]. Currently, the research has found that the freezing period promoted a decreasing of phytoplankton diversity over time [77]. However, the driving mechanism of phytoplankton population succession during the freezing period was not clear (Figure 5). Cyanobacteria had the greatest dominance during the freeze-up period; with the further reduction in temperature and the extension of the freeze-up time, the cyanobacteria entered into a dormant period [78]. Diatoms had a clear dominance during the freeze-up period, which had a direct relationship to their physiological characteristics of regulating the water, sugar, and fat in the cells to increase the ability of drought resistance [79]. During the thawing period, cyanobacteria and green algae dominate, due to maximal photosynthesis [80].

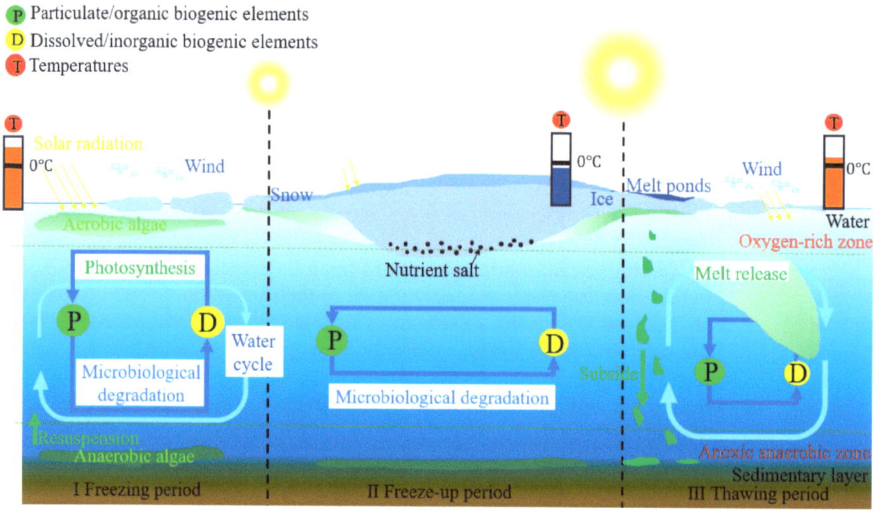

Figure 5. Mechanisms of lake-algal-bloom occurrence during freezing and thawing processes.

5. Response Mechanism of Lake Algal Bloom to Climate Warming and Prevention Strategies

5.1. Response Mechanisms of Lake Algal Blooms to Climate Warming

The response mechanism to climate change mainly manifested in the study of the effect of warming and increasing CO_2 concentration on algal blooms. Rising temperatures directly affected lake-water temperature, vertical mixing between suspended and insoluble particles, thermal stratification of water bodies, and biological-community structure [80,81]. Rising air temperature increased the concentration of nutrients and the absorption efficiency of algae by increasing water temperature and reducing the concentration of dissolved oxygen near sediments, thereby exacerbating the scale, frequency, and intensity of algal blooms [82]. The increase in the duration of thermal stratification in deep-water lakes exacerbated the phenomenon of hypoxia [83], and reducing the thickness of the mixed layer in water further promoted the expansion of cyanobacteria blooms [84]. The shortening of freezing period led to the advancement of spring-algal-bloom phenology [78]. All studies confirm that warmer temperatures lead to higher water temperatures, longer vertical stratification of the water column, thinner mixed layers, and earlier melting of the ice cap. Together, these phenomena led to earlier spring algal blooms and longer bloom durations. In addition, higher CO_2 concentrations also increased algal photosynthesis to increase phytoplankton biomass accumulation [85–87]. A portion of the carbon dioxide present in the lake was dissolved in the water, changing the pH of the water and having an effect on the nutrient-cycling process, with the water becoming less hard and the concentration of calcium ions decreasing [88]. The growth of some acid-loving cyanobacteria in this environment was promoted [89]. Furthermore, CO_2 can also improve the absorption and utilization efficiency of Microcystis for nutrients rather than other algae [90]. In summary, cyanobacteria had a more positive response to climate change.

5.2. Prevention and Control Strategies

Based on the driving factors, of occurrence principles and response mechanisms to climate change, we proposed the prevention and control strategies for lake spring cyanobacteria blooms in response to future climate change in terms of (i) improving the predictive ability of spring algal blooms in lakes, and (ii) strengthening spring-algal-bloom prevention and control in river basin lakes. Traditional ecological prediction models focus on calculating probability distributions between phytoplankton and nitrogen, and phosphorus and water-quality factors. However, this model lacks the mechanistic analysis of cyanobacterial blooms under the synergistic effect of multiple environmental factors. It also generates large errors in simulation prediction under long-term series and multi-temporal dynamic changes [84,91]. In the future, it will be necessary to establish a cyanobacterial-bloom-prediction and -warning model dominated by meteorological factors, the synergistic effects of multiple environmental factors, and the integrated effects of biochemical reaction processes. We hope to improve the ability of predicting the risk of spring algal blooms in lakes in this way [92]. The quantification of endogenous pollution in the watershed pollution load should be further strengthened, in order to develop more scientific thresholds and strategies for reducing pollutants in watershed [93,94].

5.3. A New Comprehensive Framework

Based on the comprehensive research on the effect of freeze-thaw processes on lake algal blooms, an overall framework was constructed with "principles analysis-model construction-simulation prediction-evaluation management" (Figure 6). At present, most of the existing studies used the control variable method to quantify the contribution of a single environmental factor, which made it difficult to effectively predict the evolution of algal blooms in complex scenarios in the future. There was a lack of climate-change-driven ecological and environmental risk assessment and socio-economic impact analysis of algal blooms in lakes, which restricted the prevention, control, and management of the hazards of algal blooms in lakes. Therefore, it is of great significance to construct an integrated lake-climate hydrodynamics–water-quality algal-bloom model to simulate and predict the development trend of algal bloom in lakes.

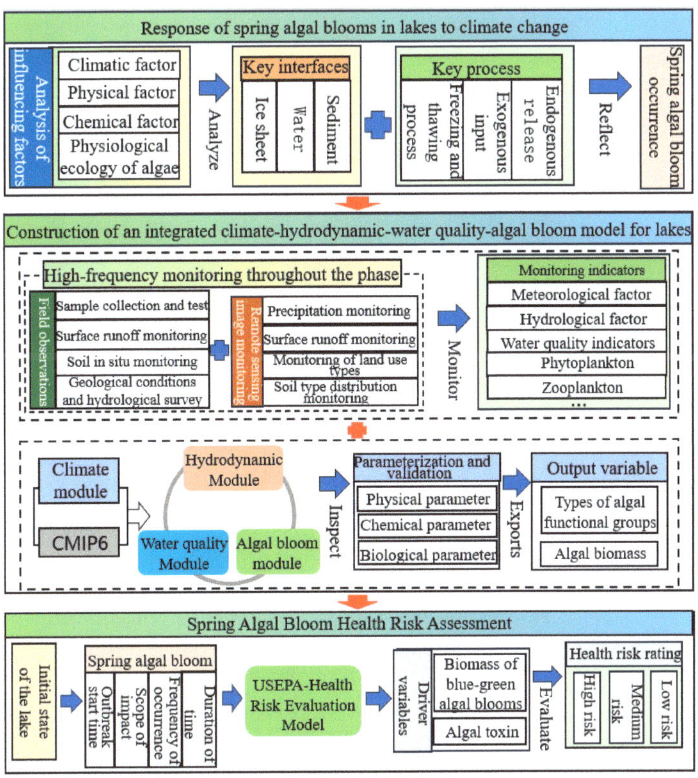

Figure 6. A framework for studying the effects of freezing and thawing on spring algal blooms in lakes under climate change.

In order to construct a comprehensive model of lake-climate hydrodynamics–water-quality algal blooms, it is necessary to first clarify the driving mechanism of freeze-thaw processes on spring algal blooms under historical conditions. Secondly, coupled with climate models, potential risks of spring algal blooms under future climate scenarios should be predicted. The key to this framework is to clarify the mechanisms of functional and representational levels of spring algal blooms in lakes, with the biomass and density of the main algal communities that trigger algal blooms as constraint indicators, effectively simulating the development and evolution process of the scale, frequency, and duration of algal blooms driven by climate conditions. The evaluation indicators for health-risk assessment are algal biomass and the accumulative-risk index. Algal toxin content takes tourism revenue, causes fishery-resource loss, requires algal-bloom-control cost, and impacts urban gross domestic product which are evaluation indicators for socio-economic development. We must minimize the disaster-risk levels caused by spring algal blooms in lakes, improve public awareness and understanding of the disaster problem, and effectively promote the synergistic effect of pollution reduction and carbon reduction.

6. Conclusions and Outlook

Based on 16 years of analysis of the literature, hot topics of research have shifted from the impact of freeze-thaw processes on lake nutrients, physicochemical properties, and hydrodynamic characteristics to the competitive driving mechanism of freeze-thaw processes on spring phytoplankton growth. The freeze-thaw process (freeze-thaw, freeze-up, thawing) directly or indirectly affects the occurrence of spring algal blooms by affecting biogeochemical process. Phosphorus conversion in lakes during freeze-thaw processes explains spring algal blooms better than nitrogen. Therefore, monitoring and controlling of phosphorus-containing substances needs to be strengthened throughout the freeze-thaw and thawing processes in lakes. It is pointed out that spring algal blooms in lakes will face problems such as expanding their scope, increasing their intensity, advancing their occurrence time, and prolonging their duration under climate change and human activities. We propose prevention and control strategies for spring algal blooms in lakes under future changing environments.

(1) Strengthening the monitoring of high-frequency lake-water-quality and -plankton data, which is necessary in order to accurately determine the limiting nutrient threshold for spring algal blooms.
(2) It is recommended to remove the snow on the surface of the ice sheet during the freeze-up period to reduce the input of external pollutants in spring.
(3) During the thawing period, attention should be paid to the release of nutrients, especially phosphorus, caused by sediment re-suspension.
(4) Installing circulating pumps in areas with high concentrations of lake pollutants during the thawing period to increase the hydrodynamic circulation in local areas.

Author Contributions: Conceptualization, Z.Z. and X.L.; methodology, Z.Z.; software, Z.Z.; validation, X.L. and Z.Z.; formal analysis, X.L.; investigation, G.Q.; resources, Y.M.; data curation, Z.Z.; writing—original draft preparation, Z.Z.; writing—review and editing, Z.Z. and X.L.; visualization, Z.Z.; supervision, X.L. and C.D.; project administration, G.Q. and Y.M.; funding acquisition, Y.W. and G.Z. All authors have read and agreed to the published version of the manuscript.

Funding: This research was supported by the National Natural Science Foundation of China (Grant NO. 42207088 and 42101051); the Postdoctoral Science Foundation of China (Grant NO. 2022M723129 and 2021M693155) and the Strategic Priority Research Program of the Chinese Academy of Sciences, China (Grant NO. XDA28020501).

Data Availability Statement: Some or all data, models, or code generated or used during the study are available from the corresponding author at request.

Conflicts of Interest: The authors declare no conflict of interest.

References

1. Sharma, S.; Blagrave, K.; Magnuson, J.J.; O'reilly, C.M.; Oliver, S.; Batt, R.D.; Magee, M.R.; Straile, D.; Weyhenmeyer, G.A.; Winslow, L.; et al. Widespread loss of lake ice around the northern hemisphere in a warming world. *Nat. Clim. Change* **2019**, *9*, 227–231. [CrossRef]
2. Wang, J.; Ouyang, W.; Liu, X.; Wang, L. Monitoring hydrological changes with satellite data: Spring thaw's effect on soil moisture and groundwater in seasonal Freezing-Thawing zones. *J. Hydrol.* **2023**, *626*, 130365. [CrossRef]
3. Zhang, X.; Li, Y.; Ren, S.; Zhang, X. Soil CO_2 emissions and water level response in an arid zone lake wetland under freeze-thaw action. *J. Hydrol.* **2023**, *625*, 130069. [CrossRef]
4. Wang, X.; Fang, C.; Song, K.; Lyu, L.; Li, Y.; Lai, F.; Lyu, Y.; Wei, X. Monitoring phycocyanin concentrations in high-latitude inland lakes using Sentinel-3 OLCI data: The case of Lake Hulun, China. *Ecol. Indic.* **2023**, *155*, 110960. [CrossRef]
5. Sharma, S.; Richardson, D.C.; Woolway, R.I.; Imrit, M.A.; Bouffard, D.; Blagrave, K.; Daly, J.; Filazzola, A.; Granin, N.; Korhonen, J.; et al. Loss of ice cover, shifting phenology, and more extreme events in northern hemisphere Lakes. *J. Geophys. Res. Biogeosci.* **2021**, *126*, 12. [CrossRef]
6. Chen, N.; Mo, Q.; Kuo, Y.M.; Su, Y.; Zhong, Y. Hydrochemical controls on reservoir nutrient and phytoplankton dynamics under storms—ScienceDirect. *Sci. Total Environ.* **2018**, *s619–620*, 301–310. [CrossRef] [PubMed]
7. Paerl, H.W.; Otten, T.G. Harmful cyanobacterial blooms: Causes, consequences, and controls. *Microb. Ecol.* **2013**, *65*, 995–1010. [CrossRef] [PubMed]
8. Bullerjahn, G.S.; McKay, R.M.L.; Bernát, G.; Prášil, O.; Vörös, L.; Pálffy, K.; Tugyi, N.; Somogyi, B. Community dynamics and function of algae and bacteria during winter in central European great lakes. *J. Great Lakes Res.* **2019**, *46*, 732–740. [CrossRef]
9. Yang, T.; Hei, P.; Song, J.; Zhang, J.; Zhu, Z.; Zhang, Y.; Yang, J.; Liu, C.; Jin, J.; Quan, J. Nitrogen variations during the ice-on season in the eutrophic lakes. *Environ. Pollut.* **2019**, *247*, 1089–1099. [CrossRef]
10. Pieters, R.; Lawrence, G.A. Effect of salt exclusion from lake ice on seasonal circulation. *Limnol. Oceanogr.* **2009**, *54*, 401–412. [CrossRef]
11. Hampton, S.E.; Moore, M.V.; Ozersky, T.; Stanley, E.H.; Polashenski, C.M.; Galloway, A.W. Heating up a cold subject: Prospects for under-ice plankton research in lakes. *J. Plankton Res.* **2015**, *37*, 277–284. [CrossRef]
12. Chen, C.; Leydesdorff, L. Patterns of connections and movements in dual-map overlays: A new method of publication portfolio analysis. *J. Assoc. Inf. Sci. Technol.* **2013**, *65*, 334–351. [CrossRef]
13. Wang, X.; Zhang, Y.; Zhang, J.; Fu, C.; Zhang, X. Progress in urban metabolism research and hotspot analysis based on citeSpace analysis—ScienceDirect. *J. Clean. Prod.* **2020**, *281*, 125224. [CrossRef]
14. Zhang, D.; Xu, J.; Zhang, Y.; Wang, J.; He, S.; Zhou, X. Study on sustainable urbanization literature based on Web of Science, scopus, and China national knowledge infrastructure: A scientometric analysis in CiteSpace. *J. Clean. Prod.* **2020**, *264*, 121537. [CrossRef]
15. Chen, C. CiteSpace II: Detecting and visualizing emerging trends and transient patterns in scientific literature. *J. Am. Soc. Inf. Sci. Technol.* **2006**, *57*, 359–377. [CrossRef]
16. Chen, C.; Ibekwe-Sanjuan, F.; Hou, J. The structure and dynamics of co-citation clusters: A multiple-perspective co-citation analysis. *J. Am. Soc. Inf. Sci. Technol.* **2014**, *61*, 1386–1409. [CrossRef]
17. Sun, K.; Deng, W.; Jia, J.; Gao, Y. Spatiotemporal patterns and drivers of phytoplankton primary productivity in China's lakes and reservoirs at a national scale. *Glob. Planet. Change* **2023**, *228*, 104215. [CrossRef]
18. Beall, B.F.N.; Twiss, M.R.; Smith, D.E.; Oyserman, B.O.; Rozmarynowycz, M.J.; Binding, C.E.; Bourbonniere, R.A.; Bullerjahn, G.S.; Palmer, M.E.; Reavie, E.D.; et al. Ice cover extent drives phytoplankton and bacterial community structure in a large north-temperate lake: Implications for a warming climate. *Environ. Microbiol.* **2015**, *18*, 1704–1719. [CrossRef]
19. Wen, Z.; Song, K.; Shang, Y.; Lyu, L.; Yang, Q.; Fang, C.; Du, J.; Li, S.; Liu, G.; Zhang, B.; et al. Variability of chlorophyll and the influence factors during winter in seasonally ice-covered lakes. *J. Environ. Manag.* **2020**, *276*, 111338. [CrossRef]
20. Wei, Z.; Yu, Y.; Yi, Y. Analysis of future nitrogen and phosphorus loading in watershed and the risk of lake blooms under the influence of complex factors: Implications for management. *J. Environ. Manag.* **2023**, *345*, 118662. [CrossRef]
21. Chen, M.; Yang, Z.; Abulaizi, M.; Hu, Y.; Tian, Y.; Hu, Y.; Yu, G.; Zhu, X.; Yu, P.; Jia, H. Soil bacterial communities in alpine wetlands in arid central Asia remain stable during the seasonal freeze-thaw period. *Ecol. Indic.* **2023**, *156*, 111164. [CrossRef]
22. Luo, D.; Jin, H.; Lü, L.; Wu, Q. Spatiotemporal characteristics of freezing and thawing of the active layer in the source areas of the Yellow River (SAYR). *Chin. Sci. Bull.* **2014**, *59*, 3034–3045. [CrossRef]
23. Han, W.; Huang, C.; Duan, H.; Gu, J.; Hou, J. Lake phenology of freeze-thaw cycles using random forest: A case study of Qinghai Lake. *Remote Sens.* **2020**, *12*, 4098. [CrossRef]
24. Cao, X.; Lu, P.; Leppäranta, M.; Arvola, L.; Huotari, J.; Shi, X.; Li, G.; Li, Z. Solar radiation transfer for an ice-covered lake in the central Asian arid climate zone. *Inland Waters* **2021**, *11*, 89–103. [CrossRef]
25. Sibert, R.J.C.; Koretsky, M.; Wyman, D.A. Culturalmeromixis: Effects of road salt on the chemical stratification of an urban kettle lake. *Chem. Geol.* **2005**, *395*, 126–137. [CrossRef]
26. Vonnahme, T.R.; Nowak, A.; Hopwood, M.J.; Meire, L.; Søgaard, D.H.; Krawczyk, D.; Kalhagen, K.; Juul-Pedersen, T. Impact of winter freshwater from tidewater glaciers on fjords in Svalbard and Greenland; A review. *Prog. Oceanogr.* **2023**, *219*, 103144. [CrossRef]

27. Barker, T.; Hatton, K.O.; Connor, L.; Moss, B. Effects of nitrate load on submerged plant biomass and species richness: Results of a mesocosm experiment. *Fundam. Appl. Limnol.* **2008**, *173*, 89–100. [CrossRef]
28. Wentzky, V.C.; Tittel, J.; Jäger, C.G.; Rinke, K. Mechanisms preventing a decrease in phytoplankton biomass after phosphorus reductions in a German drinking water reservoirresults from more than 50years of observation. *Freshw. Biol.* **2018**, *63*, 1063–1076. [CrossRef]
29. Yang, F.; Cen, R.; Feng, W.; Zhu, Q.; Leppäranta, M.; Yang, Y.; Wang, X.; Liao, H. Dynamic simulation of nutrient distribution in lakes during ice cover growth and ablation. *Chemosphere* **2021**, *281*, 130781. [CrossRef]
30. Buck, R.L. Nutrient Uptake and Water Quality in Great Salt Lake Wetland Impoundments. Ph.D. Thesis, Utah State University, Logan, UT, USA, 2022.
31. Crosman, E.T.; Horelj, D. Winter Lake Breezes near the Great Salt Lake. *Boundary-Layer Meteorol.* **2016**, *159*, 439–464. [CrossRef]
32. Ding, W.; Wang, G.; Ren, H.; Li, H.; Lü, W.; Jiang, X. Recognizing the variation of DNA-P during and after the algal bloom in lake Hulun. *Chemosphere* **2023**, *343*, 140293. [CrossRef] [PubMed]
33. Li, X.; Liu, Y.; Zhang, S.; Li, G.; Tao, Y.; Wang, S.; Yu, H.; Shi, X.; Zhao, S. Evolution characteristics and driving factors of cyanobacterial blooms in Hulun Lake from 2018 to 2022. *Water* **2023**, *15*, 3765. [CrossRef]
34. Binding, C.; Greenberg, T.; McCullough, G.; Watson, S.; Page, E. An analysis of satellite-derived chlorophyll and algal bloom indices on Lake Winnipeg. *J. Great Lakes Res.* **2018**, *44*, 436–446. [CrossRef]
35. Yezhova, Y.; Capelle, D.; Stainton, M.; Papakyriakou, T. Carbon fixation by the phytoplankton community across Lake Winnipeg. *J. Great Lakes Res.* **2021**, *47*, 703–714. [CrossRef]
36. Elliott, J.A.; Mcelarney, Y.R.; Allen, M. The past and future of phytoplankton in the UK's largest lake, Lough Neagh. *Ecol. Indic.* **2016**, *68*, 142–149. [CrossRef]
37. Ho, J.C.; Michalak, A.M.; Pahlevan, N. Widespread global increase in intense lake phytoplankton blooms since the 1980s. *Nature* **2019**, *574*, 667–671. [CrossRef]
38. Yang, J.; Wang, F.; Lv, J.; Liu, Q.; Nan, F.; Liu, X.; Xu, L.; Xie, S.; Feng, J. Interactive effects of temperature and nutrients on the phytoplankton community in an urban river in China. *Environ. Monit. Assess.* **2019**, *191*, 688. [CrossRef]
39. Vinçon-Leite, B.; Casenave, C. Modelling eutrophication in lake ecosystems: A review. *Sci. Total Environ.* **2019**, *651*, 2985–3001. [CrossRef]
40. Song, K.; Wen, Z.; Jacinthe, P.-A.; Zhao, Y.; Du, J. Dissolved carbon and CDOM in lake ice and underlying waters along a salinity gradient in shallow lakes of Northeast China. *J. Hydrol.* **2019**, *571*, 545–558. [CrossRef]
41. Liu, L.; Huang, Q.; Qin, B. Characteristics and roles of microcystis extracellular polymeric substances (EPS) in cyanobacterial blooms: A short review. *J. Freshw. Ecol.* **2018**, *33*, 183–193. [CrossRef]
42. Cavaliere, E.; Baulch, H.M. Denitrification under lake ice. *Biogeochemistry* **2018**, *137*, 285–295. [CrossRef]
43. Brooks, P.R.; Nairn, R.; Harris, M.; Jeffrey, D.; Crowe, T.P. Dublin Port and Dublin Bay: Reconnecting with nature and people. *Reg. Stud. Mar. Sci.* **2016**, *8*, 234–251. [CrossRef]
44. Huber, V.; Wagner, C.; Adrian, G.R. To bloom or not to bloom: Contrasting responses of cyanobacteria to recent heat waves explained by critical thresholds of abiotic drivers. *Oecologia* **2012**, *169*, 245–256. [CrossRef]
45. Winder, M. Lake warming mimics fertilization. *Nat. Clim. Change* **2012**, *2*, 771–772. [CrossRef]
46. Tang, X.; Wu, M.; Yang, W.; Scholz, M. Impact of simulated water level regulation on sediment nutrient release. *Water Air Soil Pollut.* **2015**, *226*, 248. [CrossRef]
47. Liu, X.Y.; Wang, C.H.; Yu, C.R.; Wen, L.; Duan, P.Y. Characteristics of sediments and nutrient release under different flow velocity. *Acta Sci. Circumstantiae* **2017**, *27*, 697–710.
48. Park, J.; Church, J.; Son, Y.; Kim, K.T.; Lee, W.H. Recent advances in ultrasonic treatment: Challenges and field applications for controlling harmful algal blooms (HABs). *Ultrason. Sonochem.* **2017**, *38*, 326–334. [CrossRef] [PubMed]
49. Yu, H.; Shi, X.; Wang, S.; Zhao, S.; Sun, B.; Liu, Y.; Yang, Z. Trophic status of a shallow lake in Inner Mongolia: Long-term, seasonal, and spatial variation. *Ecol. Indic.* **2023**, *156*, 111167. [CrossRef]
50. Knorr, S.; Weisener, C.G.; Phillips, L.A. The role of agricultural drainage, storm-events, and natural filtration on the biogeochemical cycling capacity of aquatic and sediment environments in Lake Erie's drainage basin. *Sci. Total Environ.* **2023**, *905*, 167102. [CrossRef]
51. Qin, B.; Deng, J.; Shi, K.; Wang, J.; Brookes, J.; Zhou, J.; Zhang, Y.; Zhu, G.; Paerl, H.W.; Wu, L. Extreme climate anomalies enhancing cyanobacterial blooms in eutrophic Lake Taihu, China. *Water Resour. Res.* **2021**, *57*, e2020WR029371. [CrossRef]
52. Daims, H.; Lebedeva, E.V.; Pjevac, P.; Han, P.; Herbold, C.; Albertsen, M.; Jehmlich, N.; Palatinszky, M.; Vierheilig, J.; Bulaev, A.; et al. Complete nitrification by nitrospira bacteria. *Nature* **2015**, *528*, 504–509. [CrossRef] [PubMed]
53. Peng, X.; Fuchsman, C.A.; Jayakumar, A.; Warner, M.J.; Devol, A.H.; Ward, B.B. Revisiting nitrification in the Eastern Tropical South Pacific: A focus on controls. *J. Geophys. Res. C Ocean JGR* **2016**, *121*, 1667–1684. [CrossRef]
54. Zhang, D.; Li, M.; Yang, Y.; Yu, H.; Xiao, F.; Mao, C.; Huang, J.; Yu, Y.; Wang, Y.; Wu, B.; et al. Nitrite and nitrate reduction drive sediment microbial nitrogen cycling in a eutrophic lake. *Water Res.* **2022**, *220*, 118637. [CrossRef]
55. Huang, J.; Zhao, W.; Ju, J.; Liu, S.; Ye, J.; Long, Y. The existence of ferric hydroxide links the carbon and nitrogen cycles by promoting nitrite-coupled methane anaerobic oxidation. *Water Res.* **2023**, *243*, 120192. [CrossRef]
56. Cai, Y.; Guo, L. Abundance and variation of colloidal organic phosphorus in riverine, estuarine, and coastal waters in the northern Gulf of Mexico. *Limnol. Oceanogr.* **2009**, *54*, 1393–1402. [CrossRef]

57. Lin, P.; Chen, M.; Guo, L. Speciation and transformation of phosphorus and its mixing behavior in the Bay of St. Louis estuary in the northern Gulf of Mexico. *Geochim. Cosmochim. Acta* **2012**, *87*, 283–298. [CrossRef]
58. Verhoeven, J.T.; Arheimer, B.; Yin, C.; Hefting, M.M. Regional and global concerns over wetlands and water quality. *Trends Ecol. Evol.* **2006**, *21*, 96–103. [CrossRef] [PubMed]
59. Tikhonravova, Y.; Kuznetsova, A.; Slagoda, E.; Koroleva, E. Holocene permafrost peatland evolution in drained lake basins on the Pur-Taz Interfluve, North-Western Siberia. *Quat. Int.* **2023**, *669*, 32–42. [CrossRef]
60. Kalinowska, K.; Napiórkowska-Krzebietke, A.; Bogacka-Kapusta, E.; Stawecki, K. Comparison of ice-on and ice-off abiotic and biotic parameters in three eutrophic lakes. *Ecol. Res.* **2019**, *34*, 687–698. [CrossRef]
61. Luo, H.; Cheng, Q.; Pan, X. Photochemical behaviors of mercury (Hg) species in aquatic systems: A systematic review on reaction process, mechanism, and influencing factor. *Sci. Total Environ.* **2020**, *720*, 137540. [CrossRef]
62. Kirillin, G.; Leppäranta, M.; Terzhevik, A.; Granin, N.; Bernhardt, J.; Engelhardt, C.; Efremova, T.; Golosov, S.; Palshin, N.; Sherstyankin, P.; et al. Physics of seasonally ice-covered lakes: A review. *Aquat. Sci.* **2012**, *74*, 659–682. [CrossRef]
63. Hrycik, A.R.; Stockwell, J.D. Under-ice mesocosm experiments reveal the primacy of light but the importance of zooplankton in winter phytoplankton dynamics. *Limnol. Oceanogr.* **2021**, *66*, 481–495. [CrossRef] [PubMed]
64. Kirillin, G.B.; Forrest, A.L.; Graves, K.E.; Fischer, A.; Engelhardt, C.; Laval, B.E. Axisymmetric circulation driven by marginal heating in ice-covered lakes. *Geophys. Res. Lett.* **2015**, *42*, 2893–2900. [CrossRef]
65. Lu, Y.; Yang, K.; Lu, X.; Li, Y.; Gao, S.; Mao, W.; Li, M. Response of supraglacial rivers and lakes to ice flow and surface melt on the northeast Greenland Ice Sheet during the 2017 melt season. *J. Hydrol.* **2021**, *601*, 126750. [CrossRef]
66. Huang, W.; Zhang, J.; Leppäranta, M.; Li, Z.; Cheng, B.; Lin, Z. Thermal structure and water-ice heat transfer in a shallow ice–covered thermokarst lake in central Qinghai-Tibet Plateau. *J. Hydrol.* **2019**, *578*, 124122. [CrossRef]
67. Powers, S.M.; Labou, S.G.; Baulch, H.M.; Hunt, R.J.; Lottig, N.R.; Hampton, S.E.; Stanley, E.H. Ice duration drives winter nitrate accumulation in north temperate lakes. *Limnol. Oceanogr. Lett.* **2017**, *2*, 177–186. [CrossRef]
68. Whiteford, E.J.; McGowan, S.; Barry, C.D.; Anderson, N.J. Seasonal and regional controls of phytoplankton production along a climate gradient in South-West Greenland during ice-cover and ice-free conditions. *Arct. Antarct. Alp. Res.* **2016**, *48*, 139–159. [CrossRef]
69. McMillan, M.; Nienow, P.; Shepherd, A.; Benham, T.; Sole, A. Seasonal evolution of supra-glacial lakes on the Greenland Ice Sheet. *Earth Planet. Sci. Lett.* **2007**, *262*, 484–492. [CrossRef]
70. Shan, G.; Xiang, Q.; Feng, X.; Wu, W.; Yang, L.; Zhu, L. Occurrence and sources of Per-and Polyfluoroalkyl Substances in the ice-melting lakes of Larsemann Hills, East Antarctica. *Sci. Total Environ.* **2021**, *781*, 146747. [CrossRef]
71. Hampton, S.E.; Galloway, A.W.E.; Powers, S.M.; Ozersky, T.; Woo, K.H.; Batt, R.D.; Labou, S.G.; O'Reilly, C.M.; Sharma, S.; Lottig, N.R.; et al. Ecology under lake ice. *Ecol. Lett.* **2016**, *20*, 98–111. [CrossRef]
72. Wei, Q.; Xu, Y.; Ruan, A. Spatial and temporal patterns of phytoplankton community succession and characteristics of realized niches in Lake Taihu, China. *Environ. Res.* **2024**, *243*, 117896. [CrossRef]
73. Leppäranta, M. *Freezing of Lakes and the Evolution of Their Ice Cover*; Springer Science & Business Media: Berlin/Heidelberg, Germany, 2014.
74. Lotter, A.F.; Bigler, C. Do diatoms in the Swiss Alps reflect the length of ice-cover? *Aquat. Sci.* **2000**, *62*, 125–141. [CrossRef]
75. Babanazarova, O.; Sidelev, S.; Schischeleva, S. The structure of winter phytoplankton in Lake Nero, Russia, a hypertrophic lake dominated by Planktothrix-like Cyanobacteria. *Aquat. Biosyst.* **2013**, *9*, 18. [CrossRef] [PubMed]
76. Twiss, M.R.; Ulrich, C.; Zastepa, A.; Pick, F.R. On phytoplankton growth and loss rates to microzooplankton in the epilimnion and metalimnion of Lake Ontario in mid-summer. *J. Great Lakes Res.* **2012**, *38*, 146–153. [CrossRef]
77. Jacquemin, S.J.; Doll, J.C.; Johnson, L.T.; Newell, S.E. Exploring long-term trends in microcystin toxin values associated with persistent harmful algal blooms in Grand Lake St Marys. *Harmful Algae* **2023**, *122*, 102374. [CrossRef] [PubMed]
78. Hrycik, A.R.; Isles, P.D.; Adrian, R.; Albright, M.; Bacon, L.C.; Berger, S.A.; Bhattacharya, R.; Grossart, H.P.; Hejzlar, J.; Hetherington, A.L.; et al. Earlier winter/spring runoff and snowmelt during warmer winters lead to lower summer chlorophyll-a in north temperate lakes. *Glob. Change Biol.* **2021**, *27*, 4615–4629. [CrossRef] [PubMed]
79. Kauko, H.M.; Pavlov, A.K.; Johnsen, G.; Granskog, M.A.; Peeken, I.; Assmy, P. Photoacclimation state of an Arctic Underice phytoplankton bloom. *J. Geophys. Res. Oceans* **2019**, *124*, 1750–1762. [CrossRef]
80. Lewis, K.M.; Arntsen, A.E.; Coupel, P.; Joy-Warren, H.; Lowry, K.E.; Matsuoka, A.; Mills, M.M.; Van Dijken, G.L.; Selz, V.; Arrigo, K.R. *Photoacclimation of Arctic Ocean Phytoplankton to Shifting Light and Nutrient Limitation*; John Wiley & Sons, Ltd.: Hoboken, NJ, USA, 2019.
81. Schreder, S.; Sommaruga, R.; Psenner, R.; Chimani, B.; Ganekind, M.; Koinig, K.A. Changes in air temperature, but not in precipitation, determine long-term trends in water chemistry of high mountain lakes of the Alps with and without rock glacier influence. *Sci. Total Environ.* **2023**, *905*, 167750. [CrossRef]
82. George, G. *The Impact of Climate Change on European Lakes*; Springer: Berlin/Heidelberg, Germany, 2010.
83. Smucker, N.J.; Beaulieu, J.J.; Nietch, C.T.; Young, J.L. Increasingly severe cyanobacterial blooms and deep water hypoxia coincide with warming water temperatures in reservoirs. *Glob. Change Biol.* **2021**, *27*, 2507–2519. [CrossRef]
84. Yindong, T.; Xiwen, X.; Miao, Q.; Jingjing, S.; Yiyan, Z.; Wei, Z.; Mengzhu, W.; Xuejun, W.; Yang, Z. Lake warming intensifies the seasonal pattern of internal nutrient cycling in the eutrophic lake and potential impacts on algal blooms. *Water Res.* **2021**, *188*, 116570. [CrossRef]

85. Zhang, M.; Duan, H.; Shi, X.; Yu, Y.; Kong, F. Contributions of meteorology to the phenology of cyanobacterial blooms: Implications for future climate change. *Water Res.* **2012**, *46*, 442–452. [CrossRef] [PubMed]
86. Yuan, Y.; Jiang, M.; Liu, X.; Yu, H.; Otte, M.L.; Ma, C.; Her, Y.G. Environmental variables influencing phytoplankton communities in hydrologically connected aquatic habitats in the Lake Xingkai basin. *Ecol. Indic.* **2018**, *91*, 1–12. [CrossRef]
87. Armitage, A.R.; Fourqurean, J.W. Stable isotopes reveal complex changes in trophic relationships following nutrient addition in a coastal marine ecosystem. *Estuaries Coasts* **2009**, *32*, 1152–1164. [CrossRef]
88. Thalasso, F.; Sepulveda-Jauregui, A.; Cabrol, L.; Lavergne, C.; Olgun, N.; Martinez-Cruz, K.; Aguilar-Muñoz, P.; Calle, N.; Mansilla, A.; Astorga-España, M.S. Methane and carbon dioxide cycles in lakes of the King George Island, maritime Antarctica. *Sci. Total Environ.* **2022**, *848*, 157485. [CrossRef]
89. An, S.; Niu, X.; Chen, W.; Sheng, H.; Lai, S.; Yang, Z.; Gu, X.; Zhou, S. Mechanism of matrix-bound phosphine production in response to atmospheric elevated CO_2 in paddy soils. *Environ. Pollut.* **2018**, *239*, 253–260. [CrossRef]
90. Wang, P.; Ma, J.; Wang, X.; Tan, Q. Rising atmospheric CO_2 levels result in an earlier cyanobacterial bloom-maintenance phase with higher algal biomass. *Water Res.* **2020**, *185*, 116267. [CrossRef] [PubMed]
91. Przytulska, A.; Bartosiewicz, M.; Vincent, W.F. Increased risk of cyanobacterial blooms in northern high-latitude lakes through climate warming and phosphorus enrichment. *Freshw. Biol.* **2017**, *62*, 1986–1996. [CrossRef]
92. Xiao, X.; Peng, Y.; Zhang, W.; Yang, X.; Zhang, Z.; Ren, B.; Zhu, G.; Zhou, S. Current status and prospects of algal bloom early warning technologies: A Review. *J. Environ. Manag.* **2024**, *349*, 119510. [CrossRef]
93. Bucak, T.; Trolle, D.; Tavşanoğlu, N.; Çakıroğlu, A.I.; Özen, A.; Jeppesen, E.; Beklioğlu, M. Modeling the effects of climatic and land use changes on phytoplankton and water quality of the largest Turkish freshwater lake: Lake Beyşehir. *Sci. Total Environ.* **2017**, *621*, 802. [CrossRef]
94. Pathak, D.; Hutchins, M.; Brown, L.; Loewenthal, M.; Scarlett, P.; Armstrong, L.; Nicholls, D.; Bowes, M.; Edwards, F. Hourly prediction of phytoplankton biomass and its environmental controls in lowland rivers. *Water Resour. Res.* **2021**, *57*, e2020WR028773. [CrossRef]

water

MDPI

Article

Analysis of Meteorological Element Variation Characteristics in the Heilongjiang (Amur) River Basin

Qi Yue [1,2,3], Gao Yu [1,2,3,*], Yu Miao [3,4,5] and Yang Zhou [1,2,3]

1 Institute of Groundwater in Cold Region, Heilongjiang University, Harbin 150080, China; hss_qiyue@126.com (Q.Y.); hss_zhouyang@126.com (Y.Z.)
2 School of Hydraulic and Electric Power, Heilongjiang University, Harbin 150080, China
3 International Joint Laboratory of Hydrology and Hydraulic Engineering in Cold Regions of Heilongjiang Province, Harbin 150080, China; hss_yumiao@126.com
4 Faculty of Geology and Survey, North-Eastern Federal University, 677000 Yakutsk, Russia
5 Melnikov Permafrost Institute, Siberian Branch of the Russian Academy of Science, 677000 Yakutsk, Russia
* Correspondence: gaoyu501@126.com; Tel.: +86-186-8689-5935

Abstract: Located in the Heilongjiang (Amur) River in north-east Asia, spanning four countries, plays a crucial role as an international border river, and its meteorological changes significantly impact the variation in water resources in the basin. This study utilizes daily average temperature and precipitation data from 282 meteorological stations in the Heilongjiang (Amur) River Basin and its surrounding areas for the period 1980–2022. The analysis employs spatial interpolation, change point testing, and model construction prediction methods. The results indicate a significant increasing trend in both overall temperature and precipitation changes within the Heilongjiang (Amur) River Basin. At the spatial scale, the annual warming rate increases gradually from the southeastern coastal region to the northwestern plateau region, while the rate of precipitation increase decreases from the southern area towards its surroundings. Temporally, the warming amplitude during the growing season decreases gradually from east to west, and the trend in precipitation changes during the growing season aligns with the overall annual precipitation trend. During the non-growing season, the warming trend shows a decrease in the plains and an increase in the plateau, while precipitation increase concentrates in the central and southern plains, and precipitation decrease predominantly occurs in the northwestern plateau region. Temperature and precipitation change points occurred in the years 2001 and 2012, respectively. In precipitation prediction, the Long Short-Term Memory (LSTM) model exhibits higher accuracy, with R (Pearson correlation coefficient) and NSE (Nash-Sutcliffe efficiency coefficient) values approaching 1 and lower NRSME values. This study provides a research foundation for the rational development and utilization of water resources in the Heilongjiang (Amur) River Basin and offers valuable insights for research on climate change characteristics in large transboundary river systems.

Keywords: Heilongjiang (Amur) River; temperature; precipitation; spatiotemporal distribution characteristics; abrupt change analysis; precipitation value prediction

Citation: Yue, Q.; Yu, G.; Miao, Y.; Zhou, Y. Analysis of Meteorological Element Variation Characteristics in the Heilongjiang (Amur) River Basin. *Water* **2024**, *16*, 521. https://doi.org/10.3390/w16040521

Academic Editor: Chin H Wu

Received: 24 December 2023
Revised: 30 January 2024
Accepted: 1 February 2024
Published: 6 February 2024

1. Introduction

Under global warming conditions, the accelerated melting of glaciers and snow has serious consequences for large rivers in mid-to-high latitudes, leading to increased occurrences of floods and river interruptions, as well as higher frequencies of water and drought disasters, wetland degradation, water resource depletion, and elevated basin ecosystem issues. As early as 2014, the IPCC's (The Intergovernmental Panel on Climate Change) Fifth Assessment Report pointed out that surface temperatures are continuously rising due to increased greenhouse gas concentrations, leading to changes in precipitation amount, intensity, and spatiotemporal distribution, consequently resulting in corresponding alterations in river water resource distribution and the water cycle [1]. In 2018, the IPCC released a special report on "Global Warming of 1.5 °C," indicating that the world could reach the 1.5 °C threshold at some point between 2030 and 2052 [2]. Climate change increases the frequency and intensity of extreme weather events such as extreme temperatures, heavy precipitation, and droughts, leading to different effects on the spatiotemporal characteristics of basin runoff [3]. Transboundary water resources constitute over half of the world's available freshwater resources, affecting the sustainable development of more than 148 countries and over 90% of the global population [4]. With the intensification of the global water crisis, transboundary river management issues are gradually attracting widespread attention from the international community [5,6]. Particularly for the densely populated Asian region, transboundary rivers in mid-to-high latitudes, influenced by changes in climate conditions, river supply intensity, and the constraints of water rights management across multiple countries and regions, are more prone to causing significant economic losses and adversely affecting basin water ecological security, lowering basin water resource utilization efficiency, and hindering political, economic, social, and environmental development among countries and regions within the basin [7].

The Amur River, spanning China, Russia, Mongolia, and North Korea, is one of the world's top ten longest rivers, and it is also a significant transboundary river. Its unique geopolitical features make its status and role significant, with it serving as a crucial shipping route between China and Russia and a natural resource reservoir passing through various countries. Among those factors consistently emphasized in recent World Water Development Reports by the United Nations Educational, Scientific and Cultural Organization (UNESCO), the ongoing reduction in permafrost in cold regions, particularly amid global warming and the broader context of climate change, stands out. Concomitantly, there is an accelerating trend in the rates of snow and glacier melt. Even in regions endowed with ample water resources, the intensification of seasonal water shortages is notable. This unfolding scenario magnifies the impact on water resources across diverse nations globally, presenting an urgent predicament. Consequently, there is an escalating risk to water security in transboundary water resources, posing a formidable challenge in water diplomacy and creating a complex situation for sustaining collaborative governance of water environments among nations [8]. In modern society, the impact of a series of human activities, including population growth, increasing demands for food and energy, urbanization, and industrial development, has led to an increased pressure on the development and utilization of scarce freshwater resources. Climate change further exacerbates water-related issues [9]. Changes in meteorological factors manifest in variations in temperature, precipitation patterns, and evapotranspiration, influencing the distribution of regional water resources. Large rivers in cold regions at middle and high latitudes exhibit higher sensitivity to climate change. Changes in the quantity of water resources within the basin are closely linked to water use for the production and daily living of residents in the four countries through which the river flows. Moreover, alterations in water ecology within the basin are associated with the potential degradation of wetlands and grasslands, contributing to ecological issues in the basin [10]. Therefore, conducting research on the spatiotemporal distribution characteristics of climate change in the Heilongjiang (Amur) River Basin and its meteorological abrupt change features is beneficial for coordinating the rational utilization of water resources in the Heilongjiang (Amur) River Basin and promoting collaborative development among the countries involved.

While the countries and regions through which the Heilongjiang (Amur) River Basin flows have a significant population density difference compared to more developed regions in the mid-to-low latitudes, the distribution of towns, villages, and rivers is closely related [11]. In some areas of Mongolia, Russia, and Northeast China, animal husbandry remains a predominant activity, and the livelihoods of residents in these regions are closely tied to the river [12]. The meteorological characteristics of the Heilongjiang (Amur) River differ from well-known large rivers in China, such as the Yellow River and the Yangtze River, in terms of meteorological variations, particularly in the mid-to-low latitudes [13].

In the summer of 2013, the Songhua River Basin experienced the most extreme precipitation in the region since 1984, leading to a dramatic increase in river flow and causing an unprecedented large-scale flood in the Amur River Basin. Yan Bo et al. [14] used the extreme precipitation-induced flood event in the Amur River Basin in 2013 as a research background. They analyzed nearly 60 years of precipitation data from 25 meteorological stations in the basin, calculated the trends of extreme indices using the Mann–Kendall test, and explored the spatiotemporal characteristics of extreme precipitation events through wavelet transform analysis of extreme indices. Semenov E. K. et al. [15] utilized data on basin runoff, air temperature, atmospheric pressure, and precipitation to create maps of air cyclones and pressure trends in the basin. They analyzed the causes of the 2013 flood in the Amur River Basin. Kalugin A. S. et al. [16] used ECOMAG (Ecological Model for Applied Geophysics) model to establish a model of Amur River runoff. They simulated the spatial distribution of certain features of basin hydrological cycles, such as snow accumulation, soil moisture, and evaporation, based on meteorological and water management monitoring standard data. Scholars like Li Mingliang [17] established a spatial information database for the Heilongjiang River Basin and developed the GBHM-HLJ (Geomorphology-based Hydrological Model—Heilongjiang Basin), a distributed hydrological model based on physical mechanisms. They introduced a generalized model with temperature-indexed changes in frozen soil hydraulic conductivity to simulate the impact of soil freeze–thaw cycles on water movement. Gelfan A. N. et al. [18] employed both computation and data transformation methods to build a climate prediction model, analyzing the sensitivity of temperature and precipitation changes to Amur River runoff variations in the 21st century. Zhang Wenxuan et al. [19] utilized Sentinel-1 synthetic aperture radar, conducting time series monitoring of the spatial extent of floods using Gamma and Gaussian distributions. The results indicated that cities along the middle and lower reaches of the Amur River, such as Khabarovsk and Amur, are prone to frequent flooding, and the overall flood area is increasing. Jia Lin [20] established a research framework for the basin's joint development mechanism concerning international rivers in the northeast region. It indicates that the international river development mechanism in the northeast is an integral part of regional economic cooperation in Northeast Asia. Lessons should be drawn from the development strategies in the Lancang–Mekong River Basin to encourage coordinated development of resources, economic growth, and environmental protection among basin countries. He Daming [21] and other researchers indicated that China's northeast and southwest regions are major source areas for international rivers in Asia. They proposed the development direction of comprehensive cross-border resources and environmental cooperation between land and sea. This involves leading international river development and geopolitical cooperation to advance and maintain political, economic, and technological cooperation among countries. Panova (Панова А. А.) [22] believes that governance of the ecological environment of the Amur River requires the joint formulation of relevant laws and regulations by the countries through which the basin flows. She cited cooperation agreements in the field of environmental protection jointly formulated by Russia and China. She suggested that all countries collaborate to conduct a comprehensive investigation and assessment of the Amur River Basin and, based on the findings, formulate more practical and applicable protective measures for the basin.

Due to the geopolitical characteristics of transnational basins and diverse development needs of different countries, current research has predominantly focused on analyzing specific basin segments within the countries through which the Amur River flows. This approach reflects the research characteristics of different countries involved in the basin. However, there is a lack of comprehensive analysis of meteorological elements across the entire basin. Analyzing the basin as a whole can better integrate the characteristics of meteorological factors across the basin. This study aims to analyze the climate change in both the overall basin and its sub-basins, as well as the meteorological anomalies in the entire basin and its tributaries. By elucidating the annual and seasonal climate change characteristics of the entire Amur River Basin, as well as the impact of climate change in different sub-basins on the overall basin, this research provides a basis for the development and utilization of water resources, rational allocation of basin water resources, ecological protection, and coordinated development of economic water demand and serves as a reference for the study of similar international rivers. The findings contribute to the promotion of harmonious coexistence and comprehensive sustainable development among basin countries.

Based on past research on the meteorological characteristics of the Heilongjiang (Amur) River Basin, evaluations of meteorological features have employed different methods from various perspectives. On one hand, in terms of data selection, the time series is relatively short, and the number of stations is limited, failing to comprehensively cover the meteorological characteristics of the entire Heilongjiang (Amur) River Basin [23–25]. On the other hand, many studies use meteorological elements as influencing factors, combined with multiple factors such as land use, vegetation cover, and ecosystems, to comprehensively explore the evolution patterns of basin runoff, hydrological processes, or hydrological responses under the influence of multiple factors. However, there has been no specialized investigation into the characteristics of basin meteorological elements under the background of long-term meteorological data sequences [26,27]. This study spatially divides the Heilongjiang (Amur) River Basin into eight sub-basins and seasonally divides it into the growing season and non-growing season based on the number of days with temperatures greater than or equal to 0 °C. It conducts comparative analyses of the spatiotemporal distribution characteristics and mutation features of climatic changes between the overall Heilongjiang (Amur) River Basin and its eight sub-basins. The study analyzes the climatic characteristics and change patterns of the main basin and each sub-basin, providing a research foundation for understanding the impact of the overall climate on water resource changes and the rational development and utilization of water resources in the basin. Simultaneously, it offers references for the study of climatic changes in similar large-scale transboundary river basins.

2. Materials and Methods

2.1. Study Area Profile

The Heilongjiang (Amur) River Basin is located in the northeastern part of Asia, serving as a significant boundary river between Northeast China and the Russian Far East. It is one of the world's ten major rivers alongside the Amazon River and the Yangtze River, with a basin area of 184.3×10^4 km^2, even larger than that of the Yangtze River (Figure 1) [28]. The entire basin and surrounding areas encompass 15 provincial-level administrative regions in four countries, including Heilongjiang Province, Jilin Province, the Inner Mongolia Autonomous Region, and parts of Liaoning Province in China, the Russian Far East, the eastern region of Mongolia, and parts of the two Koreas. In the northern part of the Heilongjiang (Amur) River Basin, it is separated by the Stanovoy Mountains (Outer Khingan Mountains) from the Lena River Basin. The western side runs along the Kent Mountains, and it extends eastward along the southern branch of the Greater Khingan Range, and the southern side is separated from the Yellow Sea and the Sea of Japan Basin by the Changbai Mountains and the Laoye Mountains. The basin's boundary follows the Sikhote-Alin Mountains northward until the mouth of the Heilongjiang River. These mountainous areas are the headwaters of the main and tributary rivers of the Heilongjiang (Amur) River Basin [29]. The basin is mainly characterized by mountainous and hilly terrain, with plains mainly distributed in the central and eastern regions of the basin, including the Jebusan Plain, Songnen Plain, and the plain in the middle and lower reaches of the Heilongjiang River. The Songnen Plain and the middle and lower reaches of the Heilongjiang River Plain in China have fertile soils, widely distributed black soil and black calcareous soil, and a deep history of cultivation and are important commodity grain bases for Heilongjiang Province and the country. They primarily cultivate crops such as soybeans, wheat, and sugar beets. The region is also concentrated with grasslands, supporting a developed livestock industry. Additionally, it hosts the well-known Zhalong Nature Reserve, an internationally important wetland conservation area.

The Heilongjiang (Amur) River has two sources, with the southern source being the Ergun River and the northern source being the Shilka River. The convergence of the two sources occurs near the village of Logu River, west of Mohe City in the Daxing'anling area of Heilongjiang Province, China, forming the main stream of the Heilongjiang (Amur) River, which eventually flows into the Strait of Tartary in Nikolaevsk (Temple Street), Russia. The basin is characterized by numerous rivers and a dense network of waterways, featuring seven major tributaries in addition to the main Heilongjiang River, forming a system of seven branches and one main stem. Notable lakes in the basin include the cross-border Hulun Lake on the Sino–Mongolian border and the transboundary Xingkai Lake on the Sino–Russian border. The major tributaries include the Shilka River and Ergun River in the river source area, the Zeya River (Jiqili River), Bureya River (Niuman River), and Amgun River (Xinggun River) on the Russian side, and the Songhua River on the Chinese side. Additionally, the Ussuri River, Ergun River, and the main stem of the Heilongjiang River are all international boundary rivers, constituting the world's longest boundary river at nearly 4000 km [30].

The Heilongjiang (Amur) River, as an important transboundary river spanning China, Russia, Mongolia, and North Korea, exhibits distinctive population compositions, economic structures, and development levels among the four countries. Additionally, the water use patterns, water demand, and the extent of water resource development and utilization vary [31]. The impact of meteorological element changes in this basin on the variation in water resources within the basin cannot be ignored. This information is instrumental in helping each country formulate development policies that are more tailored to the sustainable development of the local region. Using the river system as a link, it contributes to the collective maintenance of water resources, ecological environment, and the political and economic health and peaceful development among the countries within the basin. Simultaneously, it can serve as a reference for meteorological element changes in similar cold regions with large rivers, such as the Yenisei River Basin and the Ob River.

2.2. Materials

The elevation data of the Heilongjiang (Amur) River Basin in this study were derived from the Digital Elevation Model (DEM) provided by the Geospatial Information Authority of Japan, based on a spatial resolution of 30 m [32]. The boundary data for national borders were sourced from the Global Administrative District Boundaries data provided by the National Earth System Science Data Centre (http://www.geodata.cn/, accessed on 26 July 2022) [33]. The river network and lake data within the basin were obtained from the A Big Earth Data Platform for Three Poles (https://poles.tpdc.ac.cn/zh-hans/, accessed on 5 August 2022) which provides the Global River and Lake Vector Dataset (2010) [34]. Meteorological data were sourced from the National Centers for Environmental Information (NCEI), a division of the National Oceanic and Atmospheric Administration (NOAA) (https://www.ncei.noaa.gov/data/global-summary-of-the-day/archive/, accessed on 1 July 2022), specifically the daily precipitation and temperature data [35]. Based on the acquired data, temperature and precipitation data from 282 meteorological stations within the basin and surrounding areas were selected, covering a time series of daily observations ranging from 1980 to 2022.

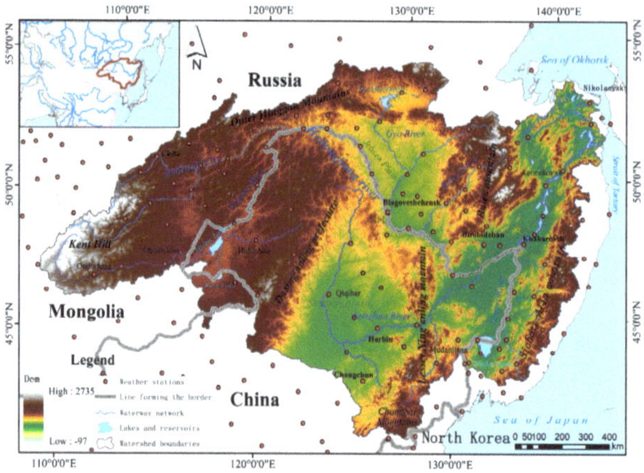

Figure 1. Geographic characteristics of the Heilongjiang (Amur) River Basin and distribution of meteorological stations.

2.3. Methodology

2.3.1. Data Preprocessing

To investigate the characteristics of meteorological element changes in the Heilongjiang (Amur) River Basin, meteorological data from 282 meteorological stations within and near the basin were selected. Due to the wide range of basin area and uneven distribution of meteorological stations within the basin, as well as the possibility of missing or omitting meteorological data in time due to various factors affecting the monitoring equipment at observation stations, there may be some interference in the analysis of later meteorological characteristics. To ensure the accuracy of meteorological feature analysis, the double mass curve method was employed to preprocess the original meteorological data from meteorological stations. The double mass curve method is a common approach for studying the consistency and variation between two parameters. It involves plotting a relationship line in a rectangular coordinate system between the continuous cumulative values of one variable and another variable over the same period. It can be used to check the consistency of hydro-meteorological elements, interpolate missing values or calibrate data, and analyze the trends and intensities of hydro-meteorological elements [36].

After calibration using the double mass curve method, we conducted a preliminary trend analysis of meteorological data using the linear trend regression test. The linear regression test is a common mathematical statistical method that can intuitively reflect the changing trend of a sequence and is widely used in the time series analysis of meteorological elements such as precipitation and temperature [37]. When analyzing the spatial distribution of meteorological elements in the basin, considering the interlaced distribution of mountains and plains in the basin and the large differences in elevation, the influence of elevation on meteorological elements should be fully taken into account, using elevation as a covariate to improve the accuracy of interpolation [38]. Combined with the local thin-plate smoothing spline function interpolation method, spatial interpolation analysis of the temporal and spatial variations in temperature and precipitation in the basin was conducted. The statistical model expression for the local thin-plate smoothing spline theory is as follows:

$$Z_j = f(m_j) + b^T y_j + e_j (j = 1, 2, \cdots, N) \quad (1)$$

where Z_j represents the dependent variable at spatial point j; m_j represents the d-dimensional spline independent variable (d = 2 in this study, representing longitude and latitude); $f(m_j)$ represents the unknown smooth function to be estimated regarding m_j; y_j represents the p-dimensional independent covariate (p = 1 in this study, representing elevation); b represents the p-dimensional coefficient vector for y_j; and e_j represents the error term.

2.3.2. Mann–Kendall Trend Test

The Mann–Kendall trend test is a common method in meteorology used to determine whether meteorological elements exhibit a certain trend of change. The advantage of this method is that sample data do not need to follow a certain distribution and are not affected by a few outliers. It has been widely used to analyze trends and step changes in the time series of elements such as precipitation, water level, and runoff [39,40]. The Mann–Kendall non-parametric test method still works well for non-normally distributed meteorological and hydrological data. In the Mann–Kendall method, when the statistic sequence curve (UF) is greater than 0, it indicates an upward trend in the time series; conversely, the opposite indicates a downward trend. If the UF is outside the significant level range, it indicates a significant change trend in the time series. If the UF and the statistic sequence reverse curve (UB) have intersections within the significant level range, the intersection is the change point [41]. The characteristics of different study areas in different time periods are different, and when it was not possible to be completely certain, further combination with other methods was needed to seek more accurate change years.

2.3.3. Pettitt Test

The Pettitt test is a non-parametric test method. It is efficient in testing continuous sequences and can identify change points in hydrological time series well. It is widely used in change point testing and has clear physical significance [42]. This test is based on the statistical function of Mann–Whitney, which assumes that two sequences $((X_1, X_2, \ldots, X_t)$ and $(X_{t+1}, X_{t+2}, \ldots, X_T))$ are from the same sequence. For continuous sequences, $U_{t,T}$ and $V_{t,T}$ are calculated as follows:

$$U_{t,T} = U_{t-1,T} + V_{t,T} \ \mathrm{t}\epsilon[2, \mathrm{T}] \tag{2}$$

$$U_{1,T} = U_{1,T} \tag{3}$$

$$V_{t,T} = \sum\nolimits_{j=1}^{T} \mathrm{sgn}(X_t - X_j) \tag{4}$$

where $U_{t,T}$ and $V_{t,T}$ are the statistic values for different time periods. When $|U_{t,T}|$ is at its maximum, the corresponding X_t is a possible change point. When the change point $U_{t,T} > 0$, the sequence has a downward change trend; otherwise, it has an upward change trend. The significance level of the potential change point is calculated as follows:

$$P_{OA}(t) = 2exp\left[-6U_{t,T}^2/\left(T^3 + T^2\right)\right] \tag{5}$$

A point is considered an effective change point when $P_{OA}(t) \leq 0.5$.

To improve the accuracy of detecting breakpoints in temperature and precipitation in the basin, the cumulative anomaly method was used in conjunction with MK and Pettitt tests to analyze the meteorological data for breakpoints. Cumulative anomaly represents the sum of all anomalies and can intuitively determine the trend of change. A larger cumulative anomaly indicates that the discrete data are greater than the mean, and the curve shows an upward trend; conversely, a smaller cumulative anomaly indicates that the discrete data are less than the mean, and the curve shows a downward trend [43]. In this study, based on the fluctuations in the cumulative anomaly curve, we judged whether there are breakpoints in the trend of meteorological elements [44]. The results obtained from the cumulative anomaly method can determine the approximate range, facilitating further accurate judgment using the M–K (Mann–Kendall) method and Pettitt test.

2.3.4. Precipitation Value Prediction

For large-scale basins in mid–high latitudes, precipitation exhibits strong non-linear characteristics in its spatiotemporal distribution. Exploring the overall patterns and trends of precipitation time series data is beneficial for precipitation forecasting, hydrological forecasting, and water resource management [45]. Commonly used methods in precipitation forecasting include physically based dynamic models and statistically based prediction models based on historical observational data [46]. Statistical prediction models, which are based on historical observational data, model long-term hydrological time series data, allowing exploration of the relationship between predictive factors and predicted precipitation [47]. In statistically based prediction models, there are single-model predictions and ensemble predictions using multiple models. With the advent and application of deep learning methods such as Artificial Neural Networks (ANNs) in recent years, these methods are gradually being applied in the field of hydro-meteorological forecasting due to their non-linear and flexible modeling characteristics.

This study compared and analyzed the prediction of precipitation in the Heilongjiang (Amur) River Basin using three models: the RNN (Recurrent Neural Networks) algorithm, STL Decomposition (Seasonal-Trend decomposition using LOESS), and the LSTM (Long Short-Term Memory) model. To compare the prediction results of different models, normalized root mean square error (NRMSE), Pearson correlation coefficient R, and the Nash–Sutcliffe efficiency coefficient (NSE) were selected as indicators to evaluate the accuracy of precipitation prediction models. Through comparative analysis of the performance of different prediction models in simulating precipitation values, the optimal model was selected to predict the main influencing factor of precipitation, namely the value of precipitation. This provides a simulated reference for predicting the occurrence of flood disasters in the areas through which the Heilongjiang (Amur) River Basin flows.

Recurrent Neural Network

An Recurrent Neural Network (RNN) is an important component of deep learning algorithms. The most significant difference differentiating it from fully connected neural networks is that the hidden layer units are not mutually independent. Hidden layer neurons are not only interrelated; the current state of the hidden layer cells is also influenced by the historical input data. This characteristic makes it very effective in extracting temporal relationships in time series data structures. An RNN is a type of neural network used for processing sequence data. At different time steps, an RNN cycles weights and connects across time steps [48].

STL Decomposition

Seasonal-Trend decomposition using LOESS (STL) is a time series decomposition algorithm based on Loess smoothing. This algorithm can decompose a time series into three components: trend component, seasonal component, and residual component.

$$Y_t = T_t + S_t + R_t \quad (6)$$

In this decomposition, Y_t represents the observed value at time t, and T_t, S_t, and R_t represent the trend component, seasonal component, and residual component at time t, respectively. The trend component describes a series of data points where the variable changes continuously over time. The seasonal component is a continuous regular pattern that repeats at fixed time intervals. The residual component represents noise or randomness, describing random fluctuations or unpredictable changes [49].

Long Short-Term Memory

A Long Short-Term Memory (LSTM) neural network is a deep neural network algorithm that transforms the hidden layer nodes into memory cells based on the Recurrent Neural Network (RNN) architecture [50]. To avoid the information loss caused by separating the input sequence variables in traditional Artificial Neural Networks (ANNs) and RNN models without considering the relationships between preceding and subsequent inputs, LSTM uses memory cells to coordinate and propagate the previous input information, continuously increasing the vector transmission process while retaining the state of the input vectors, thus providing the network with "memory function". The core functionality of LSTM lies in using finite-state storage to store and propagate neuron information [51].

Based on this, it introduces input gates, output gates, and forget gates on top of the memory cells, which are used to selectively remember and feedback the error function through the gradient descent optimization. During the forward propagation, the input gate and output gate control the activation flow into and out of the memory cells at each time step, respectively. During the backward propagation, the output gate and input gate control the error flow into and out of the memory cells at each time step, respectively. The forget gate is responsible for discarding information during the propagation process.

3. Results and Analysis

3.1. Spatial Trends of Meteorological Elements in the Heilongjiang (Amur) River Basin

3.1.1. The Spatial Characteristics of Overall Climate Change in the Heilongjiang (Amur) River Basin

Spatial interpolation analysis of temperature and precipitation data for the main body of the Heilongjiang (Amur) River Basin was conducted using the Anusplin interpolation method and linear trend analysis. The spatial distribution of annual average temperature and precipitation in the Heilongjiang (Amur) River Basin is shown in Figure 2a,b. As indicated by Figure 2a,b, the spatial distribution characteristics of the annual average temperature and precipitation in the main body of the Heilongjiang (Amur) River Basin exhibit significant spatial heterogeneity. From the southeast to the northwest of the basin, with the increase in latitude and elevation, as well as the distribution of mountainous terrain within the basin, the annual average temperature gradually decreases, and the annual average precipitation gradually decreases. The annual average temperature in the basin varies within the range of −14.37 to 6.75 °C, and the annual average precipitation varies within the range of 207.97 to 1115.05 mm. In the past 43 years, the annual average temperature in the Heilongjiang (Amur) River Basin was 0.79 °C, and the annual average precipitation was 459.98 mm. Under the combined influence of climate change and geographical factors, the meteorological characteristics of the Heilongjiang (Amur) River Basin exhibit a spatial pattern of relatively warm and humid conditions in the southeast and drier and colder conditions in the northwest of the basin.

The spatial distribution of the annual average temperature change rate in the Heilongjiang (Amur) River Basin over the past 43 years, as shown in Figure 2c, reveals a decreasing trend in temperature in the southeastern part of the basin, while the northwestern part of the basin shows a warming trend. The overall temperature change rate in the basin is 0.50 °C/10a (significant at the $\alpha = 0.01$ level). Spatially, the temperature decrease rate in the southeastern part of the basin ranges from −11.50 to −3.91°C/10a, while the temperature increase rate in the northwestern part ranges from −3.9 to 3.68°C/10a. The high-altitude areas in the northwestern part of the basin exhibit smaller temperature changes compared to the coastal areas in the northeastern part, showing a warming trend. The influence of moist and cold air from the Pacific Ocean invades the southeastern part of the basin, but the presence of mountain ranges such as the Sikhote-Alin and the Greater Khingan weakens the impact of coastal monsoons. The spatial distribution of the annual average precipitation change rate, shown in Figure 2d, indicates a decreasing trend on both the eastern and western sides of the basin, with an increasing trend in the central part. In the past 43 years, the overall annual average precipitation change rate in the basin was 15.18 mm/10a. Being spatially influenced by latitude and elevation, the precipitation change rate in the basin gradually decreases from the southern part towards the surrounding areas. Considering the combined trends of temperature and precipitation changes in the basin, there is an overall trend of "warming and humidification," but there are spatial variations within the basin due to the influence of terrain and elevation factors. Therefore, the basin as a whole was divided into eight sub-basins to further analyze the characteristics of meteorological changes in each of its sub-basins.

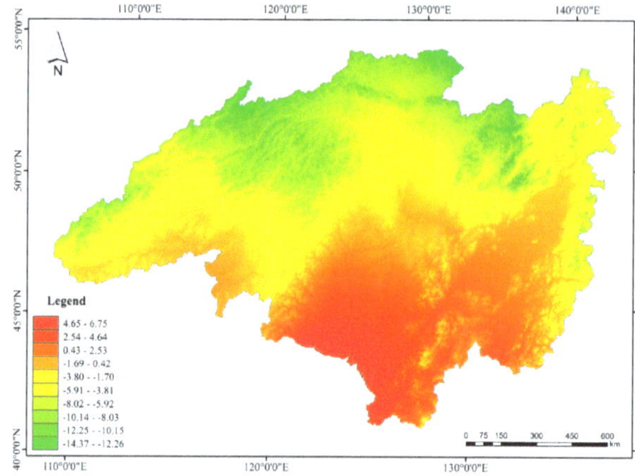

(**a**)

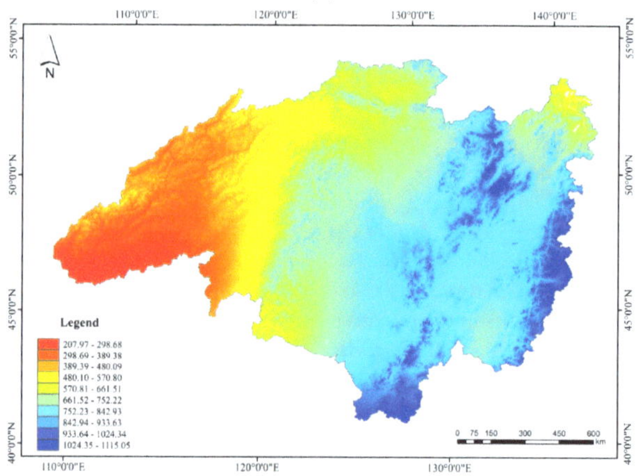

(**b**)

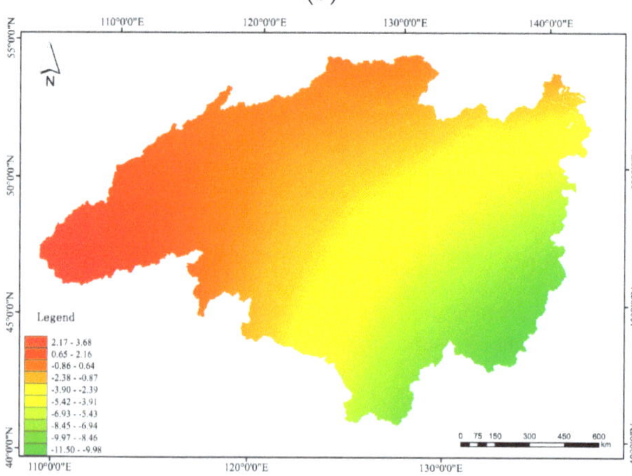

(**c**)

Figure 2. *Cont.*

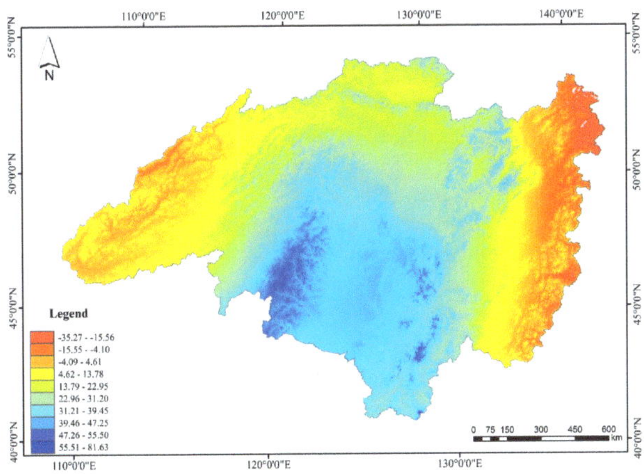

(d)

Figure 2. Characteristics of spatial distribution of temperature and precipitation changes in the Heilongjiang (Amur) River Basin, 1980–2022. (**a**) spatial distribution of mean annual temperature (in °C); (**b**) spatial distribution of mean annual precipitation (in mm); (**c**) spatial distribution of rate of change in mean annual temperature (in °C/10a); (**d**) spatial distribution of rate of change in mean annual precipitation (in mm/10a).

3.1.2. Spatial Characteristics of Climate Change in the Sub-Watersheds of the Heilongjiang (Amur) River Basin

The spatial distribution of annual average temperature and precipitation in the sub-watersheds of the Heilongjiang (Amur) River Watershed can be seen in Figure 3a,b. The mean values and trends of annual average temperature and precipitation in each sub-watershed are shown in Table 1. From Figure 3a,b and Table 1, it can be observed that there are significant differences in the annual average temperature and precipitation among the sub-watersheds of the Heilongjiang (Amur) River Watershed. Specifically, the temperature displays a distribution pattern of decreasing from the southeastern to the northwestern part of the watershed, with the lowest temperature centered in the northern part of the Shilka River, Zeya River, and Bureya River. For example, the annual average temperature in the Songhua River sub-watershed in the southeastern part is 3.54 °C, while it decreases to −3.73 °C in the Bureya River sub-watershed in the northern part, resulting in a temperature difference of 7.27 °C between the two sub-watersheds. The spatial distribution of precipitation among the sub-watersheds within the watershed is uneven, with a gradual decrease from the southeastern coastal regions to the northwestern plateau areas. The peak of precipitation is located in the southeastern coastal sub-watersheds, including the Songhua River, Wusuli River, and the main stream of the Heilongjiang River. Among them, the Songhua River sub-watershed in the southeastern part has an annual average precipitation of 641.79 mm, which is nearly twice the annual average precipitation in the Shilka River sub-watershed (339.8 mm) in the northwestern plateau region.

The spatial distribution of the annual average temperature change rates in various sub-basins of the Heilongjiang (Amur) River Basin, as shown in Figure 3c, indicates that all sub-basins are experiencing a significant warming trend (with all passing the significance level test of $\alpha = 0.01$). However, there are differences in the warming rates among the sub-basins, showing an increasing trend from the southeastern coastal basins to the northwestern plateau basins. For example, the warming rates in the Bureya River Basin, Shilka River Basin, and Jeya River Basin are all greater than 0.5 °C/10a. The Bureya River Basin has a warming rate nearly 4.5 times that of the Ussuri River Basin, which has the smallest warming rate. The spatial distribution of the annual average precipitation change rates in various sub-basins of the Heilongjiang (Amur) River Basin, as shown in Figure 3d, indicates that the precipitation change rates in the basin exhibit an increasing trend in the southern and central basins and a decreasing trend in the eastern and western basins. For instance, the Songhua River and Zeya River Basins show a significant increasing trend in precipitation, with change rates of 91.57 mm/10a and 38.52 mm/10a, respectively (both passing the significance level test of $\alpha = 0.01$). In contrast, the Wusuli River, Amgun River, and Bureya River Basins, located in the eastern part, exhibit a decreasing trend in precipitation, with change rates of −11.15 mm/10a, −26.04 mm/10a, and −8.10 mm/10a, respectively. In analyzing the spatial distribution and change rate distribution of the annual average temperature and precipitation in various sub-basins, it can be observed that the warming trend in temperature is more significant than the increasing trend in precipitation across the sub-basins.

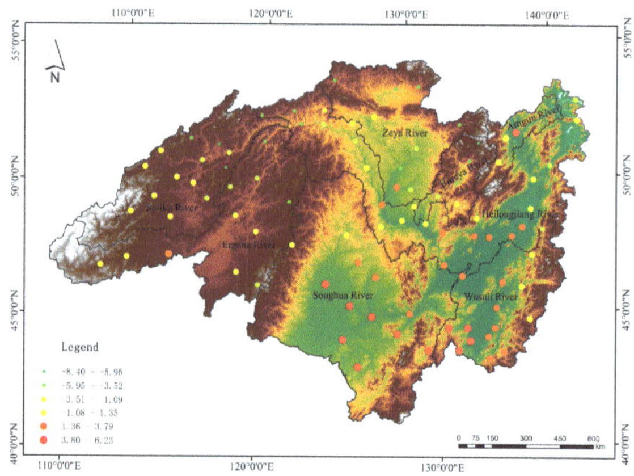

(**a**)

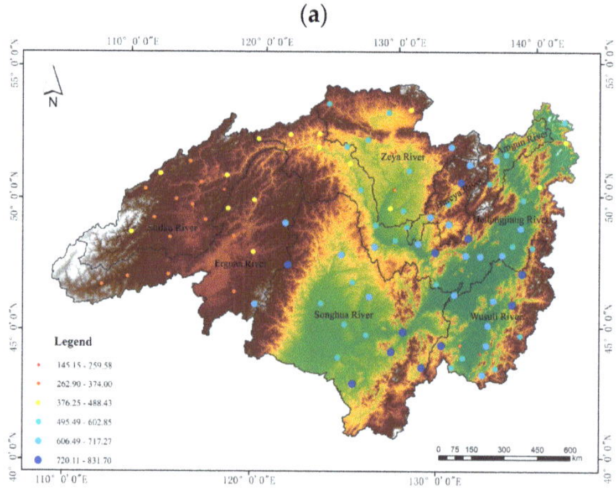

(**b**)

Figure 3. *Cont.*

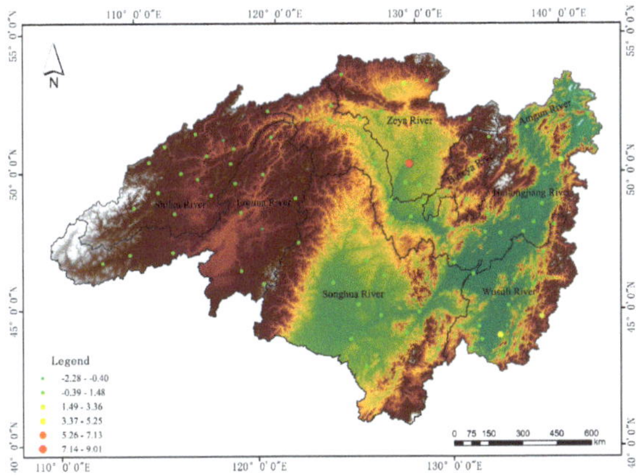

(**c**)

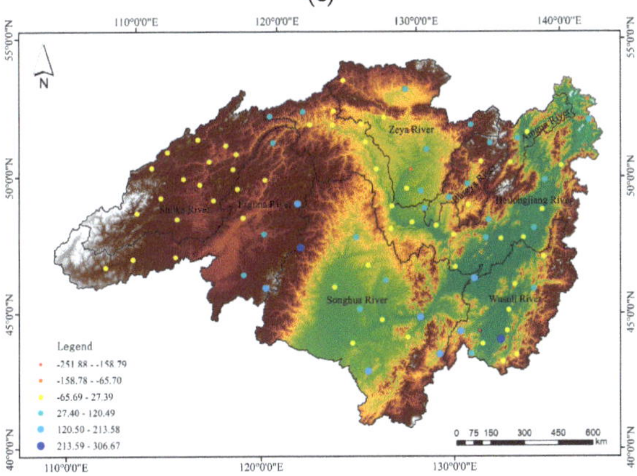

(**d**)

Figure 3. Characteristics of spatial distribution of temperature and precipitation changes in each sub-basin of the Heilongjiang (Amur) River Basin, 1980–2022. (**a**) Spatial distribution of mean annual temperature (°C) in each basin; (**b**) spatial distribution of mean annual precipitation (mm) in each basin; (**c**) spatial distribution of the rate of change in mean annual temperature in each basin (°C/10a); (**d**) spatial distribution of the rate of change in mean annual precipitation in each basin (mm/10a).

Table 1. Mean values and trends of mean annual temperature and precipitation in the basins of the Heilongjiang (Amur) River Basin.

Watershed	Mean Annual Temperature/°C	Mean Annual Precipitation/mm	Temperature Change Rate /(°C·(10a)$^{-1}$)	Temperature Change M-K Statistic Value Z	Precipitation Change Rate /(mm·(10a)$^{-1}$)	Precipitation Change M-K Statistic Value Z
Erguna River	−1.39	397.13	0.37 **	3.06	28.06	1.09
Shilka River	−1.66	339.80	0.66 **	4.33	4.33	0.15
Songhua River	3.54	641.79	0.28 **	3.20	91.57 **	2.70
Wusuli River	2.62	492.46	0.20 **	2.89	−11.15 *	−0.57
Amgun River	1.29	424.79	0.31 **	3.04	−26.04	−0.08
Bureya River	−3.73	465.09	0.89 **	4.11	−8.10	−0.21
Zeya River	−2.37	552.56	0.66 **	3.30	38.52 **	3.47
Heilongjiang River	−0.04	501.51	0.54 **	4.69	1.22	0.50
Heilongjiang (Amur) River	0.79	459.98	0.50 **	4.75	15.18	1.32

Note: * denotes passing the α = 0.05 significance level test and ** denotes passing the α = 0.01 significance level test.

3.2. Seasonal Characteristics of Meteorological Element Variation in the Heilongjiang (Amur) River Watershed

3.2.1. Seasonal Division of Sub-Watersheds

In previous studies on the seasonal characteristics of meteorological elements in regions, the seasons were often divided into spring, summer, autumn, and winter, and the characteristics of meteorological elements in each season were discussed separately. However, for the Heilongjiang (Amur) River Basin, in the middle and high latitudes, most areas within the basin experience a long and cold winter with stronger seasonality in precipitation. Taking Harbin, China, as an example, the winter season can last for more than half a year, and summer precipitation accounts for more than half of the annual total. In common seasonal divisions in high latitudes, the winter season only includes December and the following January and February, which may not adequately represent the winter conditions in the regions through which the basin flows. Additionally, most vegetation and crops require environments with temperatures above 0 °C for growth. To highlight the variations in meteorological elements in different seasons, the seasons were divided into growing seasons and non-growing seasons. Using daily temperature data from 223 meteorological stations near the Heilongjiang (Amur) River Basin from 1980 to 2022, with 0 °C as a threshold, the inflection points where the daily average temperature is continuously higher (or lower) than 0 °C were identified as the start (end) time of the growing season. The length of the growing season (from start to end) was then subjected to spatial interpolation analysis. Based on this, the annual variations in meteorological elements in the Heilongjiang (Amur) River Basin and its sub-basins were classified into growing seasons and non-growing seasons. The seasonal characteristics of meteorological elements in the basin as a whole and in each sub-basin were analyzed. The specific classification results are shown in Table 2.

Table 2. Division of growing and non-growing seasons in the watersheds of the Heilongjiang (Amur) River Basin.

Length of Growing Season	Characteristics of Changes during the Year	Season		Corresponding Watershed
		Growing Season	Non-Growing Season	
≥180 days	Temperature above 0 °C for more than 6 months	March–October	November–February	Songhua River Basin, Wusuli River Basin, Heilongjiang River Basin
120~180 days	Temperature above 0 °C for 4~6 months or above	April–September	October–March	Erguna River Basin, Zeya River Basin, Shilka River Basin
≤120 days	Temperature above 0 °C for more than 3 months	May–August	September–April	Amgun River Basin, Bureya River Basin

3.2.2. Characteristics of Meteorological Element Changes during the Growing Season in Sub-Watersheds

The length of the growing season in the Heilongjiang (Amur) River Basin is influenced by latitude, altitude, and the distribution of temperature and precipitation. The length of the growing season gradually shortens from the southeastern coastal basin to the northwestern plateau basin. The southeastern basin has a growing season length of ≥180 days, while the northwestern basin has a growing season length of ≤120 days. The remaining basins have a growing season length between 120 and 180 days (see Figure 4). The spatial distribution of the average temperature and precipitation during the growing season in the sub-watersheds of the Heilongjiang (Amur) River Watershed can be seen in Figure 4a,b. From Figure 4a,b, it can be observed that the spatial distribution of the average temperature and precipitation during the growing season is generally consistent with the spatial distribution characteristics of the annual average temperature and precipitation, with higher temperatures in the southeastern areas compared to the northwestern regions. The Songhua River sub-watershed in the central and southern part of the watershed has the highest average temperature during the growing season, reaching 14.62 °C. It is located in the Songnen Plain region with a higher vegetation coverage, which has better temperature regulation capabilities and a lower occurrence of extreme weather events compared to the Wusuli River sub-watershed, which is located at the same latitude, resulting in slightly higher temperatures during the growing season. The Amgun River and Bureya River sub-watersheds in the northern part have relatively higher temperatures during the growing season, reaching 12.11 °C. Considering the geographical features of the watersheds, these two sub-watersheds are located in the Bureya mountain range, which to some extent isolates the cold air from the Siberian Plain and the cold and wet monsoon from the Pacific, resulting in slightly higher temperatures during the growing season compared to the Shilka River sub-watershed, which is located at the same latitude and elevation. The center of precipitation during the growing season is located in the Songhua River sub-watershed, with an average precipitation of 505.45 mm. The precipitation during the growing season decreases gradually from the center to the surrounding areas within the watershed.

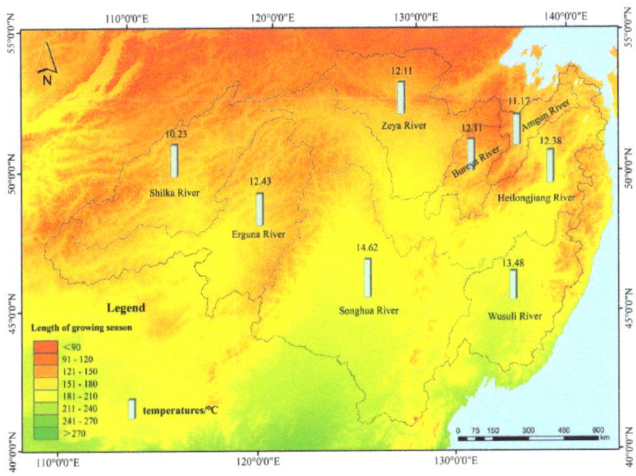

(a)

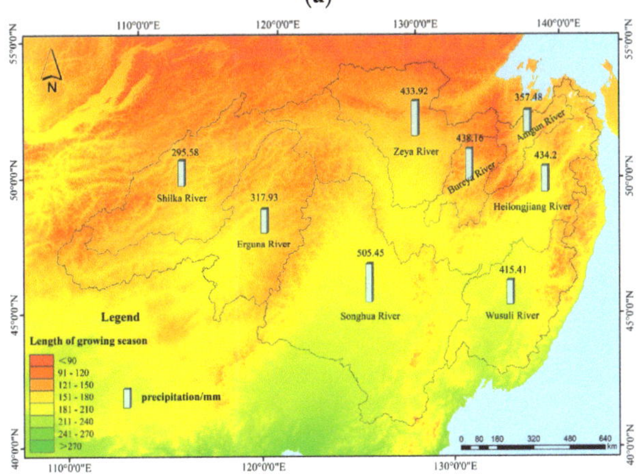

(b)

Figure 4. *Cont.*

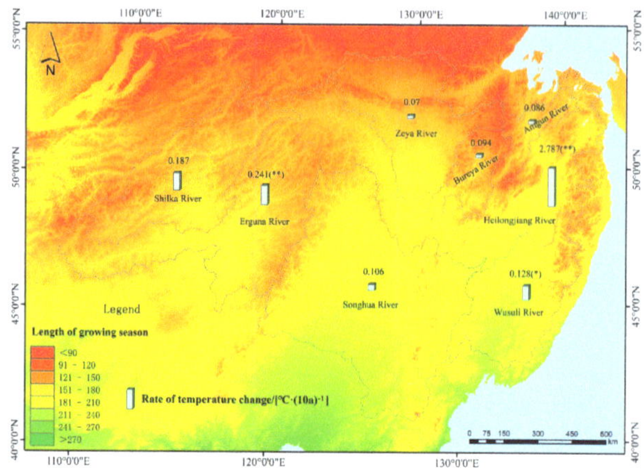

(**c**)

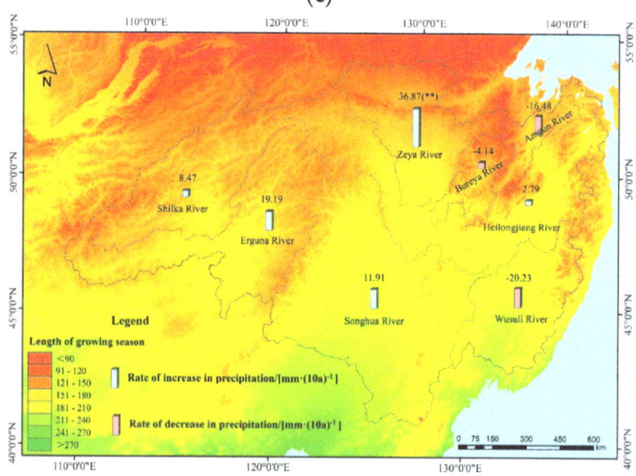

(**d**)

Figure 4. Characteristics of spatial distribution of temperature and precipitation changes during the growing season in sub-basins of the Heilongjiang (Amur) River Basin, 1980–2022. (**a**) Mean temperature (°C); (**b**) mean precipitation (mm); (**c**) mean rate of change in temperature (°C/10a); (**d**) mean rate of change in precipitation (mm/10a). (* denotes passing the α = 0.05 significance level test and ** denotes passing the α = 0.01 significance level test).

Figure 4c shows the spatial distribution of the average temperature change rate during the growing season in the sub-watersheds of the Heilongjiang (Amur) River Watershed. It indicates a significant warming trend in the average temperature during the growing season in each sub-watershed, with the warming amplitude decreasing gradually from east to west within the watershed. Specifically, the areas with relatively large warming rates are the main stream of the Heilongjiang River, Wusuli River, and the Erguna River, with warming rates of 2.787, 0.241, and 0.128 °C/10a, respectively (all passing the significance level test of α = 0.05). The Bureya River, Amgun River, and Zeya River sub-watersheds in the northern part of the watershed have smaller warming amplitudes, with rates of 0.094, 0.086, and 0.07 °C/10a, respectively. The spatial distribution of the growth season average precipitation change rate in various sub-basins of the Heilongjiang (Amur) River Basin, as depicted in Figure 4d, reveals that the growth season precipitation in the central and western sub-basins of the basin shows varying degrees of increasing trends. Among them, the Zeya River Basin exhibits the most significant increase at a rate of 36.87 mm/10a (passing the significance level test of $\alpha = 0.01$). The observed increase in the growth season precipitation in this basin may be associated with the global climate warming, leading to an augmentation of growth season precipitation. Conversely, the sub-basins located in the eastern part of the basin, including the Wusuli River Basin, Bureya River Basin, and Amgun River Basin, exhibit a decreasing trend in growth season precipitation. The reduction magnitude spatially decreases from the coastal to the inland areas. When comparing the spatial trends in the annual average precipitation and growth season precipitation in various sub-basins of the Heilongjiang (Amur) River Basin, it is evident that there is a high degree of similarity between the changes in growth season precipitation and annual precipitation in the basin. This further confirms that the variation in growth season precipitation in the Heilongjiang (Amur) River Basin plays a dominant role in its interannual precipitation changes.

3.2.3. Characteristics of Meteorological Element Changes during the Non-Growing Season in Sub-Watersheds

The spatial distribution of average temperature and precipitation during the non-growing season in the sub-watersheds of the Heilongjiang (Amur) River Watershed can be seen in Figure 5a,b. From Figure 5a,b, it can be observed that the average temperature during the non-growing season gradually decreases from the southeastern coastal regions to the northwestern sub-watersheds. The center of low temperatures is located in the Bureya River sub-watershed in the northern part, while the center of high temperatures is in the Wusuli River sub-watershed, with a temperature difference of 5.76 °C between the two. Comparing the spatial distribution characteristics of annual average precipitation and non-growing season precipitation, the spatial distribution characteristics of non-growing season precipitation are generally consistent with the distribution of annual average precipitation. The main pattern is a decrease in precipitation from the southeastern to the northwestern sub-watersheds, with the peak of precipitation located in the Songhua River sub-watershed and the center of low values in the Shilka River sub-watershed, with a difference of 72.3 mm in non-growing season average precipitation between the two.

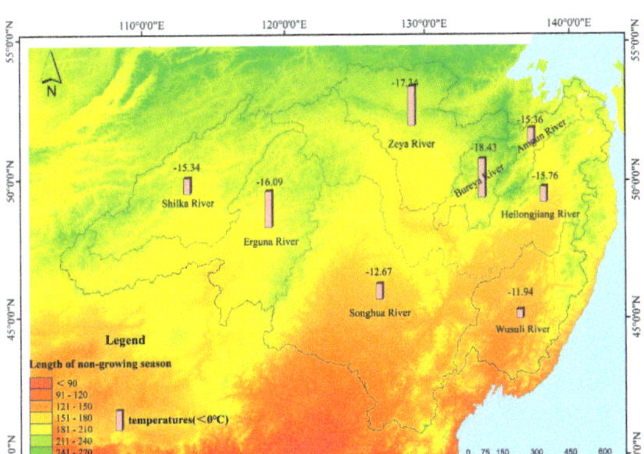

(a)

Figure 5. *Cont.*

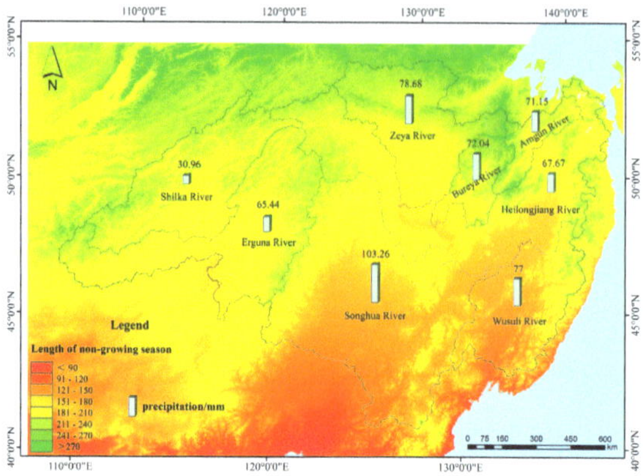

(b)

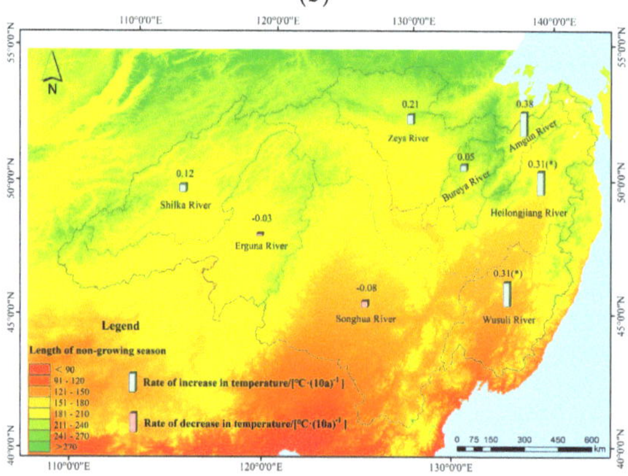

(c)

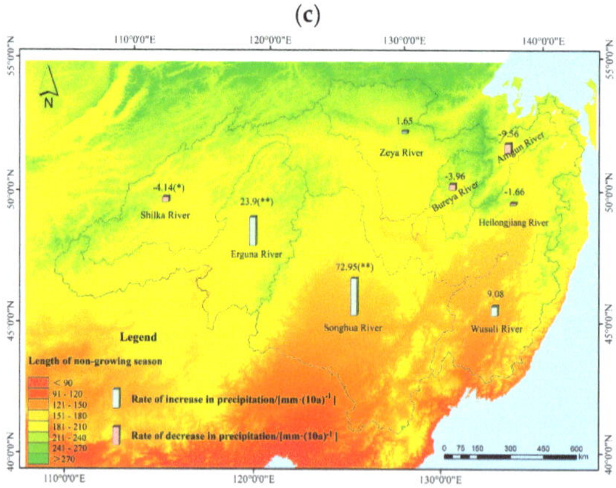

(d)

Figure 5. Characteristics of spatial distribution of temperature and precipitation changes in the non-growing season in sub-basins of the Heilongjiang (Amur) River Basin, 1980–2022. (**a**) Mean temperature (°C); (**b**) mean precipitation (mm); (**c**) mean rate of change in temperature (°C/10a); (**d**) mean rate of change in precipitation (mm/10a). (* denotes passing the α = 0.05 significance level test and ** denotes passing the α = 0.01 significance level test).

Figure 5c shows the spatial distribution of the average temperature change rate during the non-growing season in the sub-watersheds of the Heilongjiang (Amur) River Watershed. Unlike the Songhua River and Erguna River sub-watersheds, which show a weak decreasing trend, the other six sub-watersheds all exhibit a warming trend during the non-growing season, with higher warming rates in the coastal regions compared to the northwestern regions of the watershed. The Amgun River sub-watershed has the highest warming rate during the non-growing season, reaching 0.38 °C/10a, while the Bureya River sub-watershed has a relatively smaller warming rate of only 0.05 °C/10a. When considering variations in latitude and distance from the ocean, the impact of elevation on the rate of temperature changes should also be considered. When comparing the temperature changes during the growing and non-growing seasons in the sub-watersheds, it is evident that the contribution of the temperature increase during the non-growing season to interannual warming variation in the watershed is greater than that of the temperature increase during the growing season. Figure 5d illustrates the spatial distribution of the average precipitation change rate during the non-growing season in the sub-watersheds of the Heilongjiang (Amur) River Watershed. There are diverse characteristics of precipitation changes during the non-growing season in each sub-watershed. Specifically, the Songhua River sub-watershed in the southern part shows the most significant increase in precipitation, reaching 72.95 mm/10a. The Erguna River, Wusuli River, and Zeya River sub-watersheds also exhibit varying degrees of increasing trends in precipitation. On the other hand, the non-growing season precipitation in the Shilka River, Amgun River, and Bureya River sub-watersheds, which are located at higher elevations, shows a decreasing trend. In analyzing the factors influencing the increase in non-growing season precipitation in the sub-watersheds, apart from the usual monsoon effects, the significant agricultural areas in the Songhua River, Wusuli River, and Erguna River sub-watersheds should also be considered. Artificial snowfall is often used during spring planting to increase precipitation in the non-growing season.

3.3. *Analysis of Abrupt Changes in Meteorological Elements in the Heilongjiang (Amur) River Watershed*

3.3.1. Analysis of Overall Climate Change in the Heilongjiang (Amur) River Watershed

The Mann–Kendall test, cumulative departure method, and Pettitt test were used to analyze the abrupt changes in average temperature and precipitation data in the Heilongjiang (Amur) River Watershed. The results of temperature and precipitation changes in the Heilongjiang (Amur) River Watershed can be seen in Figures 6 and 7 and Table 3, with a significance level of 0.01 chosen for the Mann–Kendall test. From Figure 6a, it can be observed that the UF statistic was negative from 1983 to 1985, but there was no significant cooling trend. Since 1992 especially, the UF statistic has increased rapidly (significant at the 0.01 significance level), indicating a significant increase in the overall temperature of the watershed during this period. The UF and UB statistics intersected in 1992. It can be seen that the different methods of abrupt change analysis yielded different years of change. In the analysis of overall temperature changes, the cumulative departure method indicated an abrupt temperature change in 2001. The UF and UB curves in the Mann–Kendall test intersected in 1992. The Pettitt test identified the largest $|U|$ value at time T0, indicating an abrupt temperature change in 2001 (significant at the $p \leq 0.5$ level). Considering the results of the three methods, it can be concluded that the year with the most abrupt temperature change in the Heilongjiang (Amur) River Watershed was 2001. Analyzing Figure 7a, it can be observed that except for 1981, the UF statistic was negative from 1982 to 2018, indicating significant precipitation reductions during two periods: 1985–1990 and 2004–2012. Since 2018, the UF statistic has been positive, indicating an increasing trend in precipitation during this period. In the analysis of overall precipitation changes, the cumulative departure method indicated an abrupt change in precipitation in 2011. The UF and UB curves in the Mann–Kendall test intersected in 2018. The Pettitt test identified the largest $|U|$ value in 2011, indicating an abrupt change in precipitation in 2011 (significant at the $p \leq 0.5$ level).

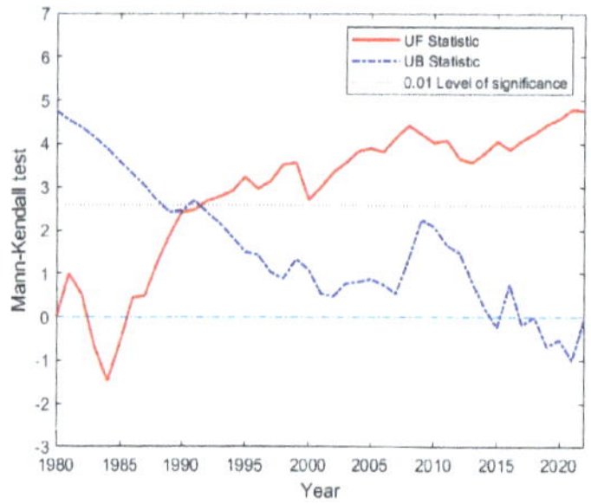

(**a**)

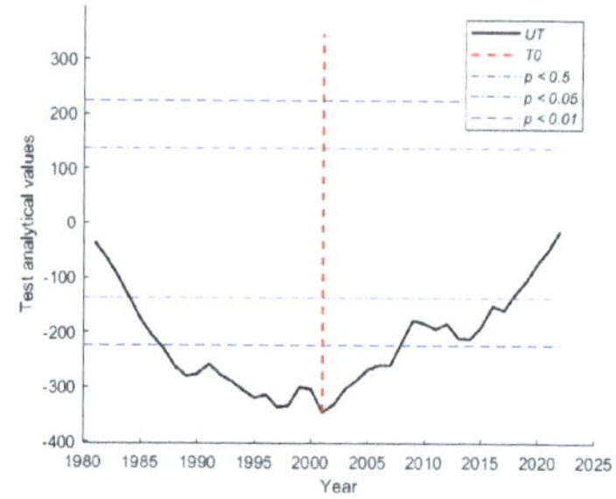

(**b**)

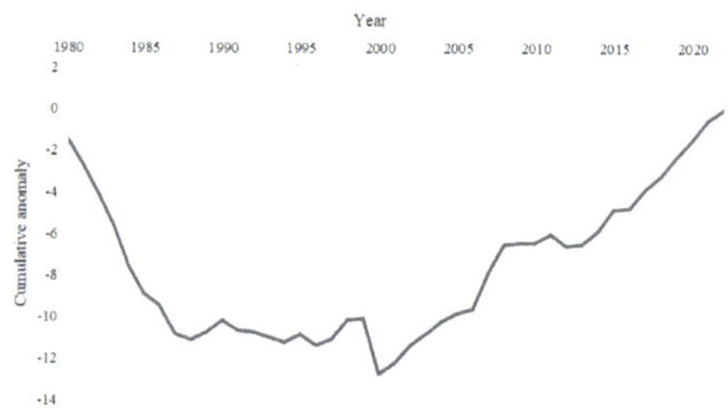

(**c**)

Figure 6. Analysis of sudden changes in mean air temperature in the Heilongjiang (Amur) River Basin, 1980–2022. (**a**) Mann–Kendall trend test; (**b**) Pettitt test; (**c**) cumulative anomaly method.

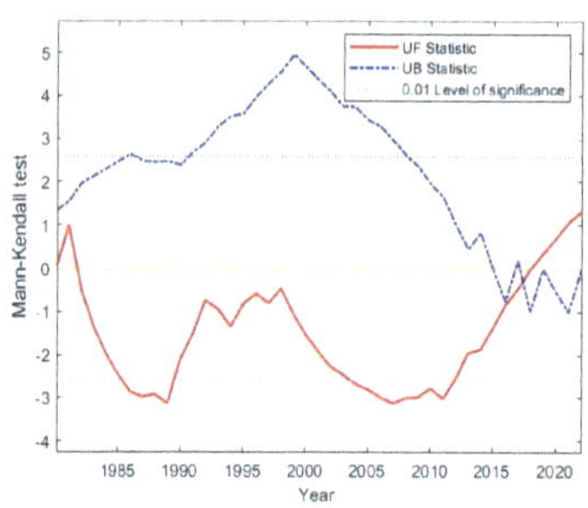

(**a**)

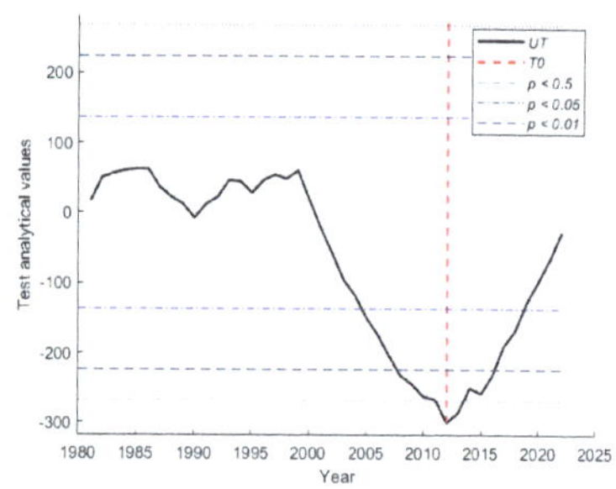

(**b**)

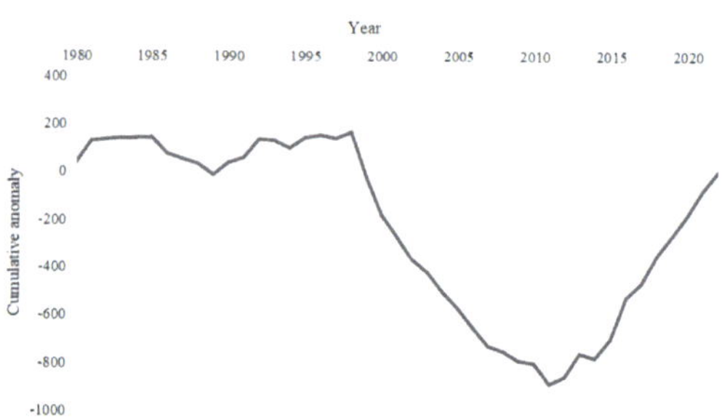

(**c**)

Figure 7. Analysis of sudden changes in mean precipitation in the Heilongjiang (Amur) River Basin, 1980–2022. (**a**) Mann–Kendall trend test; (**b**) Pettitt test; (**c**) cumulative anomaly method.

Table 3. Results of sudden change analysis of mean temperature and precipitation in the Heilongjiang (Amur) River Basin, 1980–2022.

	Year of Change			Determination of the Year of Change
	Mann–Kendall Test	Cumulative Anomaly Method	Pettitt Test	
Average temperature	1992	2001	2001	2001
Average precipitation	2018	2012	2012	2012

3.3.2. Analysis of Climate Change in Sub-Watersheds of the Heilongjiang (Amur) River Watershed

Regarding the analysis process of abrupt changes in the overall average temperature and precipitation in the watershed, the results obtained from the three methods are shown in Tables 4 and 5. From Table 4, it can be observed that the years of abrupt temperature changes in the southern sub-watersheds were slightly earlier than those in the northwestern plateau region. The Wusuli River sub-watershed experienced the earliest abrupt temperature change in 1988, while the Amgun River sub-watershed experienced the latest abrupt temperature change in 2007, likely due to topographical factors. Considering the years of abrupt temperature changes in the overall watershed, it can be seen that the larger sub-watersheds had a greater impact on the overall abrupt temperature change in the watershed. From Table 5, it can be observed that except for the Shilka River sub-watershed located in the Mongolian Plateau, the other sub-watersheds experienced an abrupt change in precipitation in 2013, leading to a large-scale flood event in the Heilongjiang (Amur) River Watershed. Specifically, the downstream area experienced a rare flood event that occurs once every 100 years. The early northward movement of the subtropical high, the eastward position and strong intensity of the East Asian trough, and frequent activity of the Northeast Cold Vortex contributed to a significant increase in precipitation during the growing season of 2013.

Table 4. Results of sudden change analysis of mean air temperature in sub-basins of Heilongjiang (Amur) River, 1980–2022.

Watershed	Year of Change			Determination of the Year of Change
	Mann–Kendall Test	Cumulative Anomaly Method	Pettitt Test	
Shilka River	1994, 2001	2001	2001	2001
Erguna River	1999	1999	2000	1999
Songhua River	1989, 2013	1989	1989	1989
Wusuli River	1988	1988	1989	1988
Amgun River	2007	2006	2007	2007
Bureya River	2004	2004	2005	2004
Zeya River	1994	1994	1995	1994
Heilongjiang River	2002, 2009	2002	2003	2002

Table 5. Results of sudden change analysis of mean precipitation in sub-basins of Heilongjiang (Amur) River, 1980–2022.

Watershed	Year of Change			Determination of the Year of Change
	Mann–Kendall Test	Cumulative Anomaly Method	Pettitt Test	
Shilka River	2013, 2018	2018	2018	2018
Erguna River	2013, 2021	2013	2013	2013
Songhua River	2013	2013	2013	2013
Wusuli River	2013	2013	2012	2013
Amgun River	2013	2013	2013	2013
Bureya River	2013, 2019	2013	2013	2013
Zeya River	2012, 2021	2013	2013	2013
Heilongjiang River	2013	2013	2013	2013

3.4. Precipitation Value Prediction for the Heilongjiang (Amur) River Basin

Using the RNN algorithm model, STL Decomposition model, and LSTM model, precipitation prediction models were established for the Heilongjiang (Amur) River Basin. A monthly precipitation dataset from 1980 to 2022, consisting of 516 months, was selected for the analysis. The precipitation amounts for the last 50 months of the dataset were predicted. The fitting and prediction results for the 50-month period are shown in Figure 8. The simulation results of the RNN model show that the predicted values exhibit a similar trend to the observed values. While most of the predicted points coincide with the observed values, there are significant errors and displacement in predicting the peaks and valleys compared to the actual situation. Regarding the simulation results of the STL model, the predicted values generally have a small deviation from the observed values. There is a slight displacement in the periods where precipitation increases or decreases. The predicted precipitation shows a gradual decrease during the growing season, and the increase in precipitation is more concentrated during the non-growing period, with the peak precipitation point being relatively stable. The LSTM model accurately predicted the periodic variations in monthly precipitation, with a trend that matches the actual data and no significant displacement. The predicted values have small errors compared to the observed values, and the peaks and valleys were accurately predicted without overestimating or underestimating. Figure 9 shows the scatter plots of the prediction results from different models. When comparing the scatter distributions of the prediction results from the three models, it can be observed that the LSTM model's predicted values are closer to the true values, indicating higher prediction accuracy. The R and NSE values of the LSTM model are 0.9734 and 0.9449, respectively. The RNN and STL models generally underestimate the precipitation compared to the observed values, with R values above 0.9 but NSE values of 0.8939 and 0.9445, respectively. The fit is slightly inferior to that of the LSTM model. The order of prediction accuracy, from high to low, is LSTM, STL, and RNN. The LSTM model has the lowest NRMSE value of 0.0628 and the highest R and NSE values among the three models. Based on these results, it can be concluded that the LSTM model offers the best prediction performance for the Heilongjiang (Amur) River Basin.

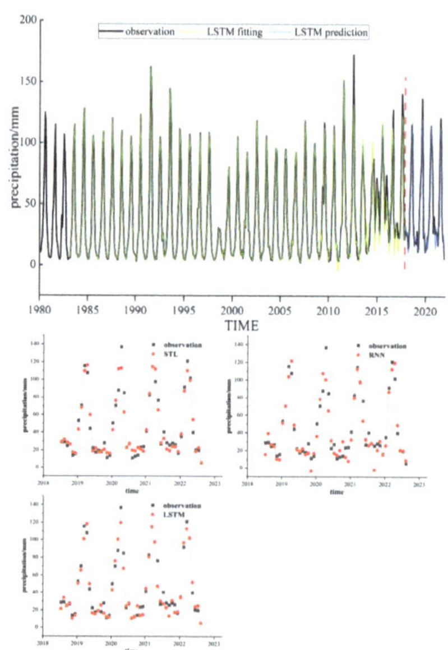

Figure 8. Precipitation prediction results of RNN, STL Decomposition, and LSTM models for Heilongjiang (Amur) River Basin. The image behind the red dashed line shows the monthly precipitation projections for the 50 months following the simulated time series.

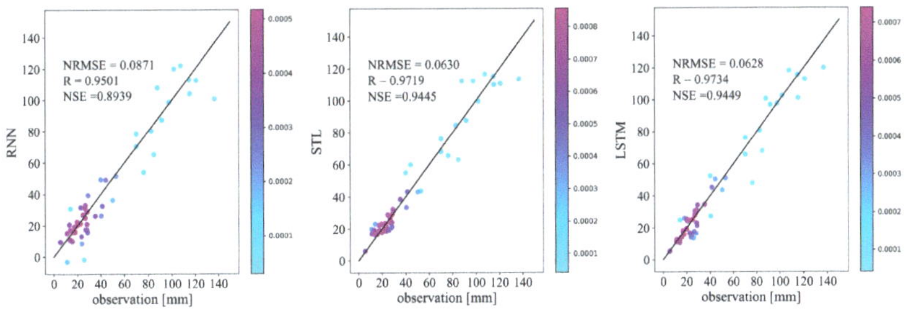

Figure 9. Scatterplot of precipitation prediction results of RNN, STL Decomposition, and LSTM model models for Heilongjiang (Amur) River Basin.

4. Discussion

(1) Over the past 43 years, the Heilongjiang (Amur) River Basin has exhibited a significant warming trend, with a warming rate of 0.50 °C/10a, surpassing the global average temperature increase rate of 0.12 °C/10a [52]. The primary factors contributing to global warming include increased population density, soil desertification, and the substantial use of carbon dioxide-producing fuels in human production and daily life, leading to elevated concentrations of greenhouse gases in the atmosphere. The retreat of glaciers and the thawing of permafrost in mid–high latitude regions further exacerbate the warming trend, creating an irreversible cycle of warming. In the Heilongjiang (Amur) River Basin, the spatial distribution of temperatures exhibits a pattern of decreasing temperatures from the southeastern to the northwestern regions. The warming amplitude gradually increases from the southeastern coastal region towards the northwestern plateau region, with higher elevations experiencing greater warming than lower elevations. During the non-growing season, the warming amplitude in the Heilongjiang (Amur) River Basin is higher than that of the growing season, indicating that the non-growing season temperature increase contributes more significantly to the interannual warming variability in the basin.

(2) Over the past 43 years, the overall annual average precipitation in the Heilongjiang (Amur) River Basin has exhibited a significant increasing trend, with an increase rate of 15.18 mm/10a. Compared to temperature changes, the causes of precipitation variations are more complex and diverse. They are influenced not only by monsoon variations but also by atmospheric circulation anomalies such as the West Pacific Subtropical High and the Northeast Cold Vortex [53]. Precipitation in the Heilongjiang (Amur) River Basin decreases sequentially from the southeastern coastal area towards the northwestern plateau region. The precipitation change rate on the higher-altitude eastern and western sides of the basin shows a decreasing trend, while it increases in the central region. Influenced by latitude and altitude, the precipitation change rate in the basin gradually decreases from the southern region towards the surrounding areas. The increase in precipitation during the growing season is greater than that during the non-growing season. This is closely related to the annual distribution and seasonal variation in precipitation in mid–high latitude regions, where the growing season (April to September) contributes to 85% of the annual precipitation. Concurrently, precipitation changes are also influenced by global temperature variations. Global warming leads to increased precipitation in continental areas, especially in mid–high latitude regions, resulting in imbalanced precipitation distribution and an increased risk of extreme weather events such as floods and droughts.

(3) Due to climate variations influenced by monsoon factors, anomalies in high-pressure systems, changes in solar radiation, and other factors, the Heilongjiang (Amur) River Basin experiences different years of abrupt changes in temperature and precipitation. The overall temperature in the basin underwent an abrupt change in 2001, while precipitation exhibited an abrupt change in 2012. The years of abrupt temperature

changes in the southern sub-basins are slightly earlier than those in the northern plateau region. Except for the Shilka River Basin located on the Mongolian Plateau, all other sub-basins experienced an abrupt change in precipitation in 2013. The different years of abrupt precipitation changes in the Shilka River Basin to some extent influence the most significant abrupt precipitation change year in the basin, leading to inconsistencies with the precipitation abrupt change years in other sub-basins.

(4) The RNN algorithm model and STL Decomposition simulation exhibited prediction trends that are generally consistent with the observed values, but they still suffered from significant simulation defects characterized by substantial displacement biases. In contrast, the LSTM model's predicted trend aligns well with the actual trend, showing no significant displacement biases. The errors between its predicted values and observed values are relatively small, and it accurately predicted peaks and troughs without overestimating peaks or underestimating troughs. The LSTM model's predicted values are closer to the true values, indicating higher prediction accuracy, with R and NSE values trending towards 1. For the Heilongjiang (Amur) River Basin, the LSTM model demonstrated the optimal predictive performance. Its predicted precipitation values can serve as crucial indicators of flood prevention, and when combined with other factors influencing flood occurrence, it can effectively provide a scientific basis for local flood control and disaster reduction efforts.

(5) In the management of the Heilongjiang (Amur) River Basin, the countries through which the basin flows should collaborate, relying on the hydrological, economic, and social links established by the basin. This collaboration should be based on a comprehensive consideration of geographical, hydrological, demographic, climatic, and climate change factors, as well as economic and social needs and other available water resources among the countries. Together, they should formulate corresponding water resource management treaties [54]. The content of such treaties should cover aspects such as transboundary water resource protection, management, and equitable utilization, ensuring the sustainable development of the basin's ecological environment, rational distribution, and utilization of water resources within the basin, and the balance of human survival and social development needs among the countries in the basin. For example, the climate in the southeastern part of the basin is characterized by a warm and humid trend. Although the climatic conditions are favorable, this region also has a relatively high population density, leading to substantial water demand. Therefore, the allocation of water resources within the basin should consider multiple factors comprehensively.

In the management of the Heilongjiang (Amur) River Basin, the countries running through the basin should join hands to formulate a water resources management treaty based on the hydrological, economic, and social ties established in the basin, taking into account the geography and hydrology, population, climate and climate change, economic and social needs, and other available water sources of the countries, and analyzing the weight of each factor [55]. The content of the treaty should cover the protection, management, and rational use and allocation of transboundary water resources in the basin, so as to ensure the sustainable development of the ecological environment in the basin, the rational allocation and use of water resources in the basin, and the balance between the needs of human survival and social development among the countries in the basin [56].

We believe that the water resources of the basin can be rationally developed and utilized while ensuring its sustainable development and ecological balance. Water resources should be proportionally allocated according to the total water resources of the basin in the four countries, taking into account the regional population density, the degree of industrialization, and the degree of agricultural agglomeration, to form a water sharing model [57]. For the Nile Basin, which is also an international river with a basin spanning 11 countries, water problems, water diplomacy challenges, and other water conflicts are more prominent [58]. With regard to the rational allocation of water resources in the basin, there are years of research results that can be used to provide a case study for the equitable allocation of water resources in the Heilongjiang (Amur) River Basin [59]. Through the development of an inclusive water sharing agreement, multiple transboundary water resource challenges, such as water security, food security, ecological security, etc., can be addressed, and joint co-operation and efforts can be promoted among the countries through which the river flows. In particular, in the management of water resources in the basin, a basin think-tank should be formed, and hydrological results should be exchanged on a regular basis, so as to make due efforts and responsibilities for the protection of water resources in the basin [60]. Last but not least, the development and protection of water resources in the basin is reflected in the surface runoff, and groundwater resources are also an important part of water resources in the basin [61]. The development, use, and protection of groundwater resources in the basin should be given the same importance as surface water. All countries should pay attention to it and share the responsibility of protecting water resources in the basin.

5. Conclusions

This study was based on daily temperature and precipitation data from 1980 to 2022 collected from 282 meteorological stations in and around the Heilongjiang (Amur) River Basin. The spatial distribution characteristics of meteorological elements for the overall basin and each sub-basin's interannual and seasonal variations were analyzed. Additionally, the meteorological variability features of the entire basin and its sub-basins were investigated. Furthermore, a predictive analysis was conducted on the precipitation for the next 50 months in the time series. The following conclusions were drawn:

(1) The southeastern part of the Heilongjiang (Amur) River Basin is relatively warm and humid, while the northwestern region is characterized by dry and cold conditions. The annual warming rate increases gradually from the southeastern coastal area towards the northwestern plateau region. Conversely, the annual increase rate of precipitation decreases from the southern region towards the surrounding areas. Over the period from 1980 to 2022, the overall trend in the Heilongjiang (Amur) River Basin indicated a "warm and humid" development pattern.

(2) There is a spatial distribution pattern in the Heilongjiang (Amur) River Basin where the annual average temperature decreases from the southeastern part to the northwestern part, while the annual average precipitation gradually decreases in the same direction. The annual warming rate increases gradually from the southeastern coastal area towards the northwestern plateau region. The rate of precipitation increase is larger in the central and southern regions, while the eastern and western regions of the basin show a decreasing trend in precipitation.

(3) The length of the growing season gradually shortens from the southeastern part to the northwestern part of the Heilongjiang (Amur) River Basin. The growing season in the northwestern part of the basin is $\leq$120 days, while in the southeastern part, it is $\geq$180 days. The remaining sub-basins have growing season lengths ranging from 120 to 180 days.

(4) The distribution of average temperature and precipitation during the growing season in each sub-basin is generally consistent with the distribution of the average annual temperature and precipitation. The rate of warming decreases gradually from east to west, and the trend of precipitation change during the growing season is consistent with the trend of precipitation change throughout the year. The average temperature during the non-growing season in each sub-basin gradually decreases from the southeastern coastal basin to the plateau basin in the northwest. The center of non-growing season precipitation remains in the central area of the Songhua River Basin, and the trend of non-growing season warming shows a decrease in the plain area and an increase in the plateau area. The increase in non-growing season precipitation is concentrated in the central and southern plain areas, while the decrease in precipitation is mainly concentrated in the northwest plateau area.

(5) Due to the influence of exceptional weather factors and the scope of the basin, different breakpoints are observed in the changes in meteorological elements. The overall temperature in the Heilongjiang (Amur) River Basin experienced a breakpoint in 2001, while precipitation underwent a breakpoint in 2012. In the southern sub-basins, the year of temperature breakpoint is slightly earlier than that in the northern plateau region, and the precipitation breakpoint is concentrated in the year 2013.
(6) The LSTM model performed better in simulating and predicting the phase changes of precipitation peaks and valleys in the basin. It has a higher prediction accuracy, with R and NSE values closer to 1. Therefore, for the Heilongjiang (Amur) River Basin, the LSTM model is considered the optimal model for precipitation prediction.

Author Contributions: Q.Y., methodology; Q.Y., software; Q.Y., validation; Q.Y., formal analysis; Q.Y., investigation; Q.Y., resources; Q.Y., data curation; Q.Y., writing—original draft preparation; Q.Y., writing—review and editing; Q.Y. and Y.Z., visualization; G.Y., supervision; Y.M. and G.Y., project administration; G.Y., funding acquisition. All authors have read and agreed to the published version of the manuscript.

Funding: This study was supported by the Strategic Priority Research Program of the Chinese Academy of Sciences (Grant No. XDA28100105); the Basic Scientific Research Fund of Heilongjiang Provincial Universities (2021-KYYWF-0049); and the Yunnan Provincial Key Laboratory of International Rivers and Transboundary Ecological Security Open Fund (2022KF03).

Data Availability Statement: Publicly available datasets were analyzed in this study. This data can be found here: (1). Available online: https://www.gsi.go.jp/, accessed on 15 September 2023. (2). Available online: http://www.geodata.cn/, accessed on 26 July 2022. (3.) Available online: https://poles.tpdc.ac.cn/zh-hans/, accessed on 5 August 2022. (4). Available online: https://poles.tpdc.ac.cn/zh-hans/, accessed on 15 September 2023. (5). Available online: https://www.ncei.noaa.gov/data/global-summary-of-the-day/archive/, accessed on 1 July 2022.

Acknowledgments: I would like to thank Gao Yu for her guidance on my thesis. I would like to thank Yu Miao and Zhou Yang for their help in writing and revising the thesis. Data support was provided by the National Earth System Science Data Centre, and A Big Earth Data Platform for Three Poles and NOAA are gratefully acknowledged.

Conflicts of Interest: The authors declare no conflict of interest.

References

1. IPCC. *Climate Change 2014: Synthesis Report*; Pachauri, R.K., Meyer, L.A., Eds.; IPCC: Geneva, Switzerland, 2014.
2. IPCC. *Global Warming of 1.5 °C. An IPCC Special Report on the Impacts of Global Warming of 1.5 °C above Pre-Industrial Levels and Related Global Greenhouse Gas Emission Pathways, in the Context of Strengthening the Global Response to the Threat of Climate Change, Sustainable Development, and Efforts to Eradicate Poverty*; Masson-Delmotte, V., Zhai, P., Pörtner, H.-O., Debra, R., Jim, S., Priyadarshi, R.S., Anna, P., Wilfran, M.-O., Clotilde, P., Roz, P., et al., Eds.; Cambridge University Press: Cambridge, UK; New York, NY, USA, 2018.
3. Hao, X.; Xia, J.; Wang, R. Influence of Climate Change on Surface Water Environment. *J. China Hydrol.* **2010**, *30*, 67–72.
4. Yao, H.; Zhou, H. A Review of Transboundary Water Resources in Central Asia. *Resour. Sci.* **2014**, *36*, 1175–1182.
5. World Economic Forum. *Global Risks 2015*, 10th ed.; World Economic Forum: Geneva, Switzerland, 2015.
6. Degefu, D.M.; He, W.; Yuan, L.; Zhao, J.H. Water allocation in transboundary river basins under water scarcity: A cooperative bargaining approach. *Water Resour. Manag.* **2016**, *30*, 4451–4466. [CrossRef]
7. Ercin, A.E.; Hoekstra, A.Y. Water footprint scenarios for 2050: A global analysis. *Environ. Int.* **2014**, *64*, 71–82. [CrossRef] [PubMed]
8. UNESCO. *UN-Water, 2023: United Nations World Water Development Report 2023: Partnerships and Cooperation for Water*; UNESCO: Paris, France, 2023.
9. UNESCO. *UN-Water, 2020: United Nations World Water Development Report 2020: Water and Climate Change*; UNESCO: Paris, France, 2020.
10. Vorosmarty, C.J.; Green, P.; Salisbury, J.; Lammers, R.B. Global water resources: Vulnerability from climate change and population growth. *Science* **2000**, *289*, 284–288. [CrossRef] [PubMed]
11. Kuang, W.; Chen, L.; Liu, J.; Xiang, W.; Chi, W.; Lu, D.; Yang, T.; Pan, T.; Liu, A. Remote sensing-based artificial surface cover classification in Asia and spatial pattern analysis. *Sci. Sin.* **2016**, *46*, 1162–1179. [CrossRef]
12. Li, S.; Liu, W. Spatial Distribution of Population in Russia and Its Evolution. *Econ. Geogr.* **2014**, *34*, 42–49.
13. Feng, Y.; He, D. Research progress on international rivers in Asia. *J. Geogr. Sci.* **2006**, *16*, 271–276. [CrossRef]

14. Yan, B.; Xu, J.; Wang, Y.; Huang, F.; Han, X.; Zhang, L.; Guo, L. Evaluation of spatiotemporal characteristics of precipitation extremes variations in amur river basin. *Appl. Ecol. Environ. Res.* **2019**, *17*, 10655–10669. [CrossRef]
15. Semenov, E.K.; Sokolikhina, N.N.; Tatarinovich, E.V.; Tudrii, K.O. Synoptic conditions of the formation of a catastrophic flood on the Amur River in 2013. *Russ. Meteorol. Hydrol.* **2014**, *39*, 521–527. [CrossRef]
16. Kalugin, A.S.; Motovilov, Y.G. Runoff formation model for the Amur River basin. *Water Resour.* **2018**, *45*, 149–159. [CrossRef]
17. Li, M.; Yang, D.; Hou, J. Distributed hydrological model of Heilongjiang River basin. *J. Hydroelectr. Eng.* **2021**, *40*, 65–75.
18. Gelfan, A.N.; Kalugin, A.S.; Motovilov, Y.G. Assessing Amur water regime variations in the XXI century with two methods used to specify climate projections in river runoff formation model. *Water Resour.* **2018**, *45*, 307–317. [CrossRef]
19. Zhang, W.; Wang, J. Flood Monitoring of Heilongjiang River Basin in China and Russia Transboundary Region based on SAR Backscattering Characteristics. *J. Geo-Inf. Sci.* **2022**, *24*, 802–813.
20. Jia, L. *Joint Development System on Northeast International Rivers of China*; China Law Society Association of Environment and Resources Law (Kunming University of Science and Technology): Kunming, China, 2009; p. 7.
21. He, D.M.; Liu, C.M.; Feng, Y.; Hu, J.M.; Ji, X.; Li, Y.G. Progress and perspective of international river researches in China. *Acta Geogr. Sin.* **2014**, *69*, 1284–1294.
22. Panova, A.A. The problem of international transboundary water use on the example of the Amur River. In Proceedings of the International Scientific and Practical Conference "Ural Mining School-Regions", Yekaterinburg, Russia, 9–18 April 2018; p. 862.
23. Yu, L.L.; Xia, Z.Q.; Li, J.K.; Cai, T. Climate change characteristics of Amur River. *Water Sci. Eng.* **2013**, *6*, 131–144.
24. Tachibana, Y.; Oshima, K.; Ogi, M. Seasonal and interannual variations of Amur River discharge and their relationships to large-scale atmospheric patterns and moisture fluxes. *J. Geophys. Res. Atmos.* **2008**, *113*. [CrossRef]
25. Mokhov, I.I. Hydrological anomalies and tendencies of change in the basin of the Amur River under global warming. *Dokl. Earth Sci.* **2014**, *455*, 459–462. [CrossRef]
26. Zhou, S.; Zhang, W.; Guo, Y. Impacts of climate and land-use changes on the hydrological processes in the Amur River Basin. *Water* **2019**, *12*, 76. [CrossRef]
27. Jia, S.; Yang, Y. Spatiotemporal Characteristics and Driving Factors of Land-Cover Change in the Heilongjiang (Amur) River Basin. *Remote Sens.* **2023**, *15*, 3730. [CrossRef]
28. Dai, C.; Li, Z.; Lin, L.; Peng, C.; Xie, Y.; Cao, W. *Probing for Water Problems in the Basin of Heilongjiang (Amur) River*; Heilongjiang Education Press: Harbin, China, 2014; pp. 10–12.
29. Guo, J. *Hydrography of the Heilongjiang Basin*; New Knowledge Publishing House: Shanghai, China, 1958.
30. Dai, C.; Wang, S.; Li, Z.; Zhang, Y.; Gao, Y.; Li, C. Review on hydrological geography in Heilongjiang River Basin. *Acta Geogr. Sin.* **2015**, *70*, 1823–1834.
31. Chun, J.; Zhang, X.; Huang, J.; Zhang, P. A gridding method of redistributing population based on POIs. *Geogr. Geo-Inf. Sci.* **2018**, *34*, 124.
32. Geospatial Information Authority of Japan. Global Map Global Version [EB/OL]. Available online: https://www.gsi.go.jp/ (accessed on 15 September 2023).
33. National Earth System Science Data Center. Global Administrative District Boundaries Data [EB/OL]. 2016. Available online: http://www.geodata.cn/ (accessed on 15 September 2023).
34. A Big Earth Data Platform for Three Poles. Global River and Lake Vector Dataset Data [EB/OL]. 2010. Available online: https://poles.tpdc.ac.cn/zh-hans/ (accessed on 15 September 2023).
35. NOAA—National Centers for Environmental Information. Global Summary of The Day Data [EB/OL]. Available online: https://www.ncei.noaa.gov/data/global-summary-of-the-day/archive/ (accessed on 5 October 2023).
36. Mu, X.M.; Zhang, X.Q.; Gao, P.; Wang, F. Theory of Double Mass Curves and Its Applications in Hydrology and Meteorology. *J. China Hydrol.* **2010**, *30*, 47–51.
37. Tu, Q.; Wang, J.; Ding, Y.; Shi, H. *Probability Statistics for Meteorological Applications*; China Meteorological Press: Beijing, China, 1984; pp. 445–470.
38. Liu, Z.H.; Li, L.T.; McVicar, T.R.; Van Niel, T.G.; Yang, Q.K.; Li, R. Introduction of the Professional Interpolation Software for Meteorology Data: Anusplin. *Meteorol. Mon.* **2008**, *34*, 92–100.
39. Wei, F. *Modern Techniques for Statistical Diagnosis and Prediction of Climate*, 2nd ed.; China Meteorological Press: Beijing, China, 2007; pp. 105–142.
40. Zhang, D.; Cong, Z.; Ni, G. Comparison of three Mann-Kendall methods based on the China's meteorological data. *Adv. Water Sci.* **2013**, *24*, 490–496.
41. Weber, K.; Stewart, M. A critical analysis of the cumulative rainfall departure concept. *Ground Water* **2004**, *42*, 935. [PubMed]
42. Zhang, Y.; Song, X. Techniques of abrupt change detection and trends analysis in hydroclimatic time-series: Advances and evaluation. *Arid Land Geogr.* **2015**, *38*, 652–665.
43. Chen, L.; Wang, Y.M.; Chang, J.X.; Wei, J. Chatacteristics and Variation Trends of Seasonal Precipitation in the Yellow River Basin. *Yellow River* **2016**, *38*, 8–12+16.
44. Nan, Z. *A New Sample Expanding Method Based on EMD and Its Application in the Field of Hydrometeorology*; Chang'an University: Xi'an, China, 2020.
45. Meng, X.; Zhang, S.; Zhang, Y. The temporal and spatial change of temperature and precipitation in Hexi Corridor in recent 57 years. *Acta Geogr. Sin.* **2012**, *67*, 1482–1492.

46. Li, K. Prediction test of nonlinea dynamic model for yearly precipitation in the northweat raid area of China. *Acta Meteorol.* **2002**, *3*, 326–332.
47. Luo, D.; Zheng, F.; Chen, Q. Prediction of Inter-annual Signal of global mean surface temperature based on deep learning approach. *Clim. Environ. Res.* **2022**, *27*, 94–104.
48. Sherstinsky, A. Fundamentals of recurrent neural network (RNN) and long short-term memory (LSTM) network. *Phys. D Nonlinear Phenom.* **2020**, *404*, 132306. [CrossRef]
49. Theodosiou, M. Forecasting monthly and quarterly time series using STL decomposition. *Int. J. Forecast.* **2011**, *27*, 1178–1195. [CrossRef]
50. Schmidhuber, J.; Hochreiter, S. Long short-term memory. *Neural Comput.* **1997**, *9*, 1735–1780.
51. Van Houdt, G.; Mosquera, C.; Nápoles, G. A review on the long short-term memory model. *Artif. Intell. Rev.* **2020**, *53*, 5929–5955. [CrossRef]
52. Zhao, D.; Gao, X.; Wu, S. Trend of climate variation in China from 1960 to 2018 based on natural regionalization. *Adv. Earth Sci.* **2020**, *35*, 750–760.
53. Liu, Y.; Zhang, F.; Gong, X. Rainfall analyses of the 2013 Heilongjiang basin-wide major floods. *Water Resour. Hydropower Northeast* **2015**, *33*, 43–44+55.
54. UN Watercourses Convention. *Convention on the Law of the Non-Navigational Uses of International Watercourses*; Adopted on 21 May 1997; United Nations: New York, NY, USA, 2014.
55. Gunasekara, N.K.; Kazama, S.; Yamazaki, D.; Oki, T. Water conflict risk due to water resource availability and unequal distribution. *Water Resour. Manag.* **2014**, *28*, 169–184. [CrossRef]
56. Kasymov, S. Water resource disputes: Conflict and cooperation in drainage basins. *Int. J. World Peace* **2011**, *28*, 81–110.
57. Talozi, S.; Altz-Stamm, A.; Hussein, H.; Reich, P. What constitutes an equitable water share? A reassessment of equitable apportionment in the Jordan–Israel water agreement 25 years later. *Water Policy* **2019**, *21*, 911–933. [CrossRef]
58. Jeuland, M.; Whittington, D. Water resources planning under climate change: Assessing the robustness of real options for the Blue Nile. *Water Resour. Res.* **2014**, *50*, 2086–2107. [CrossRef]
59. Gari, Y.; Block, P.; Steenhuis, T.S.; Mekonnen, M.; Assefa, G.; Ephrem, A.K.; Bayissa, Y.; Tilahun, S.A. Developing an Approach for Equitable and Reasonable Utilization of International Rivers: The Nile River. *Water* **2023**, *15*, 4312. [CrossRef]
60. Valipour, E.; Ketabchi, H.; Morid, S. Water Resources Allocation: Iteractions Between Equity/Justice and Allocation Strategies. *Water Resour. Manag.* **2023**, *38*, 505–535. [CrossRef]
61. Giannias, D.A.; Lekakis, J.N. Policy analysis for an amicable, efficient and sustainable inter-country fresh water resource allocation. *Ecol. Econ.* **1997**, *21*, 231–242. [CrossRef]

water

MDPI

Article

A Study on the Ice Resistance Characteristics of Ships in Rafted Ice Based on the Circumferential Crack Method

Jiayu Huang [1], Feng Diao [2,3], Shifeng Ding [1], Sen Han [1], Pentti Kujala [4,5] and Li Zhou [4,*]

1 School of Naval Architecture and Ocean Engineering, Jiangsu University of Science and Technology, Zhenjiang 212100, China; 211210101207@stu.just.edu.cn (J.H.); 15001945469@163.com (S.D.); hansen@just.edu.cn (S.H.)
2 China Ship Scientific Research Center, Wuxi 214082, China; diaofeng@702sh.com
3 Lianyungang Center, Taihu Laboratory of Deep-Sea Technological Science, Lianyungang 222000, China
4 School of Ocean and Civil Engineering, Shanghai Jiao Tong University, Shanghai 200240, China; pentti.kujala@taltech.ee
5 Estonian Maritime Academy, Tallinn University of Technology, 19086 Tallinn, Estonia
* Correspondence: zhouli209@sjtu.edu.cn

Abstract: In previous studies of ship–ice interactions, most studies focused on ship–level ice interactions, overlooking potential rafted ice conditions in extreme ice conditions. The purpose of this study is to develop a numerical model for predicting ship resistance in rafted ice regions. Numerical modeling of rafted ice was carried out using preset grid cells. By comparing the model test results, the accuracy and reliability of the numerical model are verified. On this basis, we undertook the analysis of the impacts of different ice thicknesses, ship speeds, bending strengths, and crushing strengths on the ice resistance of ships under level and rafted ice conditions. The results show that the ice resistance of ships is significantly higher than that of rafted ice under the condition of level ice; however, level ice and rafted ice have different effects on ship ice resistance. Compared with level ice, the ice resistance of ships navigating in rafted ice is more concentrated. The findings of the present research can serve as a technical reference for studies focused on predicting ship resistance in rafted ice regions.

Keywords: polar ship; rafted ice; numerical simulation; ice resistance; circumferential crack method

Citation: Huang, J.; Diao, F.; Ding, S.; Han, S.; Kujala, P.; Zhou, L. A Study on the Ice Resistance Characteristics of Ships in Rafted Ice Based on the Circumferential Crack Method. *Water* **2024**, *16*, 854. https://doi.org/10.3390/w16060854

Academic Editor: Jueyi Sui

Received: 4 February 2024
Revised: 10 March 2024
Accepted: 13 March 2024
Published: 15 March 2024

1. Introduction

Typical features of sea ice in the polar regions include brash ice, floating ice, layered ice, and rafted ice. Rafted ice is one of the specific ice formations in the polar regions, especially during the initial and final sea ice periods. The dynamic effects of the fractures, extrusion, and accumulation of sea ice cause an increase in sea ice thickness. During navigation, a ship is subjected to non-linear solid ice resistance, which significantly challenges a ship's safe navigation.

For the navigation safety design of polar ships, researchers have proposed various ship performance prediction methods under sea ice conditions, which can be divided into experimental [1–5], analytical [6–10], and numerical methods [11–15]. Experimental methods include full-scale measurements and model tests. Full-scale ship trials are challenging to replicate and involve high costs, while model tests impose strict requirements on the experimental equipment and methodology. Empirical formula methods involve theoretical analyses of ship–ice interaction processes but often simplify the ship and sea ice models, which has particular limitations for complex sea ice and ship models. In recent years, with the rapid improvement of computer performance, numerical methods have been effectively applied. Numerical methods have significant development potential compared to experimental and empirical analytical methods. Numerical methods applied to ship–ice interactions mainly include the finite element method (FEM), the discrete element method

(DEM), and the circumferential crack method (CCM). The FEM has been widely applied in the context of ship–ice interaction problems over the years.

Yu et al. [16] employed the finite element method to numerically simulate periodic ice loads in the interaction between sea ice and conical structures, and the calculated sea ice bending damage process was similar to that of the results of full-scale measurements. Feng et al. [17] used the cohesive element method to simulate the interaction between ice and structures and conducted with an analysis of parameter sensitivity. It was found that the structural response was very sensitive to changes in the fracture energy, and the stress–strain curve of the body unit had a significant effect on the simulation. This method was also used by Wang et al. [18] to simulate the continuous icebreaking process of ships at different heeling angles, and they analyzed the continuous icebreaking process of different ships with different transverse inclinations, with the results showing that the ice resistance capacity of the ship and the extension length of the sea ice crevasse increased with the increase in the ship's transverse inclination angle. Lee et al. [19] proposed a method to analyze the ice load in the frequency domain, and the trend of the overall power spectral density with the bow angle was analyzed using different regression methods (linear interpolation, support vector machine, random forest, and deep neural network), and it was found that the deep neural network method performed the best. Shi et al. [20] proposed an elastic-plastic iceberg material model with temperature gradient effect to study the dynamic collision process between a floating production storage and offloading vessel (FPSO) and an iceberg. The simulation results are compared with the design specification to verify the validity of the iceberg model, and the effects of different iceberg shapes and temperatures on the collision process are analyzed. The results show that the structural damage of a floating production storage and offloading vessel (FPSO) is affected by the structural strength, the iceberg strength, and the localized shape of the iceberg.Based on the interaction process between ships and ice as well as the theory of sea ice fracturing, Lu et al. [21] proposed an edge-crack theory model. Using the extended finite element method, the mechanism of long crack propagation between parallel ice-breaking channels was studied. The maximum distance between parallel channels without sea ice fracturing was investigated and validated against experimental results.

In the discrete element method (DEM) realm, Hanse et al. [22] employed a two-dimensional discrete circular-disc viscoelastic model to simulate broken ice and adjusted numerical model calculation parameters according to ice tank experiments. Lau et al. [23] conducted a series of numerical simulations on the interaction between ice offshore structures and ice ships using the three-dimensional block discrete element model. Liu et al. [24] calculated the impact of factors such as ship speed, ice thickness, and ship width on the ice resistance of ships using the DEM. Dong et al. [25] established an ice channel model based on the discrete features of broken ice. Using image segmentation methods to extract ice channel regions and introducing intelligent corner regression networks to accurately delineate ice channel boundaries, this method has shown good accuracy in real ice channel recognition. Xie et al. [26] simulated the ship–water interaction using a coupled CFD-DEM method and established a discretized propeller model (DPM) and a body force model (BFM). The results indicate that the BFM method can be used effectively for the assessment of the main engine power and hull profile optimization during the ship development and design stages. Regarding the circumferential crack method, Zhou et al. [27] proposed a method based on the circumferential crack approach to distinguish the forms of sea ice damage according to the ship's heel angle, and they compared the numerical simulation results with the model test results, which achieved a good consistency. Moreover, Gu et al. [28] predicted the slewing motion of a polar ship in horizontal ice, considered the effect of hull camber on different damage modes of sea ice, and analyzed the results in comparison with the results of real ruler measurements, and the two results are in good agreement.

Several scholars have also worked on rafted ice material modeling. Hopkins et al. [29] utilized the discrete element approach using circular-disk and block models to validate that the relative motion of two flat ice blocks can result in either overlapping or crushing and breaking. The former leads to rafted ice, and the latter is the initial process of ice ridge formation. After observing the natural appearance of rafted ice through experiments, Leppäranta et al. [30] found that the ice crystals at the contact point between the two layers of level ice in the rafted ice formed a granular structure, and the shear strength of rafted ice was thus lower than that of level ice. Bailey et al. [31] found that the shear force at the adhesive interface of artificially created rafted ice was approximately 30% lower than that of level ice through experiments. Parmerter et al. [32] established a numerical sea ice rafting model capable of calculating the bending stress during the ice rafting process. The results showed that the increase in the bending stress of sea ice is proportional to the square of the ice thickness.

Although these methods have been applied to study ship interactions with level ice, ice floes, broken ice, ice ridges, etc., most of the research on rafted ice has focused on its mechanical properties and physical models. There has been relatively less exploration on ship collisions with rafted ice. This paper combines a preset grid method with the circumferential crack icebreaking assumption to establish a numerical model for rafted ice. The model will be used to predict the resistance of ships in the rafted ice region and compare the numerical simulation results with the model test results. On this premise, the effects of different ice thicknesses, ship speeds, and sea ice characteristics on the level of ship ice resistance and rafted ice are studied. This study supports subsequent ship resistance predictions in rafted ice regions more effectively and holds a specific engineering application value.

2. Numerical Model

2.1. Circumferential Crack Method

When ships navigate in polar regions, the interaction between the ship and the sea ice leads to localized compression and fragmentation of the free edge of the ice when the ship's bow comes into contact with the ice. With the increase in the contact area between the ship and the sea ice, there is a corresponding increase in crushing force, resulting in circumferential cracks parallel to the contact area or radial cracks perpendicular to the contact area. Based on the physical phenomena observed in full-scale measurements and model tests, the hypothesis of circumferential crack occurrence is adopted in this study. The geometric shape of the fractured floating ice is assumed to be wedge-shaped, with the ice wedge angle denoted as θ. The icebreaking radius of the shape of the ice wedge is R, as expressed in the literature [33].

$$R = C_l \times l(1.0 + C_v \times v_{n,2}) \tag{1}$$

where C_l and C_v represent empirical parameters, $v_{n,2}$ denotes the relative normal velocity between the ship and the sea ice, l refers to the characteristic length of the ice, which can be expressed as follows:

$$l = \left[\frac{E_i h_i^3}{12(1 - v^2)\rho_w g} \right]^{1/4} \tag{2}$$

where E_i signifies the elastic modulus, h_i represents the ice thickness, v represents the poisson ratio, ρ_w signifies the density of water, and g denotes the acceleration in gravity.

This paper converts the fan-shaped ice wedge to a square through area-equivalent treatment, as shown in Figure 1. Assuming that the icebreaking radius of the ice wedge is equal to that of the side length of the square grid cell and that the areas are equal, the icebreaking angle θ of the two satisfy the following:

$$R^2 = \frac{\theta}{2} R^2 \rightarrow \theta = 2rad \tag{3}$$

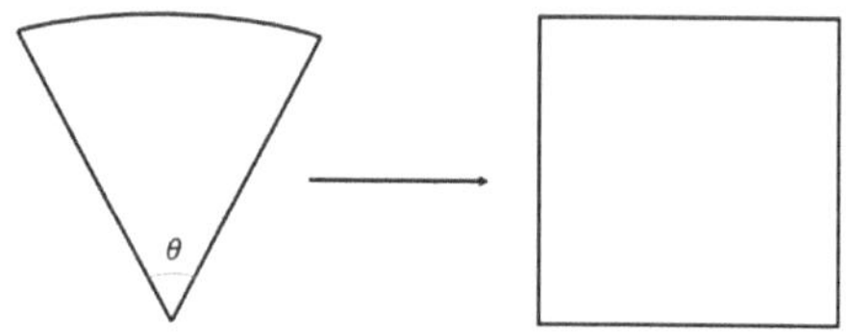

Figure 1. The principle of area equivalence.

2.2. Icebreaking Force

During polar ship icebreaking navigation, the compressive force gradually increases as the contact area between sea ice and the ship's hull increases. Before bending failure, the icebreaking force F_{cr} generated by the compression between the ship's hull and sea ice, with the force being perpendicular to the contact area, can be expressed as follows [34]:

$$F_{cr} = \sigma_c \cdot A_c \tag{4}$$

where A_c is the contact area and σ_c is the crushing strength of the ice.

2.3. Contact Area

When $L_d \cdot \tan(\varphi) \leq h_i$, the contact area between the ship and the sea ice has not reached the bottom of the sea ice at this point, resulting in a triangular contact area as follows:

$$A_c = \frac{1}{2} L_h \frac{L_d}{\cos(\varphi)} \tag{5}$$

When $L_d \cdot \tan(\varphi) \geq h_i$, the contact area between the ship and the sea ice reaches the bottom, resulting in a quadrilateral contact as follows:

$$A_c = \frac{1}{2}\left(L_h + L_h \frac{L_d - h_i/\tan(\varphi)}{L_d}\right) \frac{h_i}{\sin(\varphi)} \tag{6}$$

where A_c represents the contact area, L_h denotes the contact length, L_d refers to the contact length, h_i represents the ice thickness, and φ signifies the outward tilt angle at different ship nodes.

2.4. Ice Failure Model

During the interaction between ships and sea ice, the failure mode of sea ice is influenced by various factors, including ship angle, ice thickness, and the relative velocity between the ice and the ship. The ice failure model includes both bending failure and crushing failure in the present study. According to the research by Zhou et al. [27], different sea ice damage modes were used to distinguish the relationship between bending and crushing damage, and they found that the hull camber angle produces different sea ice damage modes and that the ship–ice friction coefficient affects the ultimate hull camber angle of the sea ice failure mode. On this basis, Gu et al. [28] assumed that the friction coefficient between the ship and the sea ice was 0.1 and calculated that the limiting angle of the ship–ice failure mode was 84.2894°, which means that, during the icebreaking voyage, if the angle between the ship and the ice is more than 84.2894°, this will lead to crushing damage, whereas if it is less than 84.2894°, this will lead to bending damage.

According to Kerr [35], the expression for the ultimate load of ice bending failure is given as follows:

$$P_f = C_f \left(\frac{\theta}{\pi}\right)^2 \sigma_f h_i^2 \tag{7}$$

where C_f signifies the empirical parameter, θ signifies the idealized ice fracturing angle, σ_f denotes the bending strength of ice, and h_i represents the ice thickness.

When ice experiences crushing failure, the localized icebreaking force acting on the hull, according to ISO/FDIS 19906-2019 [36], can be expressed as follows:

$$F_{cr} = P_G \cdot A_c \tag{8}$$

$$P_G = C_R\left[\left(\frac{h}{h_1}\right)^n\left(\frac{L}{h}\right)^m + f_{AR}\right] \tag{9}$$

where C_R represents the ice strength coefficient, h signifies the ice thickness, h_1 signifies the reference ice thickness of 1 m, m and n are the empirical coefficients, and f_{AR} represents the ice strength coefficient.

2.5. Rafted Ice Model

Currently, there are two main types of rafted ice. One type is that of finger-rafted ice, where the ice body does not move as a rigid body, accumulating internal stresses that are then released a shear force, thereby forming finger-rafted ice. Another type is that of layered rafted ice, where one ice layer fractures and climbs onto another under external dynamic forces, becoming rafted and forming a layered structure [37]. Figure 2 illustrates the phenomenon of vertical layering in the material thickness direction of consolidated ice, including the smooth level ice layer, the consolidation layer, and the submerged layer. The submerged layer is formed by immersion in water, interacts with the level of ice under the action of buoyancy, and ultimately forms the intermediate consolidation layer. Shafrova et al. [38] conducted experiments on the freezing process of first-year ice ridges and noted that various factors, such as seawater infiltration, temperature, salinity, and pressure, affect the strength of the frozen bond between ice bodies during the formation process. While studying the material properties of consolidated ice, Chen et al. [39] obtained a fragment function relationship between its compressive strength and the strain rate. The calculation formula is as follows:

$$\sigma_c' = \left\{ \begin{array}{ll} 0.37\dot{\varepsilon}^{0.2} & \dot{\varepsilon} > 4.6 \times 10^{-4} \\ 53 \ \ \dot{\varepsilon}^{0.2} & \dot{\varepsilon} \leq 4.6 \times 10^{-4} \end{array} \right\} \tag{10}$$

where σ_c' is the revised compressive strength.

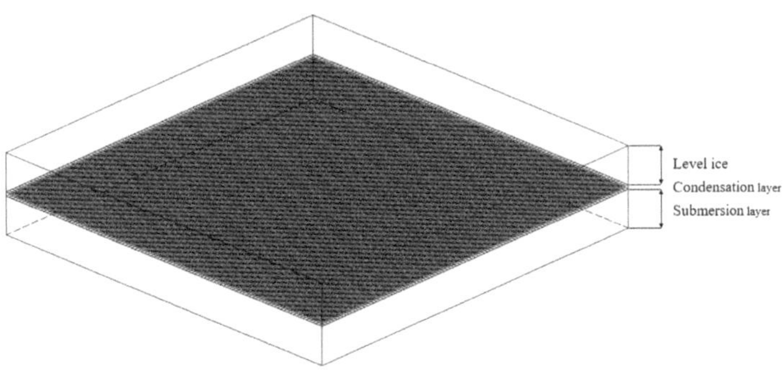

Figure 2. Rafted ice model specimen.

This study focuses on layered rafted ice, which belongs to composite ice formations. The stacking of two level ice layers primarily develops it. Currently, numerous scholars have proposed corresponding numerical models based on the characteristics exhibited by ice. Ni et al. [40] introduce cohesive elements to numerically model the intra- and interlayer structures of the rafted ice layers, respectively. By randomly deleting the cohesive elements within the model, the porosity of the natural rafted ice was successfully simulated, and numerical calculations of the collision between a ship and rafted ice were carried out, and it was found that the method was excellent in simulating the crack extension of rafted ice. In this study, based on the significant vertical layered structure of rafted ice, the solidified and submerged layers are collectively referred to as the second layer of rafted ice in the numerical model. And a correction factor is introduced to define the constitutive parameters of the ice layer, which can include the mechanical properties of the low sea ice in the condensation layer, reasonably expressing the differences between the rafted ice layers, with the formula provided below.

$$\sigma = \left\{ \begin{array}{ll} C\sigma_1 & n = 1 \\ C\sigma_2 & n = 2 \end{array} \right\} \tag{11}$$

where σ is the ice strength, C is the correction factor, and n is the number of layers of rafted ice.

In the process of ship–ice interactions, the action of icebreaking forces causes the formation of ice cracks. As these cracks spread, the ice gradually breaks and destroys. The formation of rafted ice crevasses is simulated using grid cells, and the sea ice failure model is introduced. The rafted ice is separated into isolated grid cells. The side length of the grid cell is related to the icebreaking radius R, as shown in Figure 3. In the numerical model, when the icebreaking force reaches the load-bearing limit of the grid cells, the rafted ice is destroyed. Introducing this failure model allows for a more detailed consideration of the rafted ice layer's destruction process and simulates the ice layer's fracture behavior in numerical simulations. Each grid cell represents a discrete unit of the ice layer, and by monitoring the impact of icebreaking forces on these units, it is possible to track the real-time generation and propagation of cracks, ultimately simulating the complex failure process of the rafted ice under the action of icebreaking forces. Figure 4 illustrates the computational flow of the ship icebreaking simulation, which mainly includes the numerical model and numerical process.

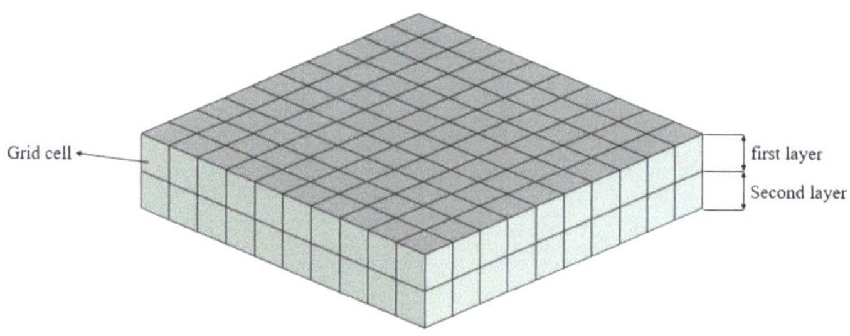

Figure 3. Numerical modeling of rafted ice.

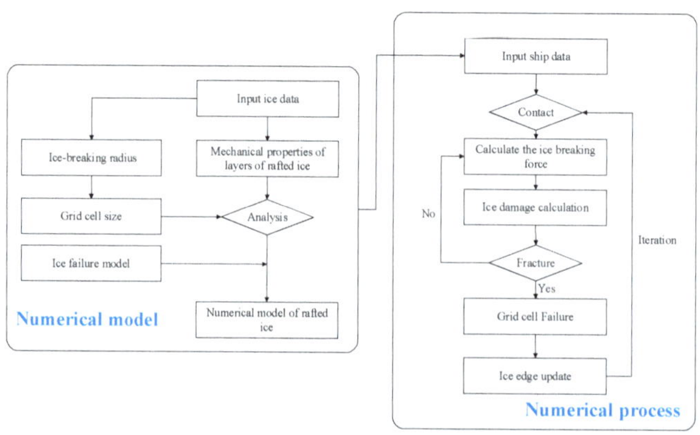

Figure 4. Computational program flowchart.

3. Model Test in Rafted Ice

3.1. Experimental Description

A relevant model test of a ship model sailing in rafted ice was performed in an outdoor ice tank of Harbin Engineering University [41]. The ice water tank is 20 m in length, 2 m in width, and 1.5 m deep, and it can naturally make different ice features in winter. The purpose of this is to conduct a towing test on a ship operating in rafted ice, and the test follows Froude similarity and Cauchy similarity using a polar ship with a model scale of 1:60. The main particulars of the full-scale and model-scale ships are listed in Table 1.

Table 1. Main particulars of the ship.

Principal Hull Data	Full-Scale	Model-Scale
Length between perpendiculars/m	122.5	2.04
Beam/m	23.32	0.38
Draught/m	7.8	0.13
Stem angle/°	20	20
Waterline angle/°	34	34

Xu et al. [41] selected three different speeds, 0.17, 0.27, and 0.37 m/s, for the ship model towing tests in the rafted ice region, and two tests were conducted for each speed, and the settings of these six ship model tests are listed in Table 2.

Table 2. Parameters of rafted ice.

Case	Towing Speed/m/s	Bending Strength/MPa	Crushing Strength/MPa
1	0.17	0.85	1.73
2		0.91	1.40
3	0.27	0.84	1.76
4		0.97	1.86
5	0.37	0.72	1.14
6		0.74	1.06

In the numerical simulation, as shown in Figure 5, the rafted ice is divided into upper and lower layers, with each being composed of numerous square ice grids. Each ice grid's length can be taken as the ship's icebreaking radius during navigation. The model is divided into upper and lower layers that accurately reproduce the construction of rafted ice, with each layer consisting of several square grids. Fine discretization of the waterline at the ship's draft ensures a reasonable simulation of the rafted ice damage and a precise description of the location of the ship–ice grid contact points. Figure 6 provides the initial top view of the ice–ship interaction.

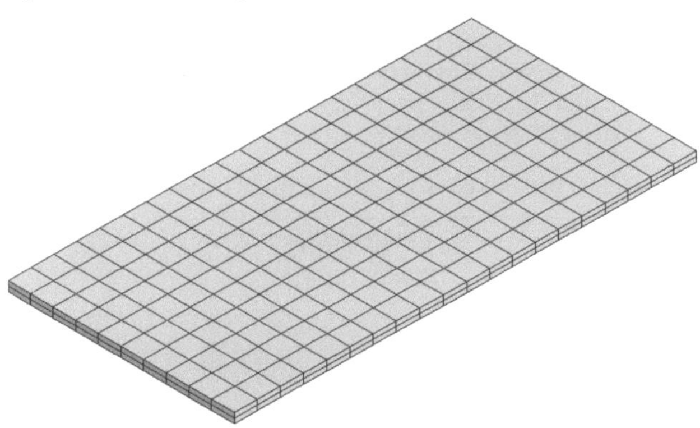

Figure 5. Numerical model of the rafted ice.

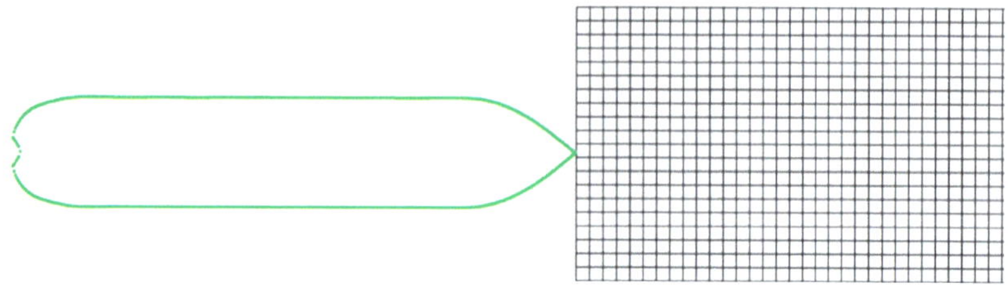

Figure 6. Initial state of the rafted ice and the polar ship.

3.2. Comparison of Model Tests

In the numerical simulation, each grid cell size represents the icebreaking radius. White grids indicate no contact between the ship and these grid cells. Red grids indicate interaction between the ship and the grid cells, while blue grids indicate grid cells that have failed. Figure 7 illustrates the interaction process between the ship and the rafted ice in the numerical simulation. With the continuous progress of the ship, the contact area between the hull and the grid cells gradually increases. When the icebreaking force exceeds the load-bearing limit of the grid cells, the grid cells fail, indicating the occurrence of fractures and a failure in the rafted ice.

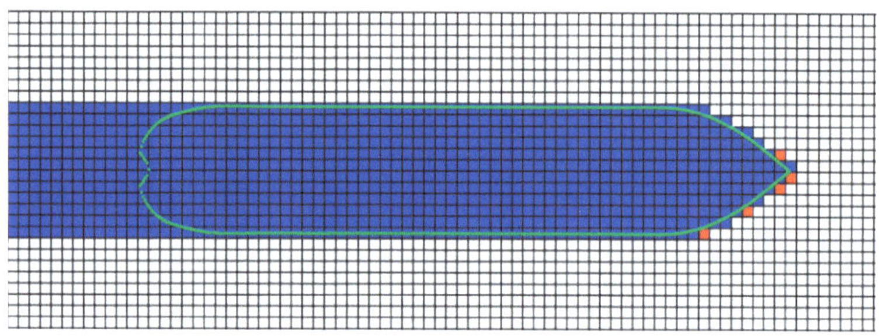

Figure 7. Numerical simulation of phenomena.

In this paper, the test conditions of Case 5 are selected to analyze the numerically simulated time history curves of total resistance, as shown in Figure 8. The simulation results show obvious periodic characteristics with relatively stable peaks. This is due to the fact that the grid cell parameters of the rafted ice are fixed in the numerical simulation, and the ice resistance value oscillates and changes within a certain amplitude when the ship reaches the icebreaking stabilization stage. The comparison between the numerical simulation results and the experimental results is presented in Table 3. It can be observed that with increasing ship speed, the resistance in the rafted ice also increases. By comparing six different experimental conditions, the error between the two ice resistance values is within 10%, indicating a good consistency and verifying the accuracy of the numerical model to some extent.

Table 3. Comparison between model test results and numerical results.

Case	Average in Experiment/N	Average in Simulation/N	Error/%
1	25.45	24.69	2.9
2	28.13	30.91	9.8
3	34.61	31.83	8.0
4	29.14	27.39	5.9
5	42.52	43.10	1.3
6	35.78	37.80	5.6

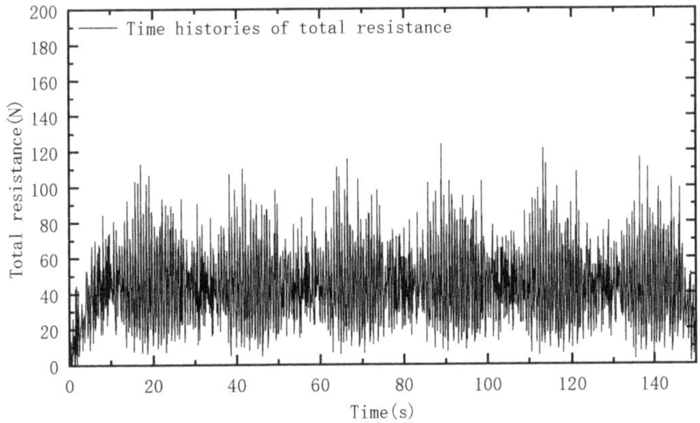

Figure 8. Time history curve of total resistance in numerical simulation.

4. Sensitivity Analysis of Parameters

The numerical model simulating the ship's motion in rafted ice established in this paper has been well validated through comparisons with previous experiments and numerical results. Additionally, this study focuses on the ice resistance characteristics of ships in the level and rafted ice regions. The model test study from the ice tank of Aalto University in Finland was chosen [42]. The model ship of MT Uikku had a scale of 1:31.6. The key parameters of both the full-scale and model-scale platforms are presented in Table 4. The ship model was towed by a trailer at a constant speed on level ice, and various experimental data were obtained by changing the ice thickness and velocity. Three test conditions were selected from this model test for comparisons, which were Case1, 2, and 3, with the parameters of the level ice being listed in Table 5.

Table 4. Main particulars of MT Uikku.

Principal Hull Data	Full-Scale	Model-Scale
Length between perpendiculars/m	150	4.75
Beam/m	21.3	0.67
Draught/m	9.5	0.3
Stem angle/°	30	30
Waterline angle/°	21	21

Table 5. Settings of level ice in model tests.

Case	Towing Speed/m/s	Towing Speed/kn	Ice Thickness/m	Bending Strength/MPa	Crushing Strength/MPa
1	0.09	0.97	0.76	0.844	2.192
2	0.09	0.97	1.03	0.669	2.485
3	0.09	0.97	0.63	1.029	5.389

Sequential ice resistance results through Case1, 2, and 3 were analyzed, as shown in Figure 9. The squares and triangles are the average and maximum ice resistance observed for Case1, 2, and 3, respectively, and the dots and pentagrams are the average and maximum ice resistance for Case1, 2, and 3 in the numerical simulation. In Case1, the numerical simulation's average value is 598.59 kN, which closely matches the measurement of 560 kN from the ice tank experiment, with an error of approximately 6.8%. In Case2, the numerical simulation's average value is 759.9 kN, while the measured value in the ice tank experiment is 830 kN, with an error of approximately 8%. In Case3, the average error between the two is 10%. By comparing the maximum ice resistance values under the three test conditions, it can be observed that only in Case2 is there a significant error between the numerical simulation and the ice tank test in the peak ice resistance. The main reason for this error is the inherent variability in ice parameters in different regions of level ice during the model test preparation process. However, the numerical simulation discretizes the ice field using grids coupled with predetermined ice parameters, which, to some extent, affects the peak ice resistance. Based on the above analysis, the numerical results are qualitatively and quantitatively consistent with the experimental data. The numerical method can also predict ice resistance for ships in level ice.

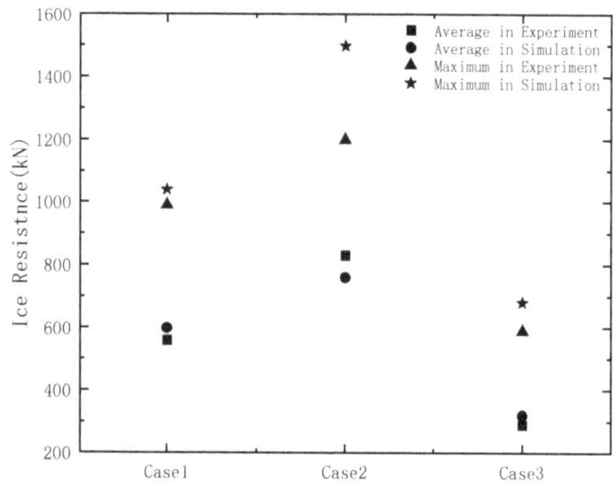

Figure 9. The numerical simulation and model test results for level ice.

Numerical simulations were conducted under Case1, 2, and 3 to compare the average ice resistance of level and rafted ice of the same ice thicknesses for the three conditions, as shown in Figure 10. The ice resistance of rafted ice is significantly lower than that of level ice at the same speed, and the difference in ice resistance gradually increases as the ice thickness increases from 0.63 m to 1.03 m. Since the structure of rafted ice is composed of two thin layers of level ice that undergo secondary freezing, its overall strength is lower than that of level ice; therefore, rafted ice is more prone to damage during ship–ice interactions.

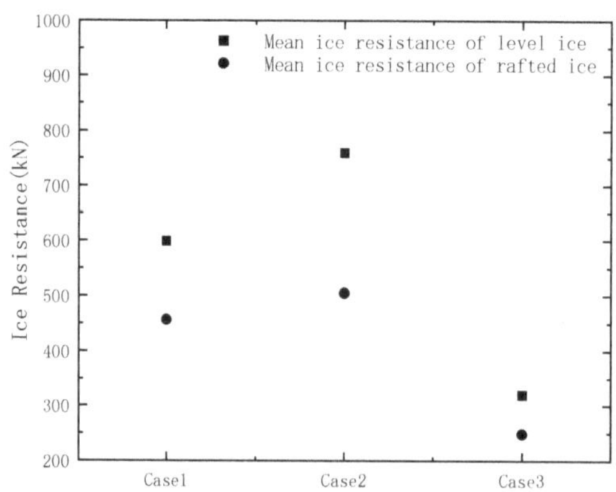

Figure 10. Comparison of level ice and rafted ice resistance under the same conditions.

Due to the different formation mechanisms and internal structures of level ice and rafted ice, there are specific differences in their resistance characteristics. As shown in Figure 11, in Case1, for example, the time history curves of ice resistance for both the level ice and the rafted ice were analyzed. It can be observed that within the first 25 s, the two trends are similar, but the resistance is lower than that of the level ice during the same period. As the icebreaker keeps moving forward, both show an increasing trend in resistance. The ice resistance exhibits periodic fluctuations once the ship enters the stable icebreaking stage. The ice resistance fluctuates with more prominent peaks when interacting with level ice. However, due to the differences in the mechanical properties of the rafted ice layers in the numerical simulation, the ultimate loads on the ice grids are inconsistent. In Figure 12a, the first rafted ice layer where the ship's bow is first contacted by the two grid cells has already failed, while in Figure 12b, the two grid cells of the second rafted ice layer at the same position are in action, and the forces generated by the different layers of ice cause the ship's ice resistance to fluctuate more significantly. The peaks of the fluctuations are more significant.

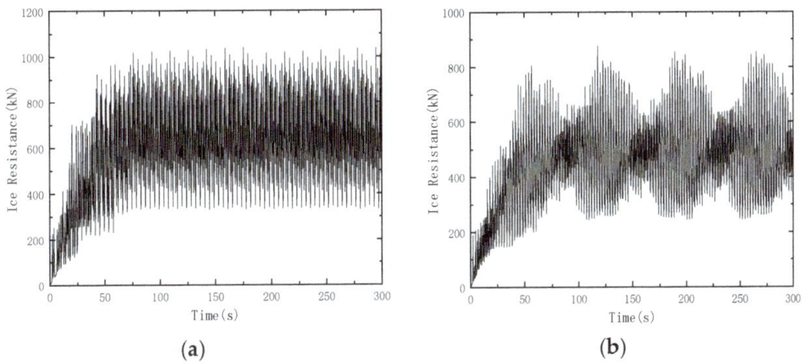

Figure 11. Time histories of ice resistance. (**a**) Time history of level ice in Case1; (**b**) time history of level rafted ice in Case1.

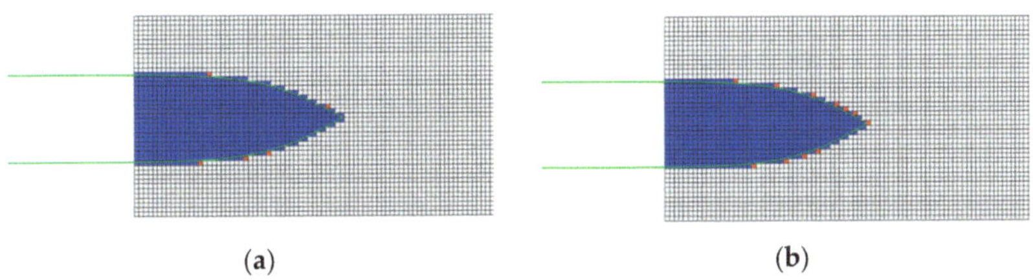

Figure 12. Interaction process of ship and rafted ice. (**a**) First layer of rafted ice (t = 110 s); (**b**) second layer of rafted ice (t = 110 s).

4.1. Influence of Ice Thickness

The ice thickness is a crucial component influencing crushed ice. Diverse ice thicknesses of 0.6 m, 0.8 m, 1.0 m, and 1.2 m were selected to simulate the icebreaking of ships in the level and rafted ice regions, and the sailing speed was 1 kn, and the sea ice parameters were referred to in Tables 5 and 6. Figure 13 compares the mean ice resistance of ships in level and rafted ice regions under varying ice thicknesses. As the ice thickness rises from 0.6 m to 1.0 m, the ship ice resistance of level ice increases from 558.02 kN to 1386.8 kN, while the ship ice resistance of the rafted ice increases from 335.20 kN to 819.14 kN. The ice resistance of ships in both the level and rafted ice regions increases with the growing ice thickness. Under the same ice thickness circumstances, the resistance of ships in the rafted ice region is relatively close to that of the level ice region at 0.6 m ice thickness, but at 1.2 m ice thickness, there is a significant difference between the two. It can be found that the ship ice resistance in level ice is more sensitive to the change in ice thickness compared to the ship ice resistance in the rafted ice area.

Figure 14 illustrates the distribution of ship ice resistance in level and rafted ice conditions under four different ice thicknesses at a certain ship speed. It can be observed that the center of distribution of ship ice resistance in both level and rafted ice increases with the growth of ice thickness. In level ice, the distribution of ship ice resistance gradually transitions from a left-skewed distribution to a right-skewed distribution, indicating that increasing ice thickness increases the probability of encountering peak values in ship ice resistance. In addition, the distribution of ship ice resistance for rafted ice increases with the increase in ice thickness, especially in the conditions of 1 m and 2 m ice thickness, but the distribution of ship ice resistance is lower compared with that of level ice under the same ice thickness conditions. Notably, under different thickness conditions, the distribution of ship ice resistance in rafted ice is more concentrated compared to that of level ice. This suggests that the internal structure of level and rafted ice has distinct influences on the distribution of ice resistance. The structure of rafted ice, being more intricate, results in a more concentrated distribution of ice resistance, potentially leading to increased fatigue effects on the ship's structure.

Table 6. The mechanical parameters for numerical simulation of rafted ice.

Case	Velocity/kn	Bending Strength/MPa	Crushing Strength/MPa	Number of Layers	Correction Factor
1	0.97	0.844	2.192	1	1.0
		0.759	1.972	2	0.9
2	0.97	0.669	2.485	1	1.0
		0.602	2.236	2	0.9
3	0.97	1.029	5.389	1	1.0
		0.926	4.850	2	0.9

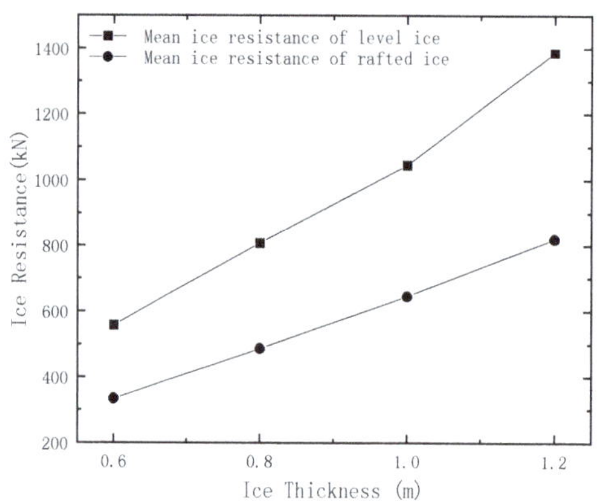

Figure 13. Mean ice resistance of level and rafted ice under different ice thicknesses.

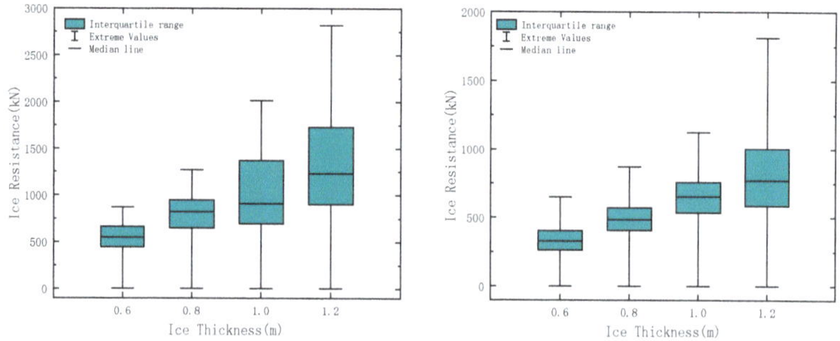

(a) (b)

Figure 14. Comparison of ship ice resistance distribution in level and rafted ice under different ice thicknesses. (**a**) Level ice; (**b**) rafted ice.

4.2. Influence of Ship Speed

This article selects speeds of 2 kn, 3 kn, 4 kn, and 5 kn to simulate ship icebreaking in the level ice and rafted ice regions. The sea ice parameters are shown in Tables 5 and 6, with an ice thickness of 0.76 m. Figure 15 illustrates the variation in ship ice resistance trends for level and rafted ice at different ship speeds. With an almost linearly increasing relationship, ship speed highly influences the ship's ice resistance in different ice conditions. A comparison between ship speeds of 2 kn and 5 kn shows that the ship ice resistance of level ice rises by 39.2%, while that of rafted ice increases by 38.4%. It can be observed that the resistance of level ice is more sensitive than that of rafted ice under the influence of ship speed.

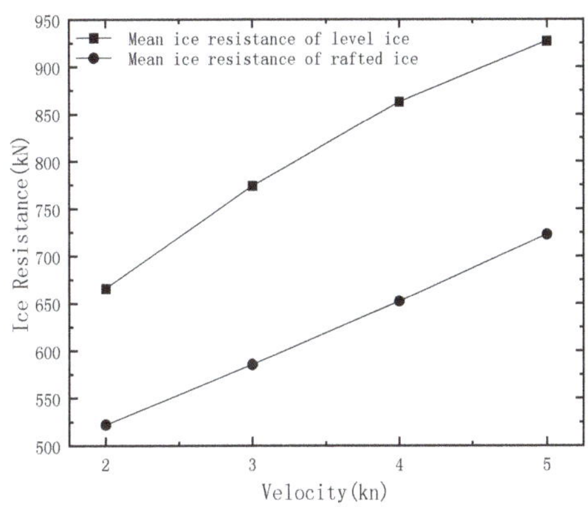

Figure 15. Mean ice resistance of the level and rafted ice under different ship speeds.

The results in Figure 16 show that the median line of ship ice resistance increases with speed in all cases except for that of the case of a speed of 4 kn in level ice. The variation in ship speed directly influences the size of the grid cells. Figure 17 illustrates the icebreaking state of vessels in level ice simultaneously under four different ship speeds. It is observed that, under the condition of a 4 kn speed in level ice, continuous crushing occurs between the ship's side and the grid cells. This leads to a more drastic variation in ship ice resistance, significantly increasing the probability of peak values and causing a more dispersed overall distribution of ice resistance. Therefore, speed not only impacts the peak magnitude of ice resistance but also significantly influences the distribution of ship ice resistance. This indicates that the change in speed may induce alterations in the interaction state between the vessel and ice, consequently affecting the overall characteristics of ice resistance.

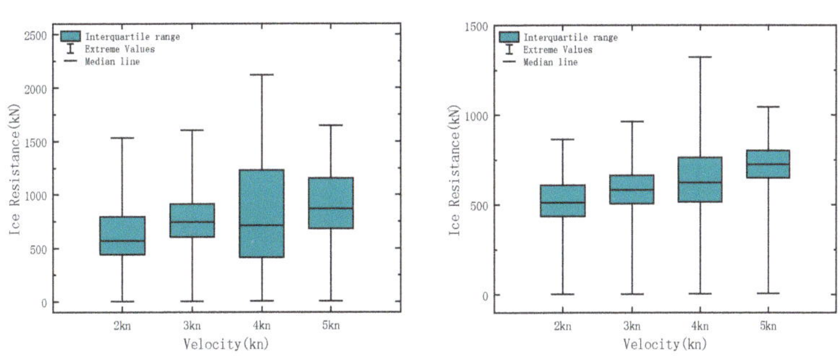

Figure 16. Comparison of ship ice resistance distribution in level and rafted ice under different ship speeds. (**a**) Level ice; (**b**) rafted ice.

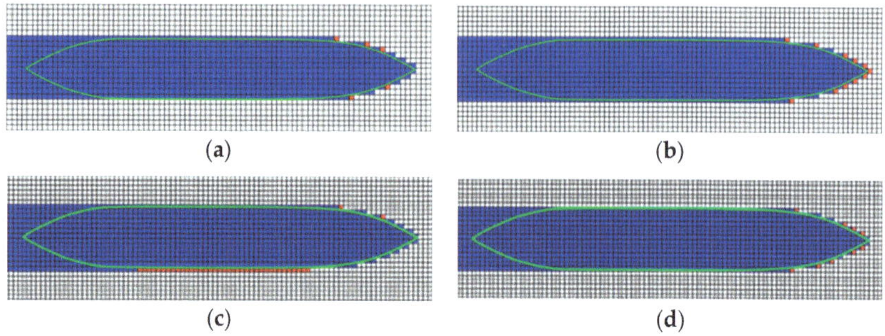

Figure 17. The icebreaking status of the ship at different speeds in a level ice scenario. (a) v = 2 kn (t = 250 s); (b) v = 3 kn (t = 250 s); (c) v = 4 kn (t = 250 s); (d) v = 5 kn (t = 250 s).

4.3. Influence of Bending Strength

In the numerical simulations, bending strength is a crucial parameter directly influencing the load-bearing capacity of each ice grid. To analyze the impact of bending strength on ship ice resistance, especially considering the environmental differences in the growth of rafted ice, which are primarily reflected in the parameter variations of the lower sea ice layer, this paper selected the correction factors of 0.6, 0.7, 0.8, and 0.9 to set the bending strength for both the overall level ice and the lower layer of the rafted ice. Figure 18 illustrates the linearly increasing trend of ship ice resistance for the level and rafted ice as the correction factor for the bending strength rise from 0.6 to 0.9. The ship ice resistance for level ice increases from 539.15 kN to 701.42 kN, while for rafted ice, it increases from 373.14 kN to 431.87 kN. It is worth noting that the ice resistance generated by the ship in level ice remains higher than that in rafted ice under the influence of bending strength.

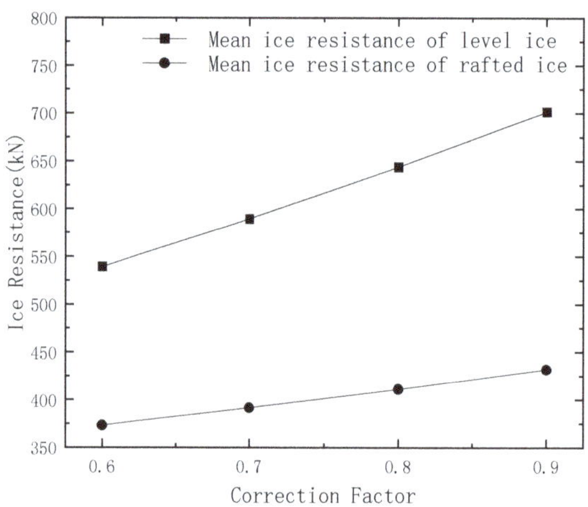

Figure 18. Mean ice resistance of the level and rafted ice regions at different bending strengths.

As shown in Figure 19, it can be observed that the median line of the ship ice resistance in level and rafted ice increases with the increasing bending strength, which implies that a higher peak in the ship ice resistance has an effect on the central tendency. Bending strength directly influences the load-bearing capacity of the ice layer, and with an increase in bending strength, the magnitude of ice resistance experienced by the vessel markedly rises. From the analysis of the overall distribution of ship ice resistance, it can be seen that the larger distance between the minima of the median line makes the ship ice resistance in level ice have a right-skewed distribution, while in the rafted ice, the median line and the minima are equally distant from each other, thereby showing a normal distribution. Despite an increase in bending strength, the overall trend changes relatively insignificantly. This suggests that bending strength does not significantly impact the distribution of ice resistance.

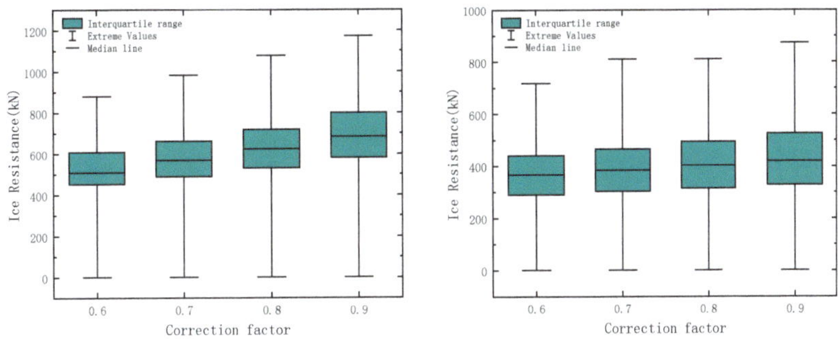

Figure 19. Comparison of ship ice resistance distribution in level and rafted ice under different bending strength. (**a**) Level ice; (**b**) rafted ice.

4.4. Influence of Crushing Strength

Crushing strength is a crucial parameter in the interaction between ships and ice. In numerical simulations, correction factors of 0.6, 0.7, 0.8, and 0.9 were selected to set the compression strength for level ice and the lower layer of rafted ice. The sailing speed is 0.97 kn, and the ice thickness is 0.76 m. Figure 20 shows that with the increase of crushing strength, the average and maximum ship ice resistance tend to decline. The icebreaking force between the ship and the ice is affected by the crushing strength of the sea ice. Analysis of individual interactions between the ship and ice grids reveals that enhancing crushing strength shortens the time required for the breaking force to reach the ice's load-bearing limit. The force remains zero until colliding with the next ice grid, leading to a decreasing trend in the mean ice resistance.

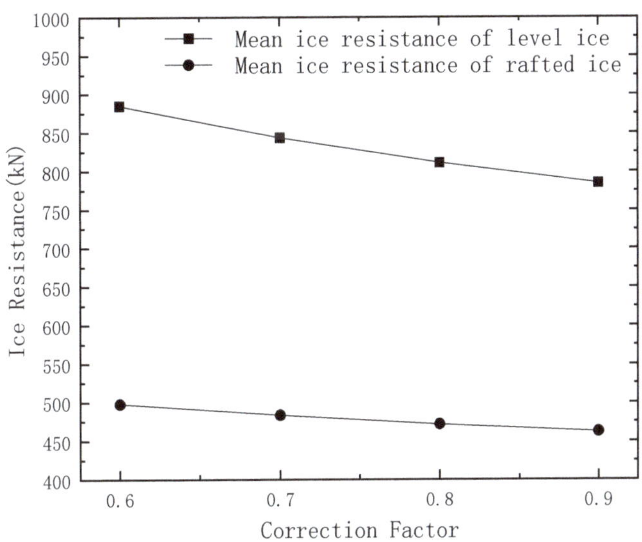

Figure 20. Mean ice resistance of the level and rafted ice regions at different crushing strengths.

As shown in Figure 21, it can be observed that the median line of ship ice resistance decreases gradually with the increase in breaking strength in both the level and rafted ice. In the level ice, the median line of ship ice resistance is further away from the end of the minima, which makes the overall left-skewed distribution. In the rafted ice, the median line of ship ice resistance is further away from the extreme value end, and the overall distribution is right-skewed. This is due to the crushing strength directly affecting the icebreaking force between the ship and the ice. The icebreaking force between the ship and the ice increases with the increase in crushing strength. This means that the time of grid cell failure is accelerated under the condition of the same carrying capacity of the grid cells in the level ice. In the process of making contact with the next grid cell, no ice resistance is generated, which leads to a decrease in the trend of the ice resistance of the ship in the level ice. However, the process of collision between the ship and the rafted ice results in the failure of the upper grid cell, which does not mean that the lower grid cell will also fail due to differences in the mechanical properties of the layers of rafted ice. This alternating action leads to a significant difference in the distribution of the overall ice resistance from that in the level ice.

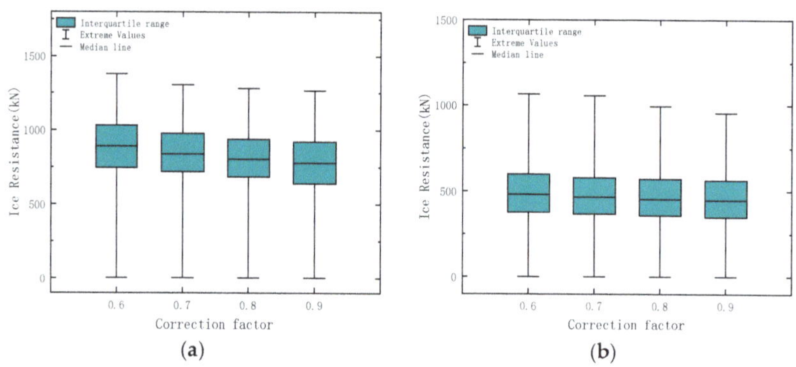

Figure 21. Comparison of ship ice resistance distribution in level and rafted ice under different crushing strengths. (**a**) Level ice; (**b**) rafted ice.

5. Conclusions

Based on the assumption of circumferential crack icebreaking, this paper employed a predefined grid method to numerically simulate the icebreaking navigation of ships in the rafted ice region. The numerical simulation results were compared with experimental data, demonstrating a good consistency. On this basis, the paper further investigates the influence of some key parameters on ship ice resistance and analyzes the resistance distribution characteristics of the level ice and rafted ice areas using probability density functions. The following conclusions are made:

1. According to the structural characteristics and mechanical properties of rafted ice, this paper adopts a new numerical method to establish a numerical model and to simulate the icebreaking process of ships in the rafted ice area, which are the keys to success.
2. Moreover, this paper utilizes the established numerical model for ship icebreaking in rafted ice areas to conduct numerical simulations and validate it against six operational conditions in an ice tank model experiment. The results demonstrate that this method can accurately predict the ice resistance experienced by ships in rafted ice areas, with the error between the simulated resistance value and the experimental value being within 10%.
3. Therefore, the model accurately predicts ship ice resistance in level ice regions through numerical case analysis. It can compare the effects of ice thickness, ship speed, bending strength, and crushing strength on ship ice resistance in both the level and rafted ice areas. Simulation results indicate that the ship ice resistance in the level and rafted ice

regions linearly increases with ice thickness, ship speed, and bending strength while linearly decreasing with crushing force. Comparing ship ice resistance in the level and rafted ice regions, the influences of ice thickness, ship speed, bending strength, and crushing strength on ship ice resistance are more sensitive in level ice than in rafted ice.

4. Numerical simulations of ships operating in level and rafted ice show that the ice resistance generated in level ice is more significant than that in rafted ice. However, this does not imply that the potential damage to the vessel caused by rafted ice can be easily overlooked. In reality, the ice resistance from the interaction between the ship and the rafted ice is more concentrated than that in the level ice. This concentration can make the ship's structure more susceptible to fatigue, increasing the risks associated with polar navigation.

The present method should be further verified with more measured data in the future. Moreover, variations in ship speed can impact the size of grid cells in the ice field, consequently influencing the distribution of ice resistance experienced by the ship, which could be studied further.

Author Contributions: Methodology, F.D. and J.H.; analysis, F.D. and J.H.; investigation, S.D.; resources, S.D. and S.H.; data curation F.D. and J.H.; software, J.H. and S.H.; writing—original draft preparation, J.H.; writing—review and editing, L.Z. and P.K.; visualization, L.Z. and P.K.; supervision, L.Z. and P.K.; project administration, L.Z. and S.D.; funding acquisition, L.Z. All authors have read and agreed to the published version of the manuscript.

Funding: This work is supported by the National Key Research and Development Program (Grant No. 2022YFE0107000), General Projects of National Natural Science Foundation of China (Grant No. 52171259), and Science and Technology Commission of Shanghai Municipality Project (22DZ1204403).

Data Availability Statement: All analyzed data in this study have been included in the manuscript.

Acknowledgments: The authors would like to thank the Jiangsu University of Science and Technology (JUST), Shanghai Jiao Tong University (SJTU), China Ship Scientific Research Center, and Tallinn University of Technology (TTU).

Conflicts of Interest: The authors declare no conflicts of interest.

References

1. Schwarz, J.; Jochmann, P.; Hoffman, L. Prediction of the icebreaking performance of the German polar research vessel. In Proceedings of the 6th STAR Symposium, Portland, OR, USA, 8–12 September 1981; pp. 239–248.
2. Ettema, R.; Sharifi, M.B.; Georgakakos, K.P.; Stern, F. Chaos in continuous-mode icebreaking. *Cold Reg. Sci. Technol.* **1991**, *19*, 131–144. [CrossRef]
3. und Polach, R.B.; Ehlers, S. Heave and pitch motions of a ship in model ice: An experimental study on ship resistance and ice breaking pattern. *Cold Reg. Sci. Technol.* **2011**, *68*, 49–59. [CrossRef]
4. Kellner, L.; Herrnring, H.; Ring, M. Review of ice load standards and comparison with measurements. In Proceedings of the International Conference on Offshore Mechanics and Arctic Engineering. American Society of Mechanical Engineers, Trondheim, Norway, 25–30 June 2017; Volume 57762, p. V008T07A028.
5. Huang, Y.; Huang, S.; Sun, J. Experiments on navigating resistance of an icebreaker in snow covered level ice. *Cold Reg. Sci. Technol.* **2018**, *152*, 1–14.
6. DuBrovin, O.V. *Calculation of Broken Ice Resistance Based on Model Testing*; University of Michigan: Ann Arbor, MI, USA, 1970.
7. Lindquist, A. Straightforward method for calculation of ice resistance of ships. In Proceedings of the POAC 89, 10th Intl Conference, Port and Ocean Engineering under Arctic Conditions, Luleaa, Sweden, 12–16 June 1989.
8. Keinonen, A.J.; Browne, R.; Revill, C. *Icebreaker Characteristics Synthesis*; Report TP 12812E; The Transportation Development Centre, Transport Canada: Ottawa, ON, USA, 1996.
9. Riska, K.; Wilhelmson, M.; Englund, K. *Performance of Merchant Vessels in Ice in the Baltic*; Technical Report 52; Winter Navigation Research Board: Helsinki, Finland, 1997.
10. Jeong, S.Y.; Lee, C.J.; Cho, S.R. Ice resistance prediction for standard icebreaker model ship. In Proceedings of the ISOPE International Ocean and Polar Engineering Conference, Beijing, China, 20–25 June 2010. ISOPE-I-10-137.
11. Silling, S.A.; Epton, M.; Weckner, O.; Askari, E. Peridynamic states and constitutive modeling. *J. Elast.* **2007**, *88*, 151–184. [CrossRef]

12. Su, B.; Skjetne, R.; Berg, T.E. Numerical assessment of a double-acting offshore vessel's performance in level ice with experimental comparison. *Cold Reg. Sci. Technol.* **2014**, *106*, 96–109. [CrossRef]
13. Sawamura, J.; Yamauchi, Y.; Anzai, K. Simulation of ice force and breaking pattern for icebreaking ship in level ice. In Proceedings of the International Conference on Offshore Mechanics and Arctic Engineering, Trondheim, Norway, 25–30 June 2017; American Society of Mechanical Engineers: New York, NY, USA; Volume 57762, p. V008T07A032.
14. Li, F.; Kotilainen, M.; Goerlandt, F.; Kujala, P. A new icebreaking pattern for the application in numerical simulation of ship performance in level ice. In Proceedings of the International Conference on Offshore Mechanics and Arctic Engineering, Madrid, Spain, 17–22 June 2018; American Society of Mechanical Engineers: New York, NY, USA, 2018; Volume 51296, p. V008T07A017.
15. Lilja, V.P.; Polojärvi, A.; Tuhkuri, J.; Paavilainen, J. Effective material properties of a finite element-discrete element model of an ice sheet. *Comput. Struct.* **2019**, *224*, 106107.
16. Yu, B.; Wu, W.; Xu, N.; Yue, Q.; Liu, S. Numerical simulation of dynamic ice force on conical structure. In Proceedings of the International Conference on Port and Ocean Engineering under Arctic Conditions, Dalian, China, 27–30 June 2007.
17. Feng, D.; Pang, S.D.; Zhang, J. Parameter sensitivity in numerical modelling of ice-structure interaction with cohesive element method. In Proceedings of the International Conference on Offshore Mechanics and Arctic Engineering, Busan, Republic of Korea, 19–24 June 2016; American Society of Mechanical Engineers: New York, NY, USA, 2016; Volume 49996, p. V008T07A012.
18. Wang, F.; Zou, Z.J.; Guo, H.P.; Ren, Y.Z. Numerical simulations of continuous icebreaking process with different heel angles in level ice. In Proceedings of the International Conference on Offshore Mechanics and Arctic Engineering, Madrid, Spain, 17–22 June 2018; American Society of Mechanical Engineers: New York, NY, USA, 2018; Volume 51296, p. V008T07A010.
19. Lee, S.J.; Jung, K.H.; Ku, N.; Lee, J. A comparison of regression models for the ice loads measured during the ice tank test. *Brodogr. Teor. Praksa Brodogr. Pomor. Teh.* **2023**, *74*, 1–15. [CrossRef]
20. Shi, C.; Hu, Z.; Luo, Y. An elastic-plastic iceberg material model considering temperature gradient effects and its application to numerical study. *J. Mar. Sci. Appl.* **2016**, *15*, 370–375. [CrossRef]
21. Lu, W.; Lubbad, R.; Shestov, A.; Løset, S. Parallel channels' fracturing mechanism during ice management operations. Part I: Theory. *Cold Reg. Sci. Technol.* **2018**, *156*, 102–116.
22. Hansen, E.H.; Løset, S. Modelling floating offshore units moored in broken ice: Model description. *Cold Reg. Sci. Technol.* **1999**, *29*, 97–106. [CrossRef]
23. Lau, M.; Lawrence, K.P.; Rothenburg, L. Discrete element analysis of ice loads on ships and structures. In Proceedings of the SNAME International Conference and Exhibition on Performance of Ships and Structures in Ice, Banff, AB, Canada, 20–23 July 2008; p. D021S008R002.
24. Liu, L.; Ji, S. Dilated-polyhedron-based DEM analysis of the ice resistance on ship hulls in escort operations in level ice. *Mar. Struct.* **2021**, *80*, 103092.
25. Dong, W.; Zhou, L.; Ding, S.; Wang, A.; Cai, J.Y. Two-staged method for ice channel identification based on image segmentation and corner point regression. *China Ocean Eng.* **2024**, *38*, 1–13.
26. Xie, C.; Zhou, L.; Ding, S.; Lu, M.; Zhou, X. Research on self-propulsion simulation of a polar ship in a brash ice channel based on body force model. *Int. J. Nav. Archit. Ocean Eng.* **2023**, *15*, 100557. [CrossRef]
27. Zhou, L.; Riska, K.; Ji, C. Simulating transverse icebreaking process considering both crushing and bending failures. *Mar. Struct.* **2017**, *54*, 167–187. [CrossRef]
28. Gu, Y.; Zhou, L.; Ding, S.; Tan, X.; Gao, J.; Zhang, M. Numerical simulation of ship maneuverability in level ice considering ice crushing failure. *Ocean Eng.* **2022**, *251*, 111110. [CrossRef]
29. Hopkins, M.A.; Hibler IIIW, D.; Flato, G.M. On the numerical simulation of the sea ice ridging process. *J. Geophys. Res. Ocean.* **1991**, *96*, 4809–4820. [CrossRef]
30. Leppäranta, M.; Hakala, R. The structure and strength of first-year ice ridges in the Baltic Sea. *Cold Reg. Sci. Technol.* **1992**, *20*, 295–311. [CrossRef]
31. Bailey, E.; Sammonds, P.R.; Feltham, D.L. The consolidation and bond strength of rafted sea ice. *Cold Reg. Sci. Technol.* **2012**, *83–84*, 37–48. [CrossRef]
32. Parmerter, R.R. A model of simple rafting in sea ice. *J. Geophys. Res.* **1975**, *80*, 1948–1952. [CrossRef]
33. Wang, S. A dynamic model for breaking pattern of level ice by conical structures. In *Acta Polytechnica Scandinavica: Me*; Finnish Acad. of Technology: Helsinki, Finland, 2001; Volume 156.
34. Su, B.; Riska, K.; Moan, T. A numerical method for the prediction of ship performance in level ice. *Cold Reg. Sci. Technol.* **2010**, *60*, 177–188. [CrossRef]
35. Kerr, A.D. The bearing capacity of floating ice plates subjected to static or quasi-static loads. *J. Glaciol.* **1976**, *17*, 229–268.
36. *ISO19906 ISO*; Petroleum and Natural Gas Industries–Arctic Offshore Structures. ISO: Geneva, Switzerland, 2010.
37. Li, Z.J.; Ding, D.; Sui, J.X. Theoretical analysis on the rafted ice thickness in Liaodong Bay. *Mar. Environ. Sci.* **1997**, *16*, 22–26.
38. Shafrova, S.; Høyland, K.V. The freeze-bond strength in first-year ice ridges. Small-scale field and laboratory experiments. *Cold Reg. Sci. Technol.* **2008**, *54*, 54–71. [CrossRef]
39. Chen, X.D. Experimental Study on Sea Ice—Water Thermodynamic Process and Characteristics of Sea Ice Uniaxial Compressive Strength. Ph.D. Thesis, Dalian University of Technology, Dalian, China, 2019.
40. Ni, B.; Wang, Y.; Xu, Y.; Chen, W. Numerical Simulation of Ship Collision with Rafted Ice Based on Cohesive Element Method. *J. Mar. Sci. Appl.* **2024**, 1–10. [CrossRef]

41. Xu, X.S. Research on Ship Resistance Performance in Rafted Ice Region. Master's Thesis Thesis, Harbin Engineering University, Harbin, China, 2022.
42. Hu, J.; Zhou, L. Further study on level ice resistance and channel resistance for an icebreaking vessel. *Int. J. Nav. Archit. Ocean Eng.* **2016**, *8*, 169–176. [CrossRef]

water

MDPI

Article

Construction of Sea Surface Temperature Forecasting Model for Bohai Sea and Yellow Sea Coastal Stations Based on Long Short-Time Memory Neural Network

Yan Jiao [1,2], Ge Li [1,2,*], Peng Zhao [1,2,*], Xue Chen [1,2], Yongzheng Cao [1,2], Guiyan Liu [1,2], Lingjuan Wu [1,2], Xin Xu [3], Di Fu [1,2], Ruoxue Xin [1,2] and Chengzhen Ji [1,2]

1 North China Sea Marine Forecasting Center of State Oceanic Administration, Qingdao 266000, China; jiaoyan87@163.com (Y.J.); jichengzhen@ncs.mnr.gov.cn (C.J.)
2 Shandong Key Laboratory of Marine Ecological Environment and Disaster Prevention and Mitigation, Qingdao 266000, China
3 North China Sea Ocean Technology Support Center of State Oceanic Administration, Qingdao 266000, China
* Correspondence: lige@ncs.mnr.gov.cn (G.L.); zhaopeng@ncs.mnr.gov.cn (P.Z.)

Abstract: In order to address the issue of large errors in predicting SST along the coast using numerical models, this study adopts LSTM, a deep learning method, to develop optimal SST prediction models. The Xiaomaidao Station is selected as an example, and then the method is then extended to 14 coastal stations along the Bohai Sea and the Yellow Sea. The results show that the SST prediction model based on LSTM effectively improves forecast accuracy. The mean absolute errors for 1–3-day SST forecasts of the optimal model at Xiaomaidao Station are 0.20 °C, 0.27 °C, and 0.31 °C, and the root mean square errors are 0.28 °C, 0.36 °C, and 0.41 °C, respectively, representing an average reduction of 78% compared to those of the numerical model. Extending this approach to other forecasting sites along the Bohai Sea and the Yellow Sea results in an average 61% reduction in forecast error when compared with the numerical model. Furthermore, it is found that utilizing an LSTM model can significantly save computational resources and improve the forecasting efficiency.

Keywords: SST; LSTM; optimal forecast model; the Bohai Sea and the Yellow Sea

Citation: Jiao, Y.; Li, G.; Zhao, P.; Chen, X.; Cao, Y.; Liu, G.; Wu, L.; Xu, X.; Fu, D.; Xin, R.; et al. Construction of Sea Surface Temperature Forecasting Model for Bohai Sea and Yellow Sea Coastal Stations Based on Long Short-Time Memory Neural Network. *Water* **2024**, *16*, 2307. https://doi.org/10.3390/w16162307

Academic Editor: Dominic E. Reeve

Received: 4 July 2024
Revised: 10 August 2024
Accepted: 14 August 2024
Published: 16 August 2024

1. Introduction

The Bohai Sea and the Yellow Sea, located along the northern coastline of China, are abundant in marine resources such as fisheries, harbors, petroleum, and tourism. They have been one of the earliest areas in China to be developed and utilized for their marine resources, playing a crucial role in local economic development [1–4]. However, changes in the marine environment can significantly impact the sustainable development of the marine economy through alterations in ocean heat conditions, dynamic processes, and ecological environments [5–8]. Therefore, it is of great significance to study changes in the offshore marine environment.

Sea surface temperature (SST) is a fundamental and crucial element of the ocean. Abnormal changes in SST can result in variations in ocean circulation patterns, fluctuations in sea levels, and changes in the ecological environment [9–14], and even lead to extreme climate events such as extensive sea ice generation or marine heat waves [15–17]. For instance, at the beginning of 2010, SST in the Bohai Sea was unusually low, leading to early and rapid development of sea ice, causing significant impact on the region. The sea ice affected 61,000 people along the Bohai coast, damaged 7157 ships, froze 296 ports and docks along the coast, and damaged 20,787,000 hectares of aquaculture. Additionally, sea ice blocked 13 offshore islands, leaving residents unable to secure daily necessities and emergency supplies. According to statistics, the direct economic loss caused by sea ice in that year reached CNY 6.318 billion [15–17]. Another instance is an unprecedented marine heat wave event in August 2016 in the East China Sea where average SST exceeded 28.7 °C—significantly higher than the climate average by 1.8 °C. The heat wave had a significant impact on marine fisheries and aquaculture. For example, approximately 950,000 mu of sea cucumber aquaculture areas along Liaoning's coast suffered economic losses totaling CNY 6.87 billion. Furthermore, the increased SST led to delayed seeding of wakame in Dalian and other coastal areas as well as dislodging a large number of seedlings from culture ropes, resulting in significant economic losses [18–20]. Therefore, understanding SST development trends and making timely accurate forecasts can provide necessary information for relevant departments to perform disaster prevention work, in order to reduce impacts caused by marine disasters effectively [18–20].

Currently, the operational prediction of SST mainly relies on two methods: numerical models and manual experience. Numerical model prediction has the advantage of including physical processes in the model, allowing for the simultaneous calculation of prediction results across the entire spatial field using large computers. The accuracy of numerical simulation prediction results is high in the vast sea area, but it is lower in coastal sea area due to factors such as local topography, boundary conditions, initial fields, and ocean currents. In contrast to numerical models, manual experience is more effective for coastal forecasting but requires more time and may result in subjective differences depending on forecasters.

In recent years, with the emergence of artificial intelligence (AI), deep learning has once again garnered attention. AI research fields primarily include intelligent robots, machine vision, image recognition, language recognition, natural language processing, and expert systems. The concept of deep learning was first proposed by Hinton et al. from the University of Toronto in 2006 [21], referring to the process of obtaining a deep network structure containing multiple levels based on sample data through specific training methods. Typical network structures used in deep learning include convolutional neural networks (CNNs), recurrent neural networks (RNNs), generative adversarial networks (GANs), and deep belief networks (DBNs). Among these structures, RNN is particularly useful for modeling sequence data where current output depends on previous outputs, which is mainly used for dealing with time series structures.

Long short-time memory (LSTM) is further developed on the basis of RNN by not only retaining its advantages but also addressing issues such as gradient disappearance or the explosion and lack of long-term memory. LSTM's ability for long-term learning makes it suitable for solving predictive problems [22]. Currently, the LSTM method has been preliminarily applied in ocean forecasting [23–25]. For instance, Gao Libin et al. established a wave height prediction model using the LSTM method [26]. The MAE reached a minimum of 0.008 m, the RMSE reached a minimum of 0.012 m, and the correlation coefficient R reached a maximum of 0.999, indicating that LSTM has a good effect in wave height prediction. Gao Song et al. utilized LSTM to forecast ocean waves and compared them with numerical model results [27], and the RMSE and MAE decreased by 18% and 22%, respectively. Zhu Guizhong et al. adopt the LSTM-RNN method to predict the monthly mean SST of the following month in the Western Pacific Ocean, achieving an MAE of 0.15 °C and RMSE of 0.19 °C, significantly improving the accuracy of existing SST prediction models [28].

In this paper, the LSTM method is utilized to replace the numerical forecast model to build the SST intelligent forecast model in the coast of the Bohai Sea and the Yellow Sea, based on the operational SST forecast requirements. The goal is to enhance the prediction accuracy of numerical models and achieve a level comparable to manual empirical prediction. Firstly, an intelligent forecasting model is constructed using the Xiaomaidao Ocean Station as a case study, with the evaluation of forecasting error for the optimal intelligent model. Subsequently, this method is extended to 14 ocean stations along the Bohai Sea and Yellow Sea to construct forecasting models and evaluate their forecasting effects. Finally, limitations of current methods are discussed, and future work prospects are considered.

2. Data and Methods

2.1. SST Observation Data

The SST data use hour-by-hour observations from 14 ocean stations along the coast of Bohai Sea and Yellow Sea from 1 August 2018 to 31 July 2021. The observational data are mainly used for constructing and testing intelligent forecasting models of SST.

In this paper, Xiaomaidao Ocean Station is taken as an example to demonstrate the building process of the intelligent SST prediction model. Built in July 1959, the Xiaomaidao Ocean Station is situated in Xiaomaidao, Laoshan District, Qingdao, China (Figure 1). It stands out as one of the few marine environmental monitoring stations with comprehensive observation and monitoring projects in China. Additionally, it is among the earliest national demonstration stations to implement automated ocean observation, and the location of the measuring point has remained unchanged since the station was built, and the surrounding environment has not changed significantly. Surrounded by the sea and connected to the land by a seawall, Xiaomaidao has a park on the island but no permanent residents. Therefore, the observation and monitoring data collected at this station are very representative and can effectively reflect the fundamental characteristics and changing patterns of the marine environment off Qingdao.

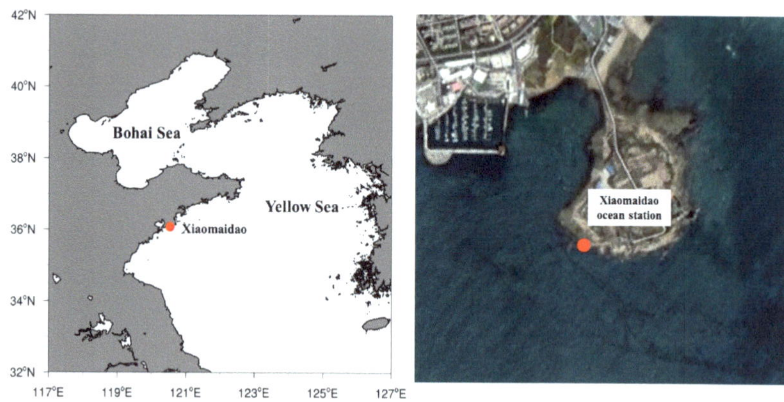

Figure 1. Location of Xiaomaidao Ocean Station.

2.2. Meteorological Forecast Data

The meteorological forecast data are sourced from the operational weather forecast system of North China Sea Marian Forecast and Hazard Mitigation Service. The system is based on the mesoscale meteorological model WRF, incorporating advanced three-dimensional variable data assimilation technology to form an atmospheric initial field to drive the regional atmospheric model. The data utilized for assimilation include conventional meteorological observation data such as GTS, buoys, and ocean stations, as well as non-conventional observation data such as satellites and aircraft. Then, combined with the parameterization scheme for weather forecasting in the Bohai Sea and the Yellow Sea, the operational model of a meteorological numerical forecast is formed, and refined numerical forecast products of meteorological elements for these regions are provided. The model has a maximum horizontal spatial resolution of 3 km, a time resolution of 1 h, and a running time of about 2 h. It can provide hourly weather forecast data for the next 7 days.

From the results of this model, hourly meteorological element data at ocean station locations were extracted including air temperature at 2 m above sea level, relative humidity at 2 m above sea level, wind speed at 10 m above sea level, wind direction at 10 m above sea level, surface heat flux, latent heat flux, etc. The data period is consistent with SST observations and covers 1 August 2018 to 31 July 2021.

2.3. SST Forecast Data

The results of SST numerical prediction are utilized to compare and verify the effect of the intelligent SST model. These predictions come from a three-dimensional temperature–salt–flow regional ocean modeling system (ROMS) operated by North China Sea Marian Forecast and Hazard Mitigation Service.

Three regional ocean models are established using multiple nesting techniques. The large area covers the entire Northwest Pacific Ocean (99°–148° E, 9° S–44° N), with a horizontal resolution of 0.1° and 25 vertical layers. The central area is the East China Sea (117°30′–135° E, 24°–41° N), with a horizontal resolution of 1/30° and 16 vertical layers. The small area covers the Yellow and Bohai Sea area (117°30′–128° E, 32°–41° N), with a horizontal resolution of 1/60° and 16 vertical layers. The output results from the Global ocean model (HYCOM + NCODA Global Analysis) are used as initial and boundary value fields for the large-area model. The simulated values from the upper-level region are used as initial and boundary value fields for both medium- and small-region models. The operation time of the models is about 0.5 h, which can provide hourly SST forecast data for the next 7 days. The construction and operation process of the model, as well as the stability test, are detailed in references [9,29–31]. The data used spans from 1 August 2020 to 31 July 2021.

2.4. Data Quality Control

In dealing with missing values and outliers in observed data, as well as default values in the numerical model, we adopt the difference method to fill gaps when there are less than or equal to 3 occurrences within 24 consecutive times. If there are more than 3 occurrences, the data for that day are not used.

2.5. LSTM Neural Network

2.5.1. Model Introduction

LSTM is a special type of RNN that is well suited for learning long time series information. Figure 2 illustrates the structural comparison between RNN and LSTM. It can be seen that, in an RNN structure, x_t represents the input information and h_t represents the output information. The traditional RNN network structure already has the capability to process time series data by transmitting processing information from previous moments to current moments and then on to subsequent moments. However, a limitation of RNN networks is that they can only receive information from adjacent sequence points, which may lead to issues such as gradient disappearing or gradient explosion when processing long sequence data.

To address this issue, LSTM replaces neural units in RNN with memory cells containing three "gates"—namely "input gates", "output gates", and "forgetting gates". The key component of LSTM is its cell state represented by a horizontal line above each memory cell—similar to a conveyor belt running through the entire chain, allowing for downward flow of information. The "input gate", "output gate", and "forget gate" play crucial roles in selectively letting information through to protect and control the state of neural units by removing or adding output information from previous moments and input information from current moments into unit states.

The formulas involved in this structural diagram are as follows:

$$i_t = \sigma(W_i \cdot [h_{t-1}, x_t] + b_i) \tag{1}$$

$$f_t = \sigma\left(W_f \cdot [h_{t-1}, x_t] + b_f\right) \tag{2}$$

$$o_t = \sigma(W_o \cdot [h_{t-1}, x_t] + b_o) \tag{3}$$

$$\widetilde{c}_t = \tanh(W_c \cdot [h_{t-1}, x_t] + b_c) \tag{4}$$

$$c_t = f_t \times c_{t-1} + i_t \times \widetilde{c}_t \tag{5}$$

$$h_t = o_t \times \tan\mathrm{h}(c_t) \quad (6)$$

In the formula, i_t, f_t, and o_t represent the "input gate", "forgetting gate", and "output gate" at time t, respectively; x_t represents the input information at time t; h_{t-1} represents the output of the previous time; W and b are the corresponding weight coefficient matrix and offset top, respectively; σ and tanh denote the Sigmoid activation coefficient and hyperbolic tangent activation function, respectively; $\widetilde{c}_t$ represents the temporary cell status; c_t represents the cell status update value at time t; and h_t is the output at time t.

After calculating the forgetting gate, input gate, and temporary cell status, the cell unit will update the cell status of the current moment. Finally, the output gate determines the output value ht of the current moment. More detailed information about LSTM can be found in reference [32].

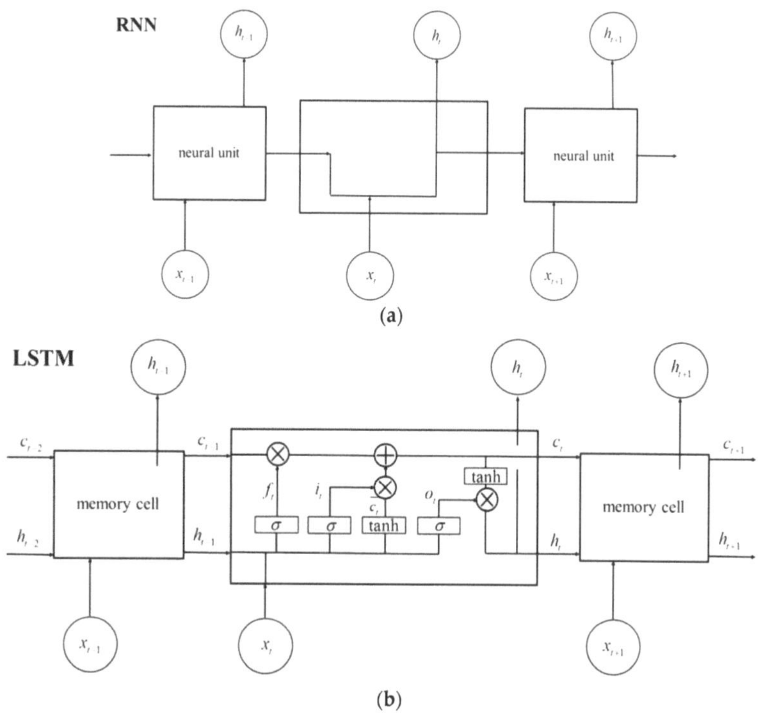

Figure 2. Structure comparison of RNN (**a**) and LSTM (**b**).

2.5.2. Model Settings

After quality control, there are 1034 days of valid data from 1 August 2018 to 31 July 2021. The data are divided into two periods: 70% (725 days) for the training model and 30% (309 days) for the testing model. The objective of this paper is to solve the problem of short-term forecasting of SST for 3 days; therefore, we set the prediction length to 72 h in order to obtain time-by-time forecasting results for SST. To achieve better training results, the parameters of the LSTM model are set as follows through control experiments: numHiddenUnits are set to 200, MaxEpochs to 50, InitialLearnRate to 0.005 s, and LearnRateDropFactor to 0.2. Finally, in order to improve the stability and accuracy of the forecast, the ensemble forecast results of 10 members are used as the final SST forecast results (Table 1).

Table 1. Parameter settings of LSTM model.

Parameters	Value
training set/%	70
test set/%	30
forecast time/hour	72
historical time/hour	72
numHiddenUnits	200
MaxEpochs	50
InitialLearnRate	0.005
LearnRateDropFactor	0.2
ensemble members	10

2.6. Test Indicators

Two indices, *MAE* and *RMSE*, were selected as indicators to measure the forecasting effect of the model using the following formula:

$$MAE = \frac{1}{m}\sum_{i=1}^{m}|YPRED_i - YTEST_i| \quad (7)$$

$$RMSE = \sqrt{\frac{1}{m}\sum_{i=1}^{m}(YPRED_i - YTEST_i)^2} \quad (8)$$

where *YPREDi* represents the model's prediction result for the *i*th sample, *YTESTi* represents the observed result for that sample, and *m* represents the number of samples used for testing.

3. Experimental Design

3.1. Experimental Scheme Setting

In this experiment, our focus is on predicting hourly changes in SST over a three-day period (represented as "Y") at Xiaomaidao Ocean Station. We will be considering factors such as hourly observation values of SST over the past three days as well as future meteorological elements potentially related to SST changes at this station (represented as "X"). It includes air temperature at 2 m above sea level, relative humidity at 2 m above sea level, wind speed at 10 m above sea level, wind direction at 10 m above sea level, surface heat flux, latent heat flux, and air–sea temperature difference [33–36]. The basic information regarding these predictive factors is shown in Table 2.

Table 2. Predictive factors for LSTM model.

ID	Variable	Abbreviation	Time Series	Data Source
①	Sea surface temperature	SST	(t − 71)~t [1]	observation
②	Air temperature	AT	(t + 1)~(t + 72)	numerical model
③	Relative humidity	RH		
④	Wind speed	WS		
⑤	Wind direction	WD		
⑥	Surface heat flux	HFx		
⑦	Latent heat flux	LH		
⑧	Air–sea temperature difference	AT-SST		observation and numerical mode

Note: [1] t is the running time of the model, and t + 1 is the first time of the forecast.

In this experiment, SST is the forecast target, and the observation data before its start time are the basic information for training; therefore, SST is a mandatory factor for each group of experiments. When designing the experiments, EXP-1 is trained with SST as the only factor, meaning that only observed values of SST in the past are used to predict future SST changes. EXP-2 to EXP-7, respectively, added one meteorological factor such as air temperature, relative humidity, wind, etc., to SST observations. Based on tests on

EXP-2 to EXP-7, the impacts of different meteorological factors on SST are discriminated. It should be noted that the wind vector consists of wind speed and direction, and the sea–air temperature difference refers to the difference between air temperature and the last observed SST. EXP-8 serves as a reserved experiment, which combines two meteorological factors that have had the greatest influence on SST selected from EXP-2 to EXP-7. In conclusion, a total of 8 groups of experiments are designed and each group undergoes 10 rounds of training, resulting in a total of 80 experiments (refer to Table 3). If it turns out that minimal error occurs in EXP-8, then more diverse combinations of factors will be adopted for further experiments.

Table 3. Experimental scheme settings.

ID	Variable Combination	Training Times
EXP-1	SST	10
EXP-2	SST, AT	10
EXP-3	SST, RH	10
EXP-4	SST, WS, WD	10
EXP-5	SST, HFx	10
EXP-6	SST, LH	10
EXP-7	SST, (AT-SST)	10
EXP-8	optimal combination of EXP-2 to EXP-7	10

3.2. Experimental Process

Figure 3 illustrates the flowchart depicting the establishment process for the LSTM method prediction model. The specific steps include (1) reading quality controlled data from the file and standardizing it; (2) training the prediction factor (XTRAIN) based on LSTM in the training set to predict the target (YTRAIN), storing the trained neural network as "NET"; (3) calling "NET", inputting the testing set's prediction factor (XTEST), and calculating the prediction target (YPRED); (4) testing YPRED against YTEST in the testing set; (5) selecting the experiment with the smallest error across all experiments as the optimal prediction model (OPM). In the daily operational forecasting, the forecast value can be obtained by simply calling NET and inputting the value of the prediction factors.

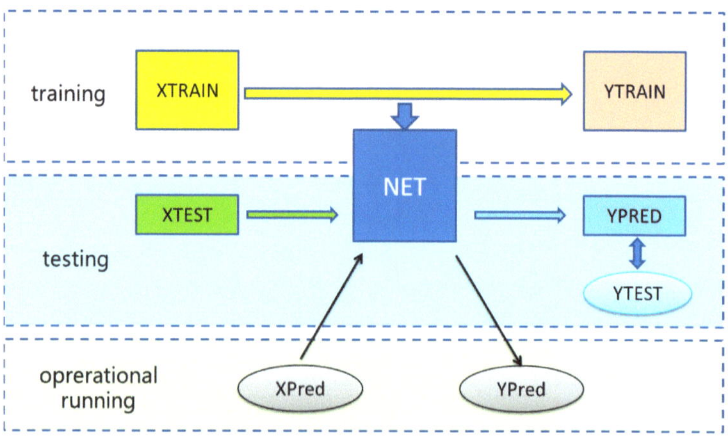

Figure 3. Flowchart of LSTM for establishing and running forecast model.

4. Results and Tests

Based on Table 3′s experimental scheme settings, the model is trained and tested using an LSTM neural network. Figure 4 displays EXP-1′s daily and hourly test results. In Figure 4a,c, blue columns represent daily MAE and RMSE of ensemble member forecasts, while red columns represent ensemble forecasts of 10 members. It is evident that ensemble prediction errors are smaller than those of individual members, indicating that ensemble prediction based on LSTM models can enhance stability and accuracy compared to single models. In Figure 4b,d, black lines depict hourly MAE and RMSE of ensemble member forecasts, with red lines representing those of ensemble forecasts—further demonstrating improved stability and accuracy.

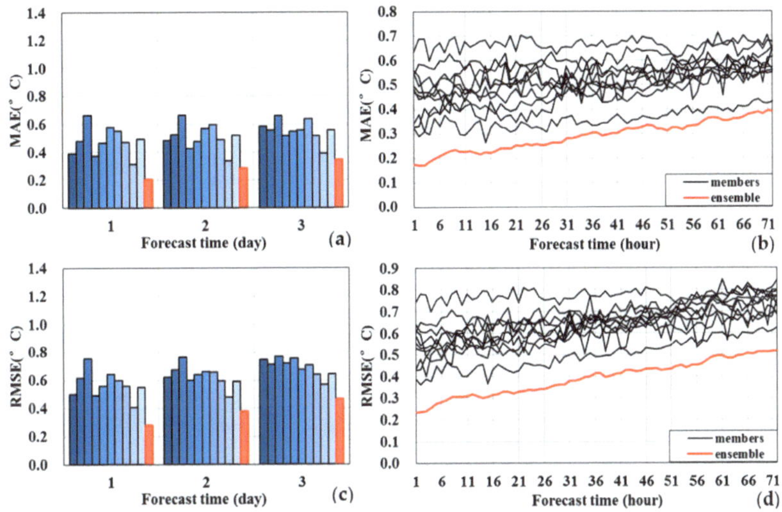

Figure 4. Results of daily and hourly forecast tests for EXP-1 with blue columns and black lines representing ensemble members, while red columns and red lines represent ensemble results of 10 members. (**a**) MAE of daily forecasts; (**b**) MAE of hourly forecasts; (**c**) RMSE of daily forecasts; (**d**) RMSE of hourly forecasts.

The same method was used to train EXP-2 through EXP-7 models; however, due to space constraints, only the best results after comparison are shown instead of listing each experiment's test results like EXP-1. Table 4 lists ensemble forecast test results for EXP-2 through EXP-7 as well as EXP-1. It is apparent that the overall effect is best for EXP-5 with errors on the second and third days smaller than those in EXP-1. The results of individual member forecasts versus ensemble forecasts of EXP-5 are depicted in Figure 5.

Based on the experimental scheme outlined above, we select the two experiments with the smallest experimental error from EXP-2 to EXP-7, combine their factors to form EXP-8, train the model, and compare its prediction effect.

Table 4. Test errors for EXP-1 to EXP-7.

	MAE			RMSE		
ID	**Day 1**	**Day 2**	**Day 3**	**Day 1**	**Day 2**	**Day 3**
EXP-1	0.20	0.28	0.34	0.28	0.38	0.46
EXP-2	0.32	0.32	0.33	0.45	0.45	0.47
EXP-3	0.46	0.49	0.53	0.62	0.66	0.72
EXP-4	0.58	0.59	0.63	0.81	0.82	0.85
EXP-5	0.24	0.27	0.31	0.32	0.36	0.41
EXP-6	0.34	0.37	0.40	0.50	0.55	0.61
EXP-7	0.29	0.31	0.33	0.46	0.49	0.53

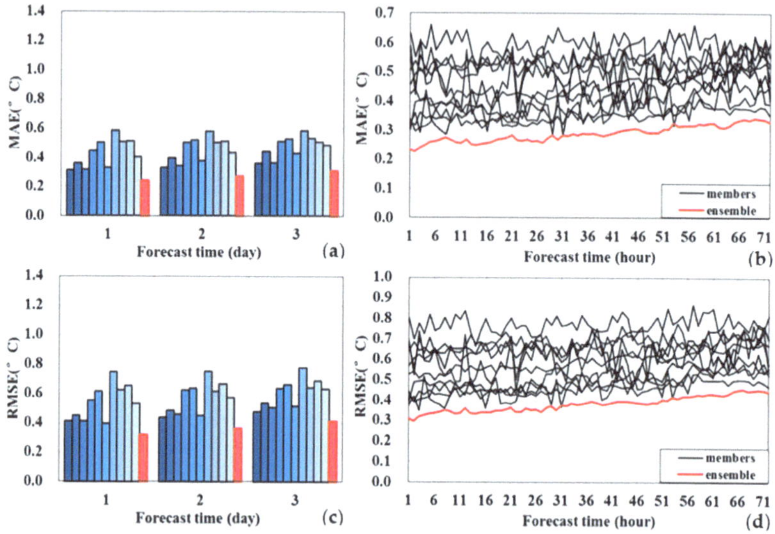

Figure 5. Results of daily and hourly forecast tests for EXP-5, with blue columns and black lines representing ensemble members, while red columns and red lines represent ensemble results of 10 members. (**a**) MAE of daily forecasts; (**b**) MAE of hourly forecasts; (**c**) RMSE of daily forecasts; (**d**) RMSE of hourly forecasts.

Figure 6 illustrates the prediction errors of eight LSTM models (green columns) alongside those of a numerical model (yellow columns). The figure indicates that EXP-1 to EXP-8 yield much smaller prediction errors compared to those of the numerical model, demonstrating clear advantages of deep learning models in coastal ocean prediction. Specifically, for 1-day SST predictions, EXP-1 performs best followed by EXP-5; for 2-day SST predictions, EXP-5 excels followed by EXP-1; meanwhile, for 3-day SST predictions, EXP-5 demonstrates superior performance followed by EXP-2. Based on these findings, the OPM of SST for Xiaomaidao Ocean Station is constructed by combining the 1-day forecast from EXP-1 with the 2–3-day forecast from EXP-5. The forecast effect is depicted in Figure 7. The MAE values for 1–3 days using the OPM are 0.20 °C, 0.27 °C, and 0.31 °C, respectively, while the RMSE values are 0.28 °C, 0.36 °C, and 0.41 °C (Figure 7a,c). In terms of hourly forecast errors, the MAEs range between 0.10 °C and 0.40 °C for forecasts from the 1st hour to the 72nd hour, with RMSEs ranging between 0.20 °C and 0.50 °C (Figure 7b,d). On average, the OPM reduces forecast errors by as much as 78% compared to those of the numerical model.

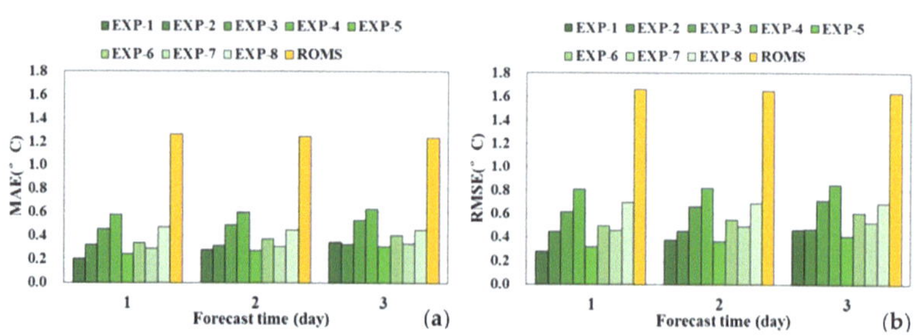

Figure 6. MAE (**a**) and RMSE (**b**) for all experimental schemes and numerical models at Xiaomaidao Ocean Station, with green columns representing EXP-1 to EXP-8, while yellow columns represent the numerical model.

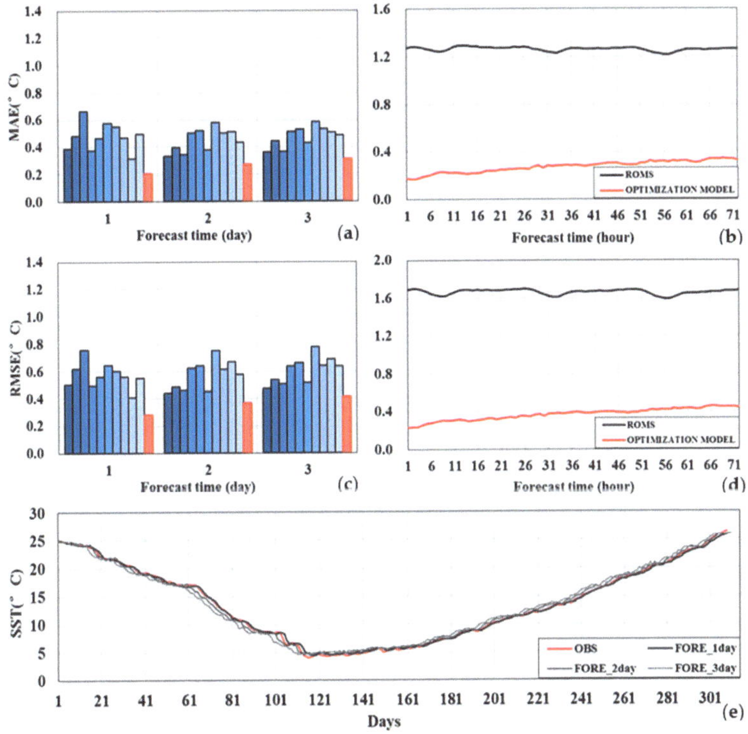

Figure 7. Results of daily and hourly forecast tests for OPM. (**a**,**c**) MAE and RMSE of daily forecasts (blue columns represent ensemble members and red columns represent ensemble results of 10 members); (**b**,**d**) MAE and RMSE of hourly forecasts (red lines represent OPM and black lines represent numerical model); (**e**) comparison of daily forecast results with observed data.

5. Model Promotion

The method used to construct the OPM of SST at Xiaomaidao Ocean Station has been extended to 14 stations along the Bohai Sea and the Yellow Sea in order to improve forecasting accuracy across a wider area.

Another example, the Xiaoshidao Ocean Station, is used to demonstrate the forecasting performance of this method. Situated in the northeast of the Shandong Peninsula and facing the Yellow Sea to the north, the Xiaoshidao Ocean Station is approximately 220 km away from Xiaomaidao. As depicted in Figure 8, the forecasting errors of eight LSTM models are significantly smaller than that of the numerical model. Among them, Exp-5 has the smallest forecast error across 1–3 days, leading us to adopt the LSTM model trained by EXP-5 as the OPM for Xiaoshidao station. The MAEs for OPM range from 0.21 °C to 0.28 °C over a span of 1–3 days, while RMSEs range from 0.30 °C to 0.40 °C, decreasing by 76% compared with those produced by the numerical model.

Figure 9 illustrates the percentage improvement/reduction of the forecast effect at 14 stations in the Bohai Sea and the Yellow Sea. The results show that the coastal SST forecast error, when utilizing the LSTM method, is reduced by an average of 61% compared to the numerical model. Despite variations in the geographical location, surrounding environment, and different impact factors, it is evident that this method can enhance prediction accuracy to a certain degree when compared with the numerical model. Furthermore, it should be noted that the OPM running time obtained through the test is less than one minute, which significantly saves computing resources and obviously improves the forecast efficiency compared with the numerical model.

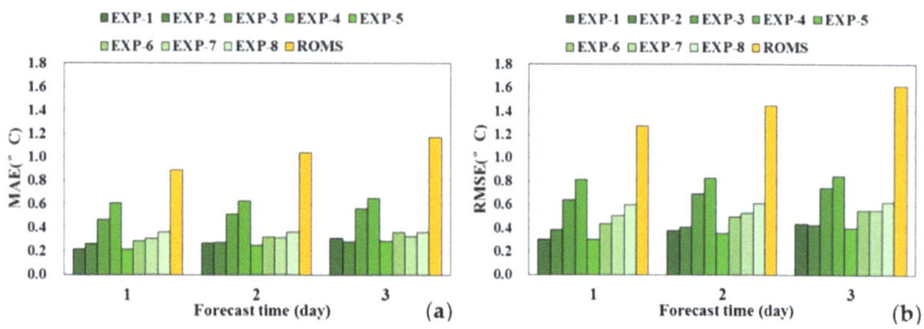

Figure 8. The same as Figure 6, but at the Xiaoshidao Ocean Station. (**a**) MAE; (**b**) RMSE.

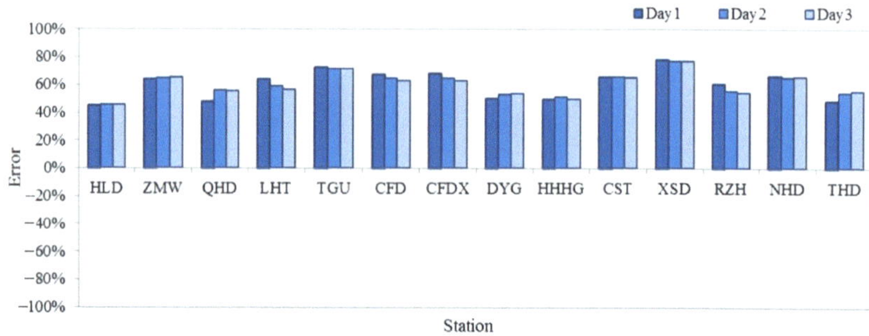

Figure 9. The percentage improvement/reduction of the forecast effect at each ocean station (positive values indicating the forecast error of the OPM lower than that of the numerical model, while negative values indicate the error of the OPM higher than that of the numerical model).

6. Summary and Discussion

In order to address large errors in predicting SST along coastlines using numerical models, this paper constructed SST prediction models for coastal stations in the Bohai Sea and the Yellow Sea based on LSTM—a type of deep learning network.

Firstly, Xiaomaidao Ocean Station was selected as an example to design an SST forecasting experiment. Factors related to SST changes—such as air temperature, wind vector, and heat flux—were extracted from the meteorological numerical model and combined with observed SST data to design different experimental schemes for LSTM model training. After testing forecast errors for each scheme, a combination yielding minimal error was selected as OPM. The 1–3-day MAEs of the OPM are 0.20 °C, 0.27 °C, and 0.31 °C, while the RMSEs are 0.28 °C, 0.36 °C, and 0.41 °C, respectively. In terms of hourly forecast errors, the MAEs range between 0.10 °C and 0.40 °C for forecasts from the 1st hour to the 72nd hour, with RMSEs ranging between 0.20 °C and 0.50 °C. When compared with the prediction results of the numerical model at the same time, it is found that the error of the OPM is reduced by an average of 78%.

The OPM construction method used for Xiaomaidao Ocean Station is extended to include 14 ocean stations along the Bohai Sea and the Yellow Sea. OPMs are constructed for each station and when compared with results from a numerical SST model for the same period, it is observed that on average, errors in predictions made by LSTM optimal models are 61% lower than those made by numerical models. This indicates that this method is universally applicable and can effectively improve coastal SST forecast accuracy. Similar studies have also been consulted. For instance, Zhang et al. developed an LSTM daily forecast model for SST in the equatorial Pacific (10° S–10° N, 120.0°–280° E) for the next 10 days, with an RMSE of 0.6 °C for the eastern equatorial Pacific and less than 0.3 °C for both central and western regions [37]. Han et al. utilized the LSTM model to predict daily SST at five buoy points in the East China Sea, with an MAE and RMSE of 0.25 °C and 0.28 °C for a one-day forecast, respectively [38]. The prediction errors of SST in these studies are similar to those found in this study, indicating that our constructed model is reasonable, reliable, and effective, especially considering the difficulty of predicting coastal SST compared to open sea SST. Furthermore, it is noted that the run time for all 14 stations using OPMs is less than one minute in total, which significantly saved computing resources and improved forecasting efficiency. Currently, this method has become a crucial reference for predicting SST in the Bohai Sea and the Yellow Sea. After an initial period of operation, it will be extended to a wider range of ocean stations in the future.

According to the OPM constructed at Xiaomaidao and Xiaoshidao ocean stations, as well as other stations, it is evident that the sea surface heat flux is the most significant factor influencing the change in SST. Following this, in terms of influence, are the sea–air temperature difference, latent heat flux, air temperature, relative humidity, and wind speed and direction. However, these factors are not orthogonal; that is, the factors affect each other. In our next step, we will consider performing the orthogonal decomposition of the influencing factors before screening them and then proceed to build a prediction model for the time series of each mode. Additionally, this study did not take into account oceanic factors such as tidal currents. Future research will consider these oceanic factors to enhance the accuracy of SST prediction. In terms of model building, we plan to integrate convolutional neural networks (CNNs) and LSTM to develop a hybrid model. The hybrid model could not only forecast the time series of SST but also incorporate linkage information between different sites.

Author Contributions: Conceptualization, G.L. (Ge Li); Methodology, Y.J.; Validation, Y.J. and Y.C.; Investigation, X.X.; Resources, G.L. (Guiyan Liu) and L.W.; Data curation, P.Z.; Writing—original draft, Y.J. and X.C.; Writing—review & editing, G.L. (Ge Li), R.X. and C.J.; Visualization, D.F.; Project administration, P.Z. All authors have read and agreed to the published version of the manuscript.

Funding: This research is supported by the Shandong Provincial Key Laboratory of Marine Ecology and Environment and Disaster Prevention and Mitigation (202104); The National Key R&D Program of China (2019YFC1408403).

Data Availability Statement: Data are contained within the article.

Acknowledgments: The authors thank the editor and anonymous reviewers for their valuable comments and suggestions to this paper.

Conflicts of Interest: The authors declare no conflicts of interest.

References

1. Zhang, Y. Development, utilization and prospect forecast of offshore oil and gas resources in Bohai Sea. *Ocean Dev. Manag.* **2003**, *20*, 6–9. (In Chinese)
2. Niu, B. Research on the Countermeasure of Shandong Province Marine Economy Sustainable Development under Resources and Environment Restraint. Master's Thesis, Ocean University of China, Qingdao, China, 2008. (In Chinese)
3. Wang, Q.; Li, Y. *Research on Marine Resources Exploitation and Environmental Problems in Bohai Sea*; China Ocean Press: Qingdao, China, 2018. (In Chinese)
4. Zhang, Y.; Han, Z.; Liu, K.; Liu, G. Study of the exploitation of marine resources—A case study of Liaoning Province. *J. Nat. Resour.* **2010**, *25*, 785–794. (In Chinese)
5. Kong, H.; Peng, B.; Liu, R. Impacts analysis of climate change on China's marine economy. *Mar. Environ. Sci.* **2018**, *37*, 9. (In Chinese)
6. Li, H.; Gao, Q. Scientific and Technological Progress, Marine Economic Development and Ecological Environmental Change. *East China Econ. Manag.* **2017**, *31*, 8. (In Chinese)
7. Zhang, B. Protection of Marine environment and sustainable development of Marine economy. In Proceedings of the 5th China Modernization Research Forum, Beijing, China, 3 August 2007. (In Chinese)
8. Liu, M. Analysis of important problems affecting the sustainable development of Marine economy in our country. *Dev. Res.* **2010**, *27*, 57–61. (In Chinese)
9. Wu, L.; Huang, J.; Ding, Y. Analysis on the causes of massive stranding of Yellow Sea green tide on Lianyungang and Rizhao coasts in 2022. *J. Oceanol. Limnol.* **2024**, *42*, 816–830. [CrossRef]
10. Wang, Y. Interference of Tropical Indian Ocean SSTA in Relationships between the Central Pacific Type ENSO and Climate Variations in the Maritime Continent Region. Master's Thesis, Nanjing University of Information Science and Technology, Nanjing, China, 2017.

11. Zhang, H.; Tao, L.; Xu, C. SST forcing off California coast and its relationship to the decadal variation of North Pacific. *Chin. J. Atmos. Sci.* **2022**, *46*, 859–872. (In Chinese)
12. Jiao, Y.; Cao, C.; Li, G.; Yuan, B.; Jiang, W.; Yu, Q. Prediction method of air temperature and sea ice in winter in the Bohai Sea and the Yellow Sea based on teleconnection. *Mar. Forecast.* **2017**, *34*, 19–24. (In Chinese)
13. Jiao, Y.; Shang, J.; Gao, S.; Wu, L.; Li, J.; Yu, Q.; Zhao, Y.; Fu, D. Influencing factors and prediction method of the green tide scale in the Yellow Sea. *Ocean Sci.* **2022**, *46*, 65–73. (In Chinese)
14. Jiao, Y.; Huang, F.; Liu, Q. Interannual Variability and Scenarios Projection of Sea Ice in Bohai Sea Part I: Variation Characteristics and Interannual Hindcast. *J. Ocean Univ. China* **2020**, *19*, 272–280. [CrossRef]
15. Guo, K.; Shang, J.; Li, G.; Cao, C.; Jiang, C.; Wang, X. Analysis of Sea Ice Conditions in the Bohai Sea and the northern Yellow Sea in 2009–2010. *Mar. Forecast* **2011**, *28*, 37–42. (In Chinese)
16. Jiang, C.; Jiang, F. Thinking on Sea Ice Disaster Monitoring, Early Warning and Disaster Prevention and Reduction in Bohai Sea. *Mar. Deve. Manag.* **2013**, *30*, 20–22. (In Chinese)
17. Li, Z. The Relationship between Sea Ice disaster and Human Activities in Bohai Sea. *Mar. Forecast* **2010**, *27*, 10–14. (In Chinese)
18. Tan, H.; Cai, R. What caused the record-breaking warming in East China Seas during August 2016? *Atmos. Sci. Lett.* **2018**, *19*, e853. [CrossRef]
19. Tan, H.; Cai, R.; Bai, D.; Karim, H.; Kareem, T. Causes of 2022 summer marine heatwave in the East China Seas. *Adv. Clim. Chang. Res.* **2023**, *14*, 633–641. [CrossRef]
20. Hu, Z.; Shan, T.; Zhang, J.; Zhang, Q.; Critchley, A.T.; Choi, H.G.; Yotsukura, N.; Liu, F.; Duan, D. Kelp aquaculture in China: A retrospective and future prospects. *Rev. Aquac.* **2021**, *13*, 1324–1351. [CrossRef]
21. Hinton, G.; Osindero, S.; Teh, Y. A fast learning algorithm for deep belief nets. *Neural Comput.* **2006**, *18*, 1527–1554. [CrossRef]
22. Yang, H. Research on Weather Foresting Based on Deep Learning. Master's Thesis, Harbin Institute of Technology, Harbin, China, 2017.
23. Gao, S.; Zhao, P.; Pan, B. A nowcasting model for the prediction of typhoon tracks based on a long short term memory neural network. *Acta Oceanol. Sin.* **2018**, *37*, 8–12. [CrossRef]
24. Jiao, Y. Analysis of Sea Ice Characteristics in Bohai Sea and Research on Multi-Time Scale Forecasting Methods. Master's Thesis, Ocean University of China, Qingdao, China, 2019.
25. Jiao, Y.; Huang, F.; Gao, S.; Liu, Q.; Ji, C.; Wang, N.; Cao, Y.; Yu, Q. Research on extended-range forecast model of sea ice in the Liaodong bay based on the long short term memory network. *Period. Ocean Univ. China* **2020**, *50*, 1–11. (In Chinese)
26. Gao, L.; Guo, M.; Zhang, S. Wave Height Forecast based on LSTM. *Fujian Comput.* **2018**, *34*, 109–111.
27. Gao, S.; Bi, F.; Huang, J. Wave Height Numerical Prediction Model Results Correction Method. Patent for invention CN112307676A, 2021. (In Chinese)
28. Zhu, G.; Hu, S. Study on sea surface temperature model based on LSTM-RNN. *J. Appl. Oceanogr.* **2019**, *38*, 7. (In Chinese)
29. Wu, L.; Wang, J.; Gao, S.; Zheng, X.; Huang, R. An analysis of dynamical factors influencing 2013 giant jellyfish bloom near Qinhuangdao in the Bohai Sea, China. *Estuar. Coast. Shelf Sci.* **2017**, *185*, 141–151. [CrossRef]
30. Wu, L.; Xu, J. Ensemble Trajectory Simulation of Large Jellyfish in the Yellow and Bohai Sea. In Proceedings of the 2016 3rd International Conference on Chemical and Biological Sciences, Amsterdam, The Netherlands, 23–25 March 2016; Volume 60. [CrossRef]
31. Wu, L.; Gao, S.; Xu, J. Development of Huanghai and Bohai Coastal Sea High Resolution and Operational Numerical TemperatureSalinity-Current Forecasting System. *J. Inst. Disaster Prev.* **2015**, *17*, 83–91. (In Chinese)
32. Christopher. Understanding. LSTM Networks. 2015. Available online: http://colah.github.io/posts/2015-08-Understanding-LSTMs/ (accessed on 15 May 2024).
33. Yi, X.; Dong, W.; Li, S. The characteristics of sea temperature variation in the Kuroshio area in the East China Sea and its causes. *Mar. Forecast* **2021**, *38*, 14. (In Chinese)
34. Lu, Y.; Liu, Y. Characteristics of Indian Ocean SST intraseasonal variation and its correlation with atmosphere. In Proceedings of the 32nd Annual Meeting of Chinese Meteorological Society-Subtropical Meteorology and Environmental Impacts, Tianjin, China, 14 October 2015. (In Chinese)
35. Wang, H. Variability of Subsurface Ocean Temperature in Tropical Pacific and Its Relationship to ENSO Cycle. Ph.D. Thesis, Institute of Oceanology, Chinese Academy of Sciences, Qingdao, China, 2009. (In Chinese)
36. Zhao, G.; Song, Y. The response of sea temperature interannual variations to sea surface wind stress and its sensitivity experiment. *J. Appl. Meteorol. Sci.* **1998**, *9*, 48–58. (In Chinese)
37. Zhang, T.; Lin, P.; Liu, H.; Zheng, W.; Wang, P.; Xu, T.; Li, Y.; Liu, J.; Chen, C. Short-Term Sea Surface Temperature Forecasts for the Equatorial Pacific Based on Long Short-Term Memory Network. *Chin. J. Atmos. Sci.* **2024**, *48*, 745–754. (In Chinese)
38. Han, Y.; Sun, K.; Yan, J.; Dong, C. Prediction of East China Sea SST based on VMD-LSTM-BLS Hybird Model. *Laser Optoelectron. Progress* **2023**, *60*, 1–9. (In Chinese)

water

MDPI

Article

Volume-Mediated Lake-Ice Phenology in Southwest Alaska Revealed through Remote Sensing and Survival Analysis

Peter B. Kirchner * and Michael P. Hannam

Southwest Alaska Network, Inventory and Monitoring, National Park Service, Anchorage, AK 99501, USA; michael_hannam@nps.gov
* Correspondence: peter_kirchner@nps.gov

Abstract: Lakes in Southwest Alaska are a critical habitat to many species and provide livelihoods to many communities through subsistence fishing, transportation, and recreation. Consistent and reliable data are rarely available for even the largest lakes in this sparsely populated region, so data-intensive methods utilizing long-term observations and physical data are not possible. To address this, we used optical remote sensing (MODIS 2002–2016) to establish a phenology record for key lakes in the region, and we modeled lake-ice formation and breakup for the years 1982–2022 using readily available temperature and solar radiation-based predictors in a survival modeling framework that accounted for years when lakes did not freeze. Results were validated with observations recorded at two lakes, and stratification measured by temperature arrays in three others. Our model provided good predictions (mean absolute error, freeze-over = 11 days, breakup = 16 days). Cumulative freeze-degree days and cumulative thaw-degree days were the strongest predictors of freeze-over and breakup, respectively. Lake volume appeared to mediate lake-ice phenology, as ice-cover duration tended to be longer and less variable in lower-volume lakes. Furthermore, most lakes < 10 km^3 showed a trend toward shorter ice seasons of −1 to −6 days/decade, while most higher-volume lakes showed undiscernible or positive trends of up to 2 days/decade. Lakes > 20 km^3 also showed a greater number of years when freeze-over was neither predicted by our model (37 times, n = 200) nor observed in the MODIS record (19 times, n = 60). While three lakes in our study did not commonly freeze throughout our study period, four additional high-volume lakes began experiencing years in which they did not freeze, starting in the late 1990s. Our study provides a novel approach to lake-ice prediction and an insight into the future of lake ice in the Boreal region.

Keywords: Southwest Alaska; lake ice; survival model; remote sensing; freeze-degree days; thaw-degree days; phenology; boreal region

Citation: Kirchner, P.B.; Hannam, M.P. Volume-Mediated Lake-Ice Phenology in Southwest Alaska Revealed through Remote Sensing and Survival Analysis. *Water* **2024**, *16*, 2309. https://doi.org/10.3390/w16162309

Academic Editors: Pentti Kujala, Zhijun Li, Fang Li, Weiping Li and Shifeng Ding

Received: 21 June 2024
Revised: 27 July 2024
Accepted: 6 August 2024
Published: 16 August 2024

1. Introduction

The seasonality of lake ice in cold region ecosystems is a key modulator of ecosystem function because it influences the physical, chemical, and biological systems of lakes through temperature and light profiles, dissolved gas concentrations, biological productivity, and human livelihoods. Studies of the Northern Hemisphere demonstrate that lake-ice cover in the past century has decreased in the order of 0.5–1 day per decade [1–4], and projections of continued warming at high latitudes indicate trends of delayed ice-cover formation and earlier breakup will continue [5–9]. Lakes are critical components of high-latitude ecosystems because they support foundational species, such as salmon, and the regional biodiversity that local populations depend on. The presence or absence of lake ice is also important to high-latitude rural communities that depend on its formation for winter travel to surrounding communities and subsistence activities such as ice fishing, hunting, trapping, and firewood or water collection [10–12].

With detailed physical data it is possible to produce models and lake-specific indices that robustly describe and predict multiple physical processes including the timing of lake stratification, and ice formation, thickening, and breakup [13–15]. Some of the most robust methods of estimating lake-ice formation rely upon quantifying the energy balance and include variables such as air and water temperatures, vapor pressure deficit, and snow properties [16–19]. Unfortunately, comprehensive observations of lake ice, or even the most rudimentary physical measurements such as local air temperature, lake-water temperature, or mean depth and volume are rarely available in the sparsely populated Arctic boreal regions. In this study, only a very limited set of measurements and observations were available for a few of the many large and biologically important lakes in Southwest Alaska. Thus, there is a need for a method to estimate historic lake ice phenology and predict how these systems might respond to a warming climate.

Because we are unable to parameterize physical models of lake ice phenology, we might look to statistical approaches for prediction. While several studies have successfully used traditional statistical approaches to evaluate lake ice phenology [20], traditional statistical methods such as multiple regression are not well-suited to time-to-event data such as lake-ice formation or breakup because years in which freeze-over does not occur are very informative but cannot be easily encoded as a date [21]. Linear mixed-effects models have been successfully used to study ice-off dates in alpine lakes [22] where non-freezing years are not a concern, and circular regression has shown promise in addressing the cyclical nature of lake-ice dynamics [23], but neither of these approaches effectively deals with non-freezing years. Alternatively, survival models which were originally developed through medical research for understanding the effects of interventions on patient survival were created specifically for analyzing time-to-events that sometimes do not occur. They have found increasing use in vegetation phenology research and provide an effective method for modeling phenology [24–26]. Furthermore, these recent implementations allow for the incorporation of daily meteorological data rather than being limited to seasonal climate summaries as predictors.

In this manuscript we (i) establish a 16-year satellite observation record of lake-ice phenology for 15 large freshwater lakes and two clusters of smaller lakes in Southwest Alaska for the water years 2002–2016, (ii) develop a daily survival model to predict lake-ice phenology using spatially-interpolated observations of antecedent thaw-degree days, cumulative thaw-degree days, cumulative freeze-degree days, and downwelling shortwave radiation, (iii) use this model to hindcast and forecast results for the water years 1981–2022, (iv) validate the model with in situ observations from five lakes, and (iv.) evaluate these results using a local 75-year fall–winter temperature record to gain a long term perspective.

2. Methods

2.1. Study Area

The seventeen lakes and lake clusters we studied were located in Southwest Alaska between 57–61° N latitude and 149–156° W longitude and range in surface area from Lake Iliamna at 2637 km^2 to two clusters of smaller lakes, each typically $\leq$5 km^2 (Figure 1). Several of these lakes are important spawning habitats for the world's largest sockeye salmon fishery, and most are within the boundaries of two US National Parks (Lake Clark National Park and Preserve, and Katmai National Park and Preserve), and two National Wildlife Refuges (NWRs) (Kenai NWR and Becharof NWR). The lake settings and morphology vary in elevation, size, surface, orientation, and surrounding terrain, which affect the solar influx, wind, and meso-climate that influence lake-ice formation and phenology (Table 1). In general, the largest lakes were formed by retreating ice sheets and debris damming, the medium-sized lakes were formed by mountain glaciers, whereas, the small-clustered lakes were formed by thermokarst activity and high latitude (freeze–thaw) hydrology, and soil formation processes on a glaciated landscape. Some of the lakes (e.g., Lake Clark, Chakachamna, Telaquana, and Twin) are located in faulted mountain valleys where glaciers follow natural weakness and are consequently long and narrow. Others (e.g., Illiamna, Naknek, and Becharof) are found in flatter terrain and are formed by retreating ice fields or piedmont glaciers with morainal damming, resulting in broader, more rounded shapes (Figure 1).

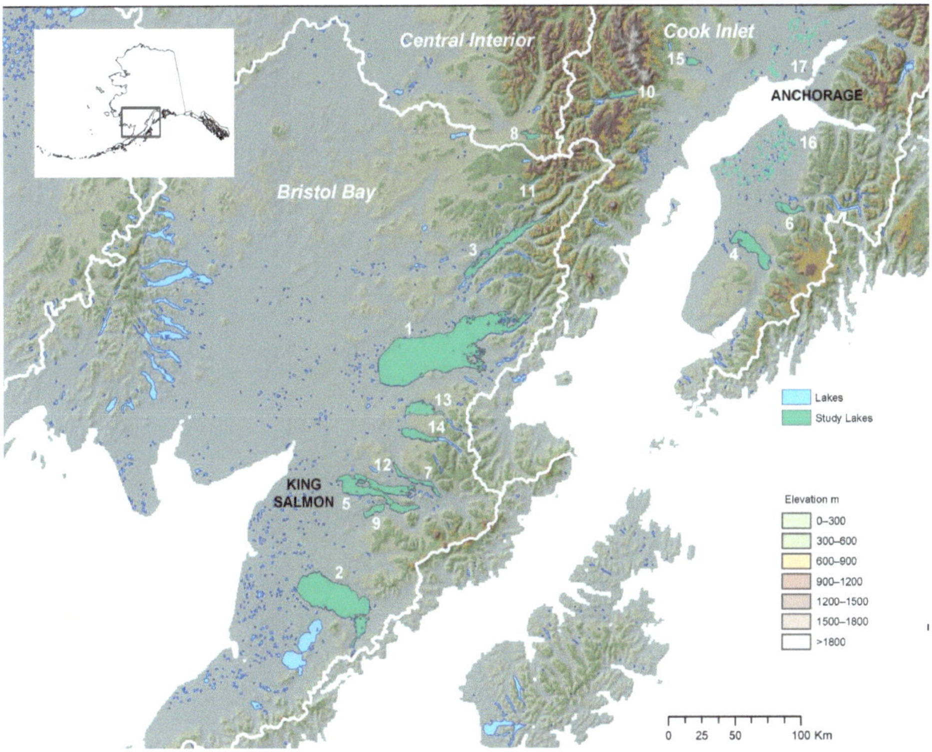

Figure 1. Location of study area, lake number refer to lake names and metrics listed in Table 1.

To evaluate the impact of lake volume on lake-ice phenology, we established a relative volume ranking. While only five of the study lakes had bathymetric data, 12 of the 15 lakes (excluding clusters) had a common metric of maximum depth available (Table 1). To harmonize these data, we calculated surface areas using USGS digital raster graphics [27], and obtained the maximum depth data from NPS measurements, or the most up-to-date source available [28–30]. We then used a simple cone-volume derivation that has been shown to perform reasonably well for small-to-medium-sized lakes to establish a volume ranking [31].

2.2. Ice Phenology Observations

Lake-ice phenology was estimated through supervised classification using the MODIS Rapid Response System Land Surface Reflectance composites of true color (bands 1-4-3), corrected near-infrared (NIR, Bands 7-2-1), and corrected shortwave infrared (SWIR, Bands 3-6-7) at an image resolution of 250 m with a temporal frequency of up to 1-day (Figure 2) [32]. The presence of snow-covered ice and clear water on lake surfaces of daily MODIS data was interpreted, and the percentage of ice cover was estimated using a 1 km^2 grid, where the percentage of lake ice equaled the percentage of visually counted grid cells with $\geq$50% ice cover. At the time of our analysis, MODIS Rapid Response System data were not available for 2009, and these images were evaluated using previous ocular estimates of the percentage of frozen surface area; a subset comparison of the two methods showed <10% error between them.

Table 1. Lake Characteristics.

Lake [A]	Area km^2	Lake Elev. m	Max. Depth m [B]	Conical Volume est. km^3	Known Volumes km^3	No-Freeze Years n = 15 MODIS [C]	No-Freeze Years n = 40 Model [D]	Duration Trend Days/Year [E]	MAD Days [F]	RMAD [G]
1 Iliamna Lake	2637	11	301	264.58	115.3	3	6	−0.5 * [−0.8,−0.2]	40	0.43
2 Becharof Lake	1195	4	92	36.65	44.0	5	11	0.2 * [0.0,0.6]	31	0.57
3 Lake Clark	336	75	266	29.79	32.3	3	3	0.2 [−0.8,0.8]	46	0.63
4 Tustumena Lake	298	34	290	28.81	-	4	11	0.2 [−0.4,0.9]	80	1.35
5 Naknek Lake	458	13	160	24.43	-	4	6	0.2 [−0.1,0.5]	33	0.44
6 Skilak Lake	99	58	174	5.74	7.2	2	2	−0.1 [−0.1,−0.4]	53	0.69
7 Lake Grosvenor	73	35	112	2.73	-	3	3	−0.2 [−0.2,−0.4]	30	0.40
8 Telaquana Lake	48	376	132	2.11	2.9	0	0	−0.5 * [−0.7,−0.1]	23	0.21
9 Lake Brooks	75	20	82	2.05	-	1	2	0.0 [−0.3,0.4]	25	0.28
10 Chakachamna	74	346	80	1.97	-	1	0	−0.1 [−0.8,0.4]	15	0.11
11 Twin Lakes	27	601	103	0.93	-	0	0	−0.5 * [−0.8,−0.2]	33	0.20
12 Lake Coville	33	35	62	0.68	-	0	0	−0.6 * [−0.7,−0.3]	31	0.23
13 Kukaklek Lake	173	247	-	-	-	0	0	−0.6 * [−0.7,−0.4]	20	0.23
14 Nonvianuk Lake	133	191	-	-	-	0	0	−0.6 * [−0.7,−0.4]	34	0.24
15 Beluga Lake	44	75	-	-	-	0	0	−0.3 * [−0.5,−0.2]	10	0.06
16 Northern Kenai	88	60	-	-	-	0	0	−0.3 * [−0.5,−0.1]	16	0.10
17 Lower Susitna	61	39	-	-	-	0	0	−0.6 * [−0.9,−0.3]	15	0.09

Notes: [A] Lake number corresponds with Figure 1 ordered by relative volume, where max depth was available; [B] maximum lake-depth estimates, ***lakes in cryptodepressions***; [C] number of ***no-freeze observations*** in 15-year MODIS record; [D] number of years of ***<50% probability of freeze*** in 40-year model run; [E] positive and ***negative*** trends in duration of freeze (* credibly different from 0) [with upper and lower terciles]; [F] MAD (variation) of MODIS-observed duration of ice cover; [G] RMAD (relative variation) of MODIS-observed duration of ice cover.

Figure 2. MODIS image classification for Lake Iliamna 25 February 2015. True color (bands 1, 4, 3) (**left**), NIR and SWIR (bands 7, 2, 1) (**middle**), and estimation of percentage of ice cover using grid mask (**right**).

The metrics used in this analysis were freeze-over (>90%), breakup (<90%), and duration (days between freeze-over and breakup) (Figure 3). If the scene was not interpretable due to cloud cover or a missing image, the metric was reported as the median day between the first and last interpretable images, and the number of days between interpretable images was reported as uncertainty [33]. Data are reported in day of water year (DWY) and labeled for the year in which it ends (e.g., water year 2016 is from 1 October 2015 to 30 September 2016). To summarize the variability in observations, we report the mean absolute deviation ($MAD = n^{-1}\sum_i^n|\overline{x} - x_i|$) or, when more informative, we present the relative mean absolute deviation ($RMAD = MAD/\overline{x}$) (Table 1).

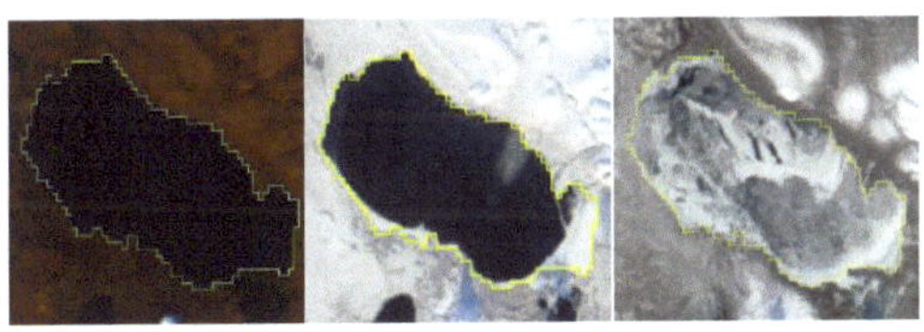

Figure 3. Examples of metric thresholds from Lake Becharof. MODIS true color images with no ice cover on October 11, 14% on December 23, and 92% (freeze-over) on February 15.

2.3. Meteorologic Data

The meteorologic data used for modeling were extracted from the gridded daily weather and climatology variables of Daymet [34], available from the Oak Ridge National Laboratory Distributed Active Archive Center. We extracted data at each lake and lake cluster centroid for the water years of 1981–2022. The variables were daily minimum and maximum air temperature, and incident shortwave radiation; mean temperatures were considered the mean of the daily minimum and maximum temperatures.

$$\left(\overline{T}_{DWY,y,i} = 0.5\left(T^{max}_{DWY,i,y} + T^{min}_{DWY,i,y}\right)\right. \tag{1}$$

Antecedent thaw degree days for water year y at each lake i were calculated as the sum of the daily mean temperature if greater than 0 °C for an accumulation period of June 1st (DWY = 245) through September 30th (DWY = 365) of the previous water year.

$$antTDD_{y,i} = \sum_{DWY=245}^{365} \max\left(0, \overline{T}_{DWY,y-1,i}\right) \tag{2}$$

Accumulated climate variables were calculated at daily time steps for each day d of water year y and lake i as the sum of the daily climate value over each day from the first day (s) of the accumulation period to day d. For modeling freeze-over, the accumulation period began on October 1st, and for modelling breakup the accumulation period began on the date of freeze-over. In this way, accumulated freeze-degree days (FDD) were calculated:

$$FDD_{d,w,i} = \sum_{DWY=s}^{d} \min\left(0, \overline{T}_{DWY,y,i}\right) \tag{3}$$

accumulated thaw-degree days TDD were calculated:

$$TDD_{d,y,i} = \sum_{DWY=s}^{d} \max\left(0, \overline{T}_{DWY,y,i}\right) \tag{4}$$

and accumulated solar radiation SWR was calculated:

$$SWR_{d,y,i} = \sum_{DWY=s}^{d} Rad_{DWY,y,i} \tag{5}$$

where *Rad* is the incident solar radiation.

For Southwest Alaska, 1 October was chosen for accumulating *FDD* because little or no accumulation of *FDD* occurred before that date in the lakes studied. Specifically, 12 of 17 lakes accumulated no *FDD* before 1 October during the satellite observation period. For the five lakes where *FDD* did accumulate before 1 October, the accumulated *FDD* was less than 6% of those accumulated by the date of freeze-over, and we did not expect this minimal *FDD* accumulation to be influential (Appendix C). The accumulation period for *antTDD* captured >90% of the interannual variability (Figure A5), suggesting this approach will adequately capture the influence of seasonally accumulated heat content.

2.4. Survival Model

Survival models have been used to represent time-to-event data, allowing for events that never happen. Thus, survival models are well suited to modeling lake-ice processes, where ice formation (in this case >90% ice cover) may not always occur. Our model uses a daily, discrete-time, survival modelling approach, borrowing from work in plant phenology [24–26]. By modelling the state of lake ice at a daily time step, we were able to use daily weather data to predict freeze-over and breakup, and generate more precise predictions.

The daily ice-cover state $F_{\{d,y,i\}}$ for day d of water year y at lake i was encoded as a Bernoulli variable, where ice cover < 90%: 0 and ice cover > 90%: 1. We modeled the daily probability of a state change, given the previous day's state, as a linear function with a logit link.

$$\Pr\left(F_{\{d,y,i\}} = 1 \middle| F_{\{d-1,y,i\}} = 0\right) = \text{logit}^{-1}\left(\beta_{\{0,i\}} + \beta_{\{1,i\}} \cdot FDD_{\{d,y,i\}} + \ldots\right) \tag{6}$$

$$\Pr\left(F_{\{d,y,i\}} = 0 \middle| F_{\{d-1,y,i\}} = 1\right) = \text{logit}^{-1}\left(\gamma_{\{0,i\}} + \gamma_{\{1,i\}} \cdot TDD_{\{d,y,i\}} + \ldots\right) \tag{7}$$

Our model allowed for lakes to have differing baseline probabilities of a state-change, and differing sensitivities to meteorologic covariates by incorporating random effects for intercepts and covariate coefficients at the individual lake level. We made inferences using a Bayesian framework with JAGS version 4.3.1 [35], and the jagsui [36], and tidybayes [37] packages in R version 4.2.3 [38]. Models were fit using Markov Chain Monte Carlo (MCMC). Three MCMC chains were burned in until the Gelman–Rubin statistic [39] was less than 1.1, before sampling the posterior. We present posterior medians and credible intervals (0.95 and 0.66) for meteorologic sensitivity estimates and model predictions. Meteorologic sensitivity is presented as the change in the log of the odds ratio of the hazard resulting from a 1-*sd* change in the meteorologic variable. As such, positive estimates indicate an increased probability of a state change with an increase in the meteorologic variable. Note that cumulative freeze-degree days become more negative as they accumulate, so a negative effect of freeze-degree days indicates an increase in the probability of a state change with colder temperatures. For the predicted date of freeze-over and breakup, we present the median of the survival distribution function, which is the first day that the cumulative probability of a state change exceeds 0.5 (i.e., the first day on which the lake is more likely to have frozen than not).

We assessed model fit using the r^2 metric recommended by Gellman et al., 2019 [40]:

$$r^2 = var(\hat{F}) / (var(\hat{F}) + var(E)) \quad (8)$$

where $\hat{F}$ is a vector of model predictions and E is a vector of model residuals. This formulation corresponds to the marginal r^2 presented by Nakagawa et al., 2012 [41] because random effects are included in the explained variance. To understand overall model fit, we present an omnibus r^2 calculated from the full dataset. Because we are primarily interested in how our models captured the year-to-year variation rather than lake-to-lake variation, we also calculated and reported r^2 separately for each lake using lake-specific predictions and residuals from the full model. To further describe model fit we present the mean absolute error ($MAE = n^{-1}\Sigma_i^n \left|\hat{f} - f_i\right|$). A detailed model description is presented in Appendix A.

3. Results

3.1. Satellite Observations

Over the 15 years of MODIS satellite observations used to calibrate our model, freeze-over (>90% ice cover) typically occurred the earliest (mid-November) in the Northern Kenai, Lower Susitna, and Beluga lakes, and the latest (late January to early February) in Becharof, Lake Clark, and Tustumena. Across all lakes, the median observed start date of freeze-over was December 20, with a *MAD* of 33 days. Breakup (<90% ice cover) typically occurred the earliest (late April to early May) in Becharof, Lake Clark, and Tustumena, and the latest (early June) in the highest elevation lakes—Twin Lakes and Telaquana (Figure 4). Breakup generally occurred more rapidly and the timing of breakup did not vary as greatly; the median start date of breakup was May 4, with a *MAD* of 19 days. The median duration of MODIS-observed ice cover > 90% was 131 days, with a *MAD* of 53 days. Several lakes experienced years in which >90% ice cover was not observed (Table 1).

Year-to-year variation within a given lake ranged from 7 to 27 days *MAD* for lake-ice formation, 10 to 80 days *MAD* for duration, and 4 to 22 days *MAD* for breakup. Scaling variation by ice-season duration revealed even greater differences. The most variable lake by this measure (Tustumena Lake) had an expected year-to-year difference in ice-season duration that exceeded its average ice-season duration by 140% (*RMAD* = 1.4), whereas the most consistent lake (Beluga Lake) could be expected to vary by only 6% of its average duration (*RMAD* = 0.06, Table 1).

In some years, multiple lakes did not freeze-over, most notably eight lakes did not freeze-over in 2003 and 2016: Iliamna, Becharof, Lake Clark, Tustumena, Naknek, Skilak, Grosvenor, and Chakachamna in 2003, and Brooks in 2016. In 2015, four lakes did not freeze-over: Iliamna, Becharof, Naknek, and Grosvenor. In 2014, two did not freeze-over: Becharof and Tustumena. In 2013, a year with sustained wind events, two did not freeze-over: Lake Clark and Tustumena (Table 1 and Figure 4). The lack of freeze-over in some years, and the interannual variability in freeze-over, duration, and breakup reflect the dynamic climate of the Southwest Alaska region during fall and early winter, which oscillates between warm and cold temperatures over several weeks and, since the 1980s, has experienced warm fall/winter temperature anomalies. The long-term variability of this cold accumulation period can be seen in the record of in situ observations from King Salmon (Figure 5) [42].

3.2. Survival Model

Our survival model for freeze-over generally fit better than our model for breakup (Figures A1 and A2). Our model described 87% of the variation in freeze-over and 52% of the variation in breakup across all lakes and years. The year-to-year variance explained by the model varied substantially among lakes: from highs of 97% and 85% to lows of 47% and 24% for freeze-over and breakup, respectively. Mean absolute error varied greatly among lakes, from 5 days to 36 days with an average of 11 days for freeze-over, from 4 to 46 days with an average of 16 days for breakup, and from 6 to 41 days with an average of 19 days for duration of ice cover (Table 1, and Figures 6, A1 and A2).

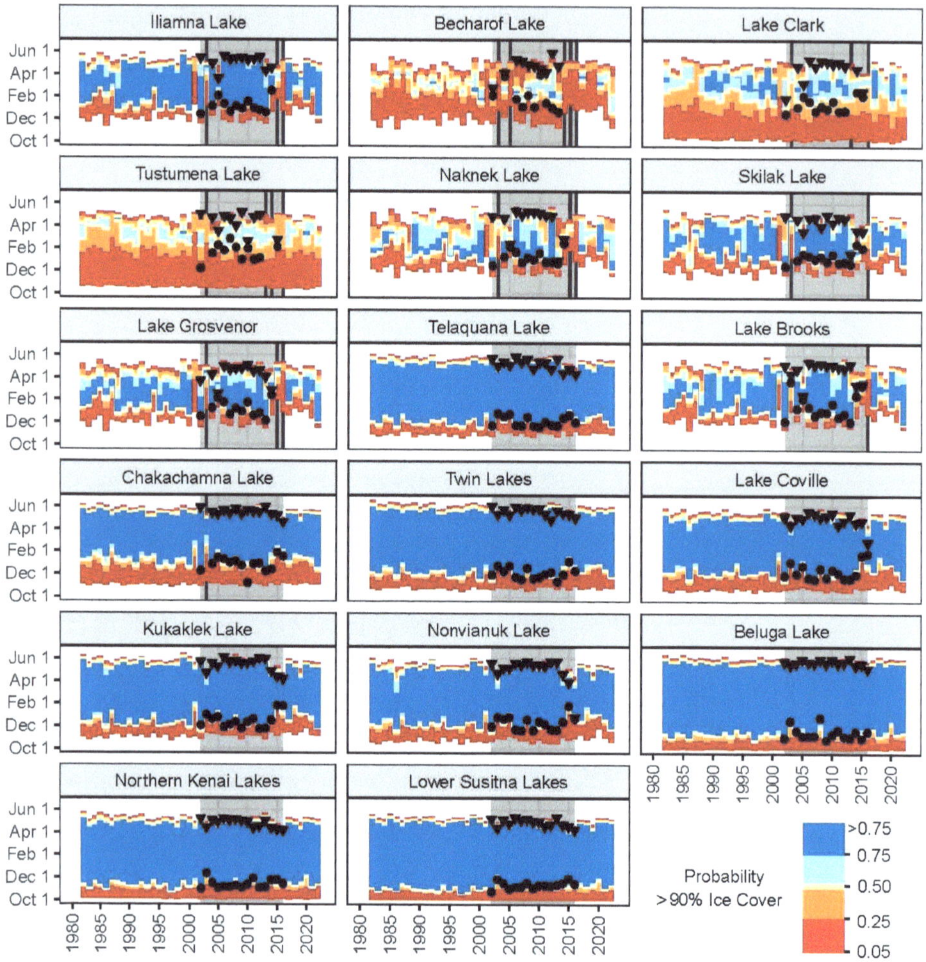

Figure 4. Modeled and observed ice cover. Modeled probability of ice cover > 90% by lake, for period of study, with observed MODIS dates of 90% freeze-up (black circles) and breakup (black triangles). The period of interpreted satellite observation is spanned by the light grey background, and years in which the lakes did not freeze are marked by dark grey vertical background bars. The highest-volume lakes are at the top of the figure.

Chilling, as measured by *FDD*, generally increased the freeze hazard (probability of freeze-over happening today if it has not happened already), but sensitivity to chilling varied greatly among lakes (Figure 7). We found evidence that greater antecedent summer heat loading (*antTDD*) decreased the freeze hazard for some lakes, but the effect was small, uncertain, and not consistently important across all lakes. Breakup hazard (probability of breakup happening today if it has not happened already) was increased at all lakes by warming (*TDD*) and cumulative ice-season solar radiation (*SWR*). Continued chilling during the ice season (*FDD*) decreased the breakup hazard for some lakes. Sensitivity to continued chilling was generally more pronounced in larger-volume lakes (Figures 7 and 8).

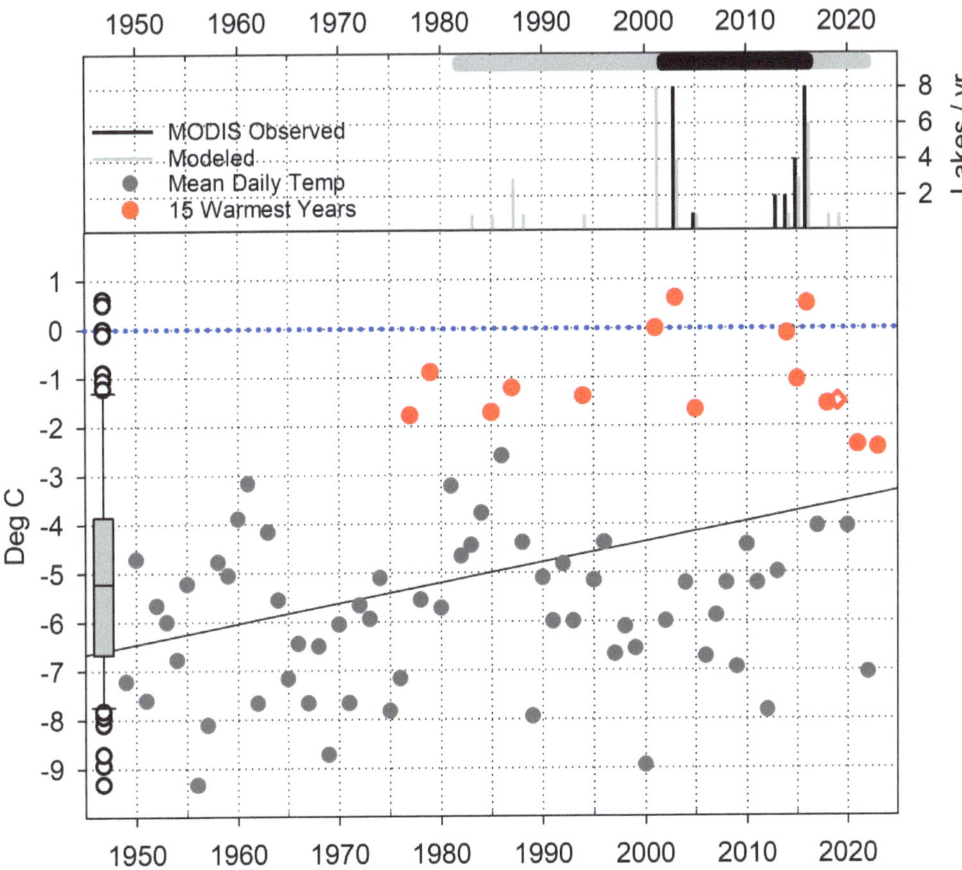

Figure 5. 1949 to 2022 regional record of King Salmon airport daily mean air temperatures during the fall/winter cold-accumulation period. Lower panel shows the 15 warmest periods (red), where 4 show a mean above or near 0 °C (dotted blue line) since the early 2000s (2019 had insufficient data, and the diamond represents nearby station). Upper panels' vertical bars show the number of lakes observed by MODIS (black) and modeled (grey) for each year that lakes did not fully freeze-over; the horizontal bars show the years of observation and modeling in this study. The boxplot on the left shows the high variance of data and predominance of years during the observation and modeling periods relative to all years.

The *RMAD* of the model-predicted ice-season duration for each lake agreed well with the *RMAD* of MODIS observations (Spearman rank correlation, $\rho = 0.92$), suggesting that the meteorologic variables that drove our model predictions could account for the observed patterns of lake variability. Our model correctly predicted 14 of 25 events when the lakes did not freeze, with one falsely predicted event. Years in which lakes did not freeze were also predicted in the anomalously warm years of the 1980s for the Becharof Lake (2), Lake Clark (2), and Tustumena Lake (6), but no non-freeze years for any other lake before 2000 (Figure 4).

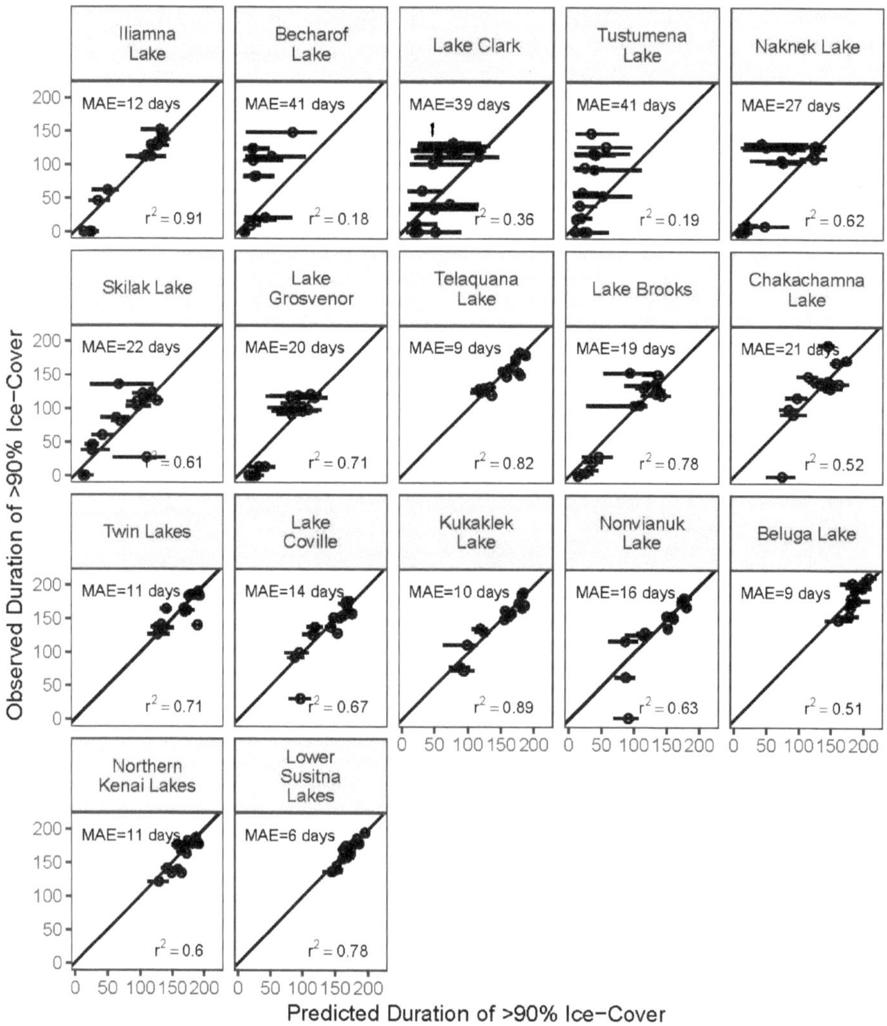

Figure 6. Model performance at each study lake. Observed versus predicted values (posterior median +/− 95% credible intervals) for the duration of ice cover, 1:1 line in black. Mean Absolute Error and r^2 of model predictions are shown for each lake. Higher-volume lakes are at the top of figure.

We estimated decreasing trends in ice-cover duration in most, but not all, study lakes, ranging from −6 to + 2 days per decade. Over the 40 years represented in our model, nine lakes had a credibly negative trend, one lake had a credibly positive trend, and five lakes had no discernable trend. All but one lake showing a decreasing trend were <20 km^3, whereas all but one lake with a positive or undiscernible trend were >20 km^3 (Table 1).

3.3. Lake Volumes

Estimated lake volumes spanned four orders of magnitude, from 0.68 to >264 km^3, where five of the lakes were >24 km^3 (three with known volumes) and the remainder were under 7 km^3. When compared to the five known volumes calculated using bathymetric measurements, we underestimated volumes in four of the lakes by 8 to 37%, and overestimated the largest lake (Lake Iliamna) by 56% (Table 1). While the volume calculations were not as precise as those made using bathymetry, considering the large differences in lake volume, this did provide a basis for an ordinal comparison that would otherwise not have been possible.

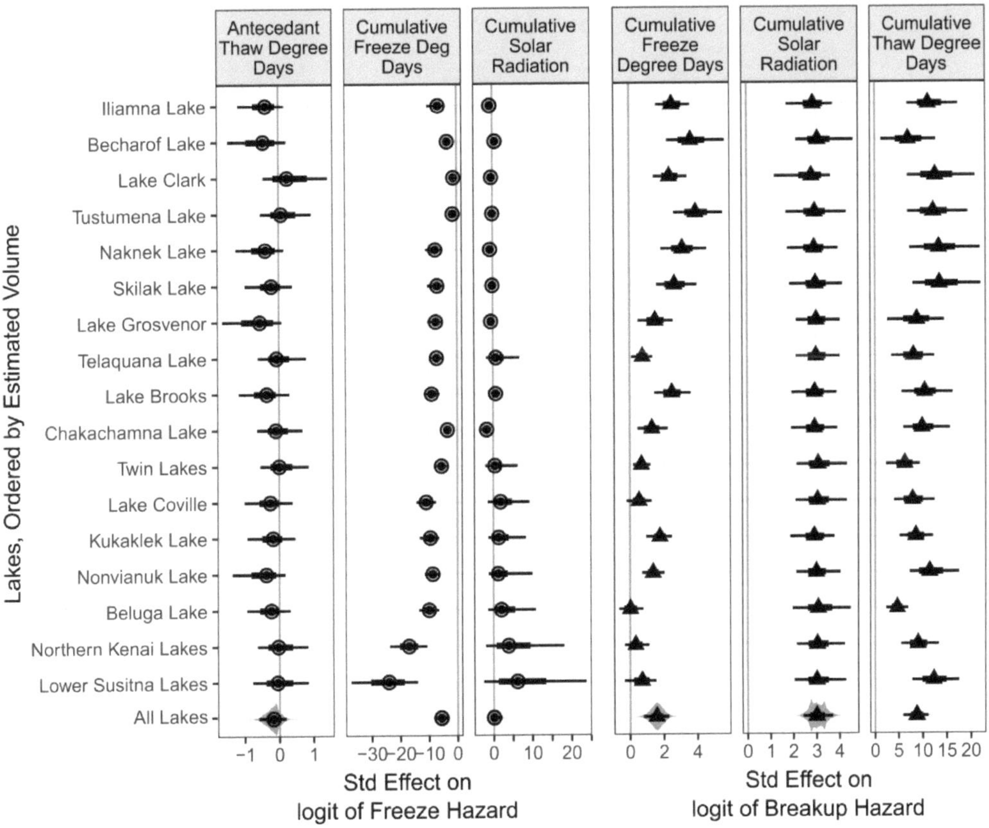

Figure 7. Meteorologic variables' effects on freeze-start (circles, left 3 panels) and breakup (triangles, right 3 panels). Each panel shows the effect of a 1-*sd* change in the predictor variable on the log of the odds-ratio of the hazard, where the freeze hazard is the probability of 90% or more ice cover occurring today, given less than 90% ice cover yesterday, and the breakup hazard is the probability of less than 90% ice cover today, given greater than 90% ice cover yesterday. Model coefficients are estimated by their posterior medians (points) +/− 95% credible intervals (thin bars) and 66% credible intervals (thick bars). Greater magnitude coefficients indicate a stronger relationship or greater apparent sensitivity to a meteorologic variable. Higher-volume lakes are at the top.

Lakes with larger estimated volumes were later to freeze-over (Spearman's rank correlation, $\rho = -0.82$), had a shorter duration of ice cover ($\rho = 0.82$), and were less likely to freeze in years with fewer *FDD*s (Table 1, and Figures 4 and 8). Relative variability in ice-season duration was greater in larger-volume lakes ($\rho = 0.84$), driven both by the tendency for large-volume lakes to be more variable in absolute terms ($\rho = 0.59$) and for larger-volume lakes to have shorter ice-season durations. Model-estimated sensitivity to chilling decreased with lake volume ($\rho = -0.80$).

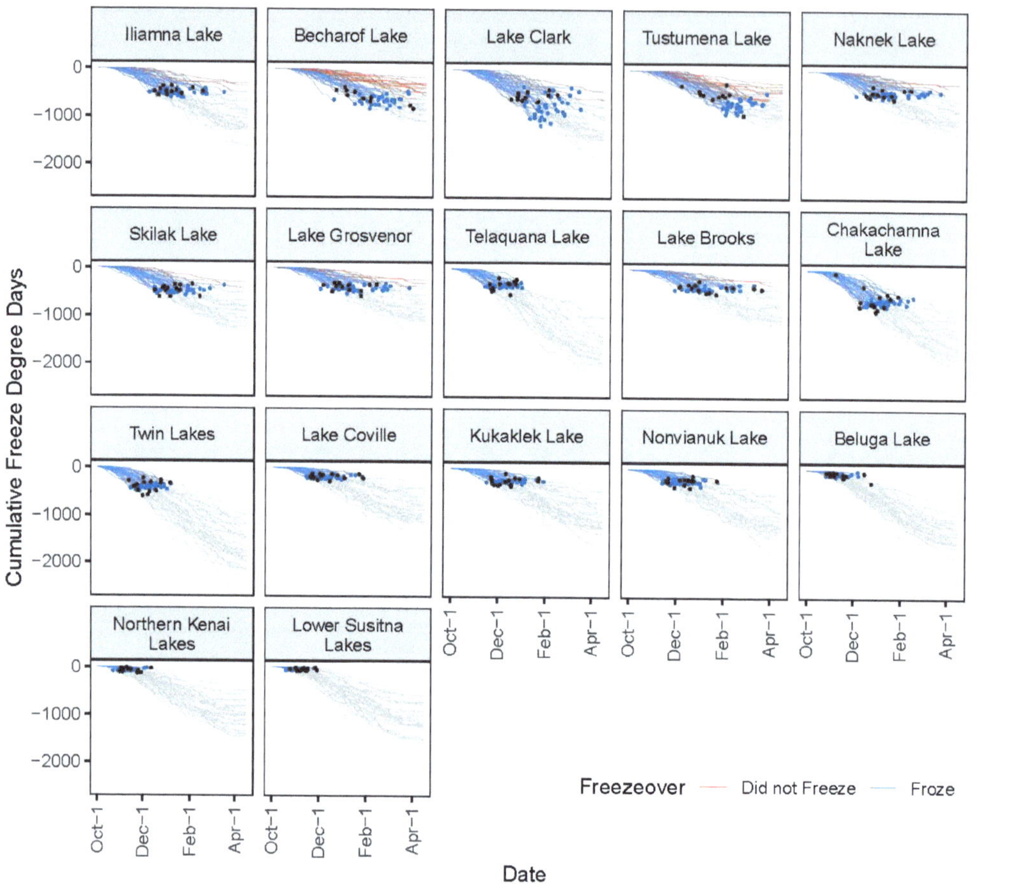

Figure 8. Freeze degree day (*FDD*) accumulation. Cumulative *FDD* before freeze-over shown in blue, after freeze-over shown in grey, date of freeze-over indicated by dot. The years in which our model indicated that lakes did not freeze by March-30 are shown in red. Higher volume lakes are at the top of figure.

3.4. Validation against In Situ Observations

Because previous studies have found good agreement between in situ observations of lake-ice formation and melt with optical (MODIS and Sentinel) satellite remote-sensing observations, we do not present a separate validation of the satellite-observed ice season dates. Zhang et al. [43] found a *MAE* of 6–8 days in over 400 observations of lake-ice formation and breakup in the state of Maine in lakes that varied in size, depth, latitude, and elevation, whereas Tuttle et al. [44] found *MAE* of approximately 2–5 days for the ice breakup dates of a lake in Svalbard, Greenland, over a 15-year observation period.

To validate our model-predicted lake-ice seasons, we looked to available in situ observations. While there were no formal systematic observations of the percentage of ice cover for our study lakes, such as regular air photos or observers following a consistent protocol, there were three independent data sets for five lakes that were available to validate our results with. These were: (i) continuous observations of water temperature from sensor arrays deployed in three lakes, and (ii) two different long-term visual observations of ice cover for the purposes of navigating boats, snowmachines, or aircraft on two lakes.

We estimated winter stratification over sensor arrays deployed in the deepest parts of three lakes. Winter stratification can be a good indicator of ice formation in lakes that are dimictic (i.e., mix twice a year by turning over the higher-density water to depth as it heats or cools to approximately 4 °C). However, these are not always representative of ice >90% cover because the relatively small area of water directly over these locations may not be ice-covered, and the deepest parts of lakes tend to be the last to freeze. The arrays deployed in Lake Clark (100 m), Naknek (70 m), and Brooks Lake (50 m) were used to collect temperature profiles at 10 m increments over the years of 2006–2022 [45]. To estimate the duration of winter stratification, we measured the duration of time that shallow waters were cooler than deep waters by more than a mean daily temperature difference of 0.5 and 1.5 °C for the shallower (50 m) and deeper (70 and 100 m) lakes, respectively [8]. All three lakes failed to stratify in the winter of 2016, corresponding to the failure to achieve freeze-over as seen in the MODIS observations and indicated by our model results. For other years, the durations of ice cover were highly correlated (Pearson's product moment correlation, $r = 0.83$–0.89, (Table 2)). Furthermore, for the years in which lake ice did not form—2013 on Lake Clark and 2015 on Naknek Lake—the overlapping temperature array data showed very short periods of stratification, which were likely not long enough for ice to form at >90% over the larger lake surface or be observed in the MODIS record.

Table 2. Model compared to in situ observations.

Lake	Validation Data [A]	Years [B]	Model Correlation	Mean Duration Obs.	Mean Duration Model	SD Obs.	SD Model	n Years	Conical Volume Estimate	Ranking by Proxy Volume
Lake Clark	T, Array, 100 m	2006–2022	0.84	60	60	47	27	16	29.79	3
Naknek Lake	T, Array, 70 m	2008–2022	0.89	87	74	45	47	14	24.43	5
Telaquana Lake	Ice obs.	2002–2020 *	0.85	150	157	21	21	12	2.73	7
Lake Brooks	T, Array, 50 m	2010–2022	0.83	73	77	46	44	12	2.05	9
Twin Lakes	Ice obs.	1982–1996 *	0.95	177	172	17	12	14	0.93	11

Notes: [A] Type of in-situ observations (and array depth) used for validation. [B] Asterisk indicates data had a gap in one or more years of ice observation.

The second set of validation data were in situ ice-cover observations for two lakes: Twin Lakes from 1982 to 1996, and Telaquana Lake from 2002 to 2020. These data were from historical observations used to record the approximate dates when the lakes would safely support foot traffic and aircraft, or were no longer navigable by boat. We again found strong correlations between the ice-cover duration derived from these observations and our modeled ice-cover duration (Twin Lakes: $r = 0.95$, Telaquana: $r = 0.85$) (Table 2).

3.5. Long-Term Climate Record

Cumulative FDD in the late fall through early winter was the strongest predictor of freeze-over, and subsequently ice-cover duration, in our model (Figure 7). To consider this in the context of the regional climate record, we used the most regionally representative long-term (74 years) record of the mean daily temperatures available. We evaluated mean daily air temperatures from October 1 through February 7, with fewer than seven missing days per period, which represent the beginning and midpoint of our seasonal modeling period (Figure 5). One of the warmer years on record, 2019, was excluded due to a lack of daily observations. These data show that the 14 warmest periods between 1 October and 7 February from 1949 to 2023 have occurred since 1977. The water years of 2014, 2015, 2016, and 2018 were the 4th, 6th, 2nd, and 9th hottest years, respectively.

4. Discussion

The ecological and economic importance of lake-ice cover is widely acknowledged, but the increasing temperatures of Earth's boreal regions have changed, and will continue to change, the patterns of lake-ice formation, duration, and breakup. However, data-intensive methods that require long-term observations and physical data are possible for only a small fraction of lakes globally. To address this, we developed a survival model to predict the historic timing and duration (phenology) of lake ice using readily accessible meteorological data and satellite-observed lake-ice cover, and hindcast predictions to explore changes over a 40-year period and evaluate them in a historic 74-year regional context.

The role of cumulative cold content through water and air temperatures has been successfully used to predict thermal stratification, mixing, and ultimately lake-ice formation by Ashton [46,47]. Likewise, cumulative heat content and "lake heatwaves" have been shown to impede the formation of lake ice [14,15,48]. Furthermore, geographic location, depth, and volume have been found to be important indicators of lake-ice dynamics [49]. Our model was developed to capture these dynamics over relevant time periods in the simplest manner possible by using geographically located predictor variables of antecedent thaw-degree days (*antTDD)*, cumulative freezing-degree days (*FDD*), cumulative thaw-degree days (*TDD*), and cumulative downwelling shortwave radiation (*SWR*). The survival model and data we used revealed three broad patterns of variability, threshold response, and trend in the phenology of lake ice.

4.1. Variability

The lakes in this study vary considerably in their ice phenology, where lower-volume lakes (<20 km^3) freeze-over consistently and for longer durations, and higher-volume lakes (>20 km^3) freeze-over less consistently for shorter and more variable durations. The consequences of such interannual variability are best expressed when scaled by the average duration of ice cover. This relative variability revealed even more profound variation: the most variable lake (Tustumena Lake) had an expected year-to-year difference in ice-season duration that exceeded its average ice season by 140%, whereas the most consistent lake (Beluga Lake) could be expected to vary by only 6% of its average duration. Relative variability in ice-season duration was greater in larger-volume lakes, driven both by the tendency for large-volume lakes to be more variable in absolute terms and for larger-volume lakes to have shorter ice-season durations.

Interannual variation in duration of ice cover was driven more by variation in freeze-over than breakup, where freeze-over was driven primarily by chilling (accumulated *FDD*). Sensitivity to chilling decreased with lake volume, therefore large-volume lakes require a greater accumulation of *FDD* before freeze-over occurs, explaining their shorter ice-season durations (Figures 4 and 8). More stable, lower-volume lakes required substantially less chilling, and reliably accumulated enough *FDD* to freeze early in the fall–winter season. These patterns were further affected by the lake's location, (e.g., higher elevation and shading by surrounding terrain). Lastly, while the highest-elevation lakes (e.g., Twin, Telaquana, Chackachamna, Kukaklik, and Nonvianuk) reliably froze from year to year, there were also differences among them due to the climactic influence of the warmer Gulf of Alaska (Chackachamna) and colder interior (Telaquana) regions (Figure 1).

4.2. Threshold Response

While Lake Clark, Becharof Lake, and Tustumena Lake demonstrated non-freezing years over the duration of our study, this appears to be a new phenomenon for the other large lakes. Illiamna Lake, Nakenek Lake, Skilak Lake, Lake Grosvenor, and Brooks Lake showed little appreciable probability of a non-freezing winter in any year before 2000, but all were predicted and/or observed to not freeze in multiple winters post-2000 (Figure 9). This, coupled with fact that 12 of the 15 warmest October–February cold accumulation periods since 1949 occurred during our modeled period (1981–2021) suggests that we have reached a threshold where the expectation of several of the larger lakes freezing is substantially diminished (Figures 5 and 9).

Model estimates and MODIS observations agreed on multiple years that several lakes did not freeze. Of these years, 2016 is most notable, when eight of the nine largest lakes failed to freeze-over and the mean daily temperatures were above 0 °C (Figure 4). Notably, the only large lake that did freeze in 2016 (Telaquana Lake) is the highest-elevation large lake (376 m) located in a shaded mountain valley. The 2001 and 2003 water years were also above 0 °C, with multiple lakes neither observed nor modelled to freeze-over (Figures 5 and 9).

The past 45 years have shown the greatest climate variability in fall/winter temperatures, with 15 of the warmest and some of the coldest years occurring since 1977 (Figure 5). These warm periods also corresponded with other climate indicators in the region, including the highest summer sea-surface temperatures of the century in the Bering Sea (2003–2005 and 2014–2020) [50], warmer waters in the Gulf of Alaska starting in 1976 [51,52], and anomalously warm waters in the Gulf for the years 2003, 2005, and 2014–2016 [53]. Benson et al. [54] and Robertson et al. [55] also recognized that large-scale atmospheric and oceanic conditions like the El Niño Southern Oscillation and the Pacific Decadal Oscillation are associated with higher winter and spring temperatures since the late 1980s, and suggest these regional climate drivers are no longer stationary [56] (Benson et al., 2011). Lastly, Dauginis [57] found consistency between declining sea ice, lake ice, and snow-on trends in Southwest Alaska.

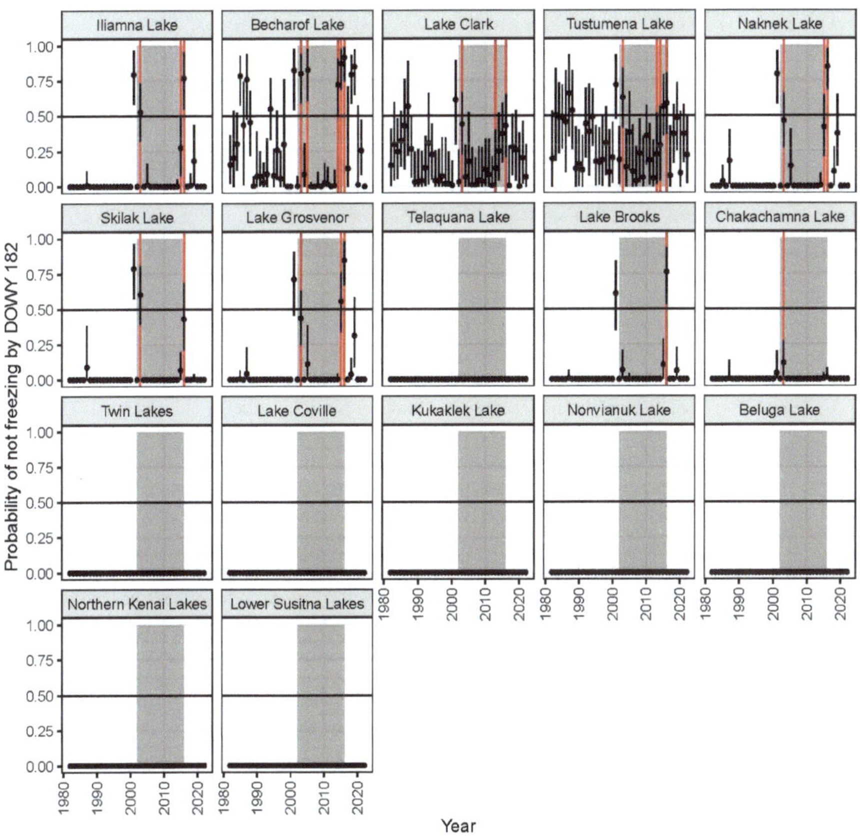

Figure 9. Modeled probability of ice cover exceeding 90%. Vertical red lines show the 0.5 probability of each lake reaching the 90% ice cover by March 30, (DWY 18). Black points and error bars show the posterior median (+/− 95% credible intervals) and the grey rectangles show the years of MODIS observations. Higher-volume lakes are at the top of figure.

There have only been a few detailed studies that have predicted future changes in the phenology of lake-ice cover. Dibke et al. [58] simulated lake-ice response to future climate in 2040–2079 using the CGCM3 Global Climate Model and the upper-level emission scenario (SRES A2). Their results propose that freeze-over will be later by 5–20 days, and breakup will be earlier by 10–30 days by the mid-to-late century. Lakes in Pacific coastal areas of North America saw the largest projected changes, while lakes in the Alaskan interior were less affected. More recently, Sharma et al. [9] suggested that, with continued climate warming, lakes in Southwest Alaska will experience an exponential decrease in reliable ice cover under a variety of climate scenarios.

4.3. Trend

The majority of lower-volume lakes (<10 km^3) show negative trends in their period of seasonal ice cover in the order of 1–6 days per decade over our modeled period. In contrast, all but one of the lakes > 20 km^3, show undiscernable or positive trends in ice-cover duration of up to two days per decade over our modeled period (Table 1).

Studies of phenology trends in northern-hemisphere lake-ice phenology have found a wide range of temporal trends that are generally negative but vary widely [9,56,57,59,60]. Our study estimated both positive and negative trends and may further explain the broad range of observed and modeled trends. The estimated trend directions were a function of (i.) the time window that we analyzed as there was a warm period near the start of our study period, and (ii.) lakes were differentially impacted by the early warm period depending on their individual characteristics. The non-negative trends were associated with a shorter duration of ice cover and a higher mean absolute error in the model fit. The trend in duration of ice cover in lakes with shorter periods of ice cover were impacted by the warm fall/winters of 85 and 87, resulting in years with short or no ice cover at the beginning of our study period, and thus a non-negative trend (Figures 4 and 9). The years 1977 and 1979 were also anomalously warm, but were not covered by our modeling period. The lakes < 20 km^3 in volume did not show modeled years with a short duration of ice cover in the 1980s, thus showed negative trends in duration of ice cover (Table 1).

4.4. Future Efforts

While our model accurately predicts many of the years and dates in which freeze-over and breakup occur, there are several instances where it does not. In 2013, Tustumena and Lake Clark did not freeze-over, but our model predicted they would (Figure 9). Temperature profile data available for Lake Clark in 2013 showed only eight days of winter stratification. This suggests in some years, especially on high-volume lakes with greater heat-storage capacities, other processes such as wind, seiche, and (temperature) profile mixing inhibit ice formation and exacerbate the ice breakup process due to open water and thinner ice cover.

Data collected from a meterology station near Tustumena suggest wind processes affected the ice-formation processes in 2013 when the mean October wind speed was 23% higher than the mean wind speed for October in 2001–2016 (Figure 10). Antecdotal evidence also suggests wind and waves on Lake Becharof may be a factor as well. Parts of Southwest Alaska can see particularly large atmospheric pressure gradients due to their position between the relatively shallow and cold waters of the Bristol Bay (Bering Sea) and the warmer waters of the Gulf of Alaska (Pacific Ocean). Local winds are frequently observed >30 mps (58 knots) for extended periods of time and, because Becharof has relatively flat surrounding terrain and a prevailing wind orientation, this can result in significant wave action, with wave heights of several meters on larger lakes [61]. A sustained wind can also create a seiche (water piling up downwind and later rebounding), which inhibits seasonal temperature stratification in addition to sustaining mixing on larger lakes. With sufficient data it is possible to calculate and model wind (and temperature stratification) effects on lakes of various sizes. However, given the limited availability of bathymetric profiles and local observations of wind direction and velocity, in addition to the complex topography surrounding these lakes, it is presently not possible for us to include wind in our model.

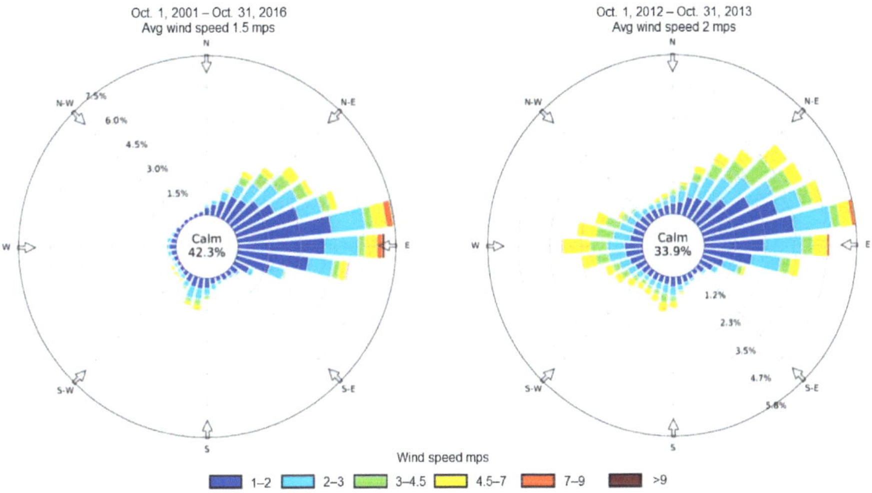

Figure 10. Wind rose comparisons. In 2013, lake ice was prevented from forming on Lake Clark and nearby Tustumena Lake. Winds observed from Soldotna, close to Lake Tustemena, demonstrate the average annual wind direction and intensity for October 2002–2016 water years (**left**) and the 2013 water year (**right**) when the average wind speed was 29% higher. (Data from Western Regional Climate Center).

Other important variables known to affect the formation of lake-ice cover are snow cover and water chemistry. While it is possible to estimate presence or absence of snow cover using satellite data [62], we found negligeable improvement when including these data in earlier versions of our model. This suggests snowpack properties such as depth, density, and cold content, in addition to the interaction with atmospheric boundary layer conditions (e.g., vapor pressure gradients), are more important than just the presence or absence of snow [18]. Water turbidity and chemistry can also affect the process of heating, cooling, and stratification in lakes, which also affects turnover and the formation of lake ice. While several of the lakes in our study are crypto depressions (depths below sea level) (Table 1), none of them has tidal influx or evidence of seawater intrusion. Some lakes have a significant influx of glacial sediment impacting light absorption and solar heating (e.g., Lake Clark, Tustumena, Skilak, Telaquana, Chakachamna, and Twin Lakes), and others have water chemistries impacted by volcanism (Becharof) [63], but we do not suspect the addition of water quality or chemistry variables would have significantly changed our modeling results.

Because this simple model does not use difficult-to-obtain forcing variables to predict forcing variables such as wind, snow, or water quality, it should be possible to generally predict future conditions with a limited set of variables. For example, multiple Global Climate Models (GCM) that have shown good capability in predicting temperature and the first-order influences on downwelling shortwave can be statically applied. There are also efficient machine learning approaches to evaluate remote sensing data at scale that could be used for calibration [43]. This approach would be particularly useful for larger-volume lakes at lower elevations that appear to be most susceptible to small changes in accumulated FDD. However, there are tradeoffs to using this simple approach as it may miss delays in ice formation or years in which lakes do not freeze-over, as seen in our hindcast predictions when compared to MODIS observations. In cases where capturing this variability is important, additional input variables for wind and snow depth would likely improve the model skill in predicting years with intermittent forcing events of high wind, as seen for some lakes in 2013 or years when snow cover departs from the mean.

5. Conclusions

While some communities, villages, and individuals have observations of lake-ice formation and breakup at specific locations, there are very few observational datasets that describe the phenology of an entire lake, much less a robust method of predicting when ice-free conditions may occur. The types of in situ observations required for building physically based models are neither available nor financially feasible for most Alaskan stakeholders. Our open-source model addresses this by demonstrating the potential to obtain accurate probabilities of past lake-ice phenology using readily available satellite data and a few meteorologic parameters that are publicly available. Further improvements to this model could be made at a regional and local level by (i) adding wind data as a variable to the model, (ii) introducing computationally derived lake-volume estimates as a variable, and (iii) incorporating additional satellite observations and uncertainties into the modeling process. This model could also be adapted as a short- or long-term predictive tool using input variables derived from near-real time data of local weather conditions or GCMs.

Author Contributions: Conceptualization, P.B.K. and M.P.H.; methodology, M.P.H. and P.B.K.; software, M.P.H.; validation, P.B.K. and M.P.H.; formal analysis, P.B.K. and M.P.H.; investigation, P.B.K.; data Curation, P.B.K.; writing—original draft preparation, P.K; writing—review and editing, P.B.K. and M.P.H.; project administration, P.B.K. All authors have read and agreed to the published version of the manuscript.

Funding: Funding was provided by the National Park Service, Inventory and Monitoring Program, Southwest Alaska Network.

Data Availability Statement: All data used in this analysis are cited in the text and sources referenced. These data are publicly available, except for GIS bathymetry data which are pending publication and available upon request. https://irma.nps.gov/DataStore/Reference/Profile/2303056. DOI: 10.57830/.

Acknowledgments: This work would not have been possible without the careful data management and image interpretation provided by the National Park Service staff including Michael Verrier, Timothy Shepherd, Chuck Lindsay, and others. Many thanks to Jim Sickman for the helpful comments and valued mentorship over the years. We also thank the reviewers who provided thoughtful insights into improving this manuscript.

Conflicts of Interest: The authors declare no conflicts of interest.

Appendix A

Detailed Model Description

We modeled lake-ice processes using a daily, discrete-time, survival modelling approach, borrowing from recent work in plant phenology. Survival models have a long history in medical research. They represent time-to-event data, allowing for events that never happen, making them well suited to modelling a process like lake-ice formation. By modelling the state of lake ice at a daily time step, we can use daily-resolved weather data to predict freeze-over and thaw-start and generate more precise predictions.

We modeled the state of lake-ice cover ($f_{d,y,i}$) at each lake (i in 1…17) and lake clusters on each day d of each water year y as a Bernoulli distributed random variable, with 0 representing ice cover < 90% and 1 representing ice cover > 90%:

$$f_{d,y,i} \sim \text{Bernoulli}\left(p_{d,y,i}\right) \tag{A1}$$

where $p_{d,y,i}$ is the probability of ice cover exceeding 90%. We decompose this probability into the sum of two conditional probabilities, the freeze-over hazard ($h^{\text{fr}}_{d,y,i}$) and the breakup hazard ($h^{\text{br}}_{d,y,i}$). The freeze-over hazard is the probability of ice cover exceeding 90% on day d, given that it did not by day $d-1$, and the breakup hazard is the probability of ice cover dropping below 90% on day d, given that it exceeded 90% on day $d-1$.

$$h^{\text{fr}}_{d,y,i} = \Pr\left(F_{d,y,i} = 1 \middle| F_{d-1,y,i} = 0\right)$$
$$h^{\text{br}}_{d,y,i} = \Pr\left(F_{d,y,i} = 0 \middle| F_{d-1,y,i} = 1\right) \quad \text{(A2)}$$

$$p_{d,y,i} = h^{\text{fr}}_{d,y,i}\left(1 - y_{d-1,y,i}\right) + \left(1 - h^{\text{br}}_{d,y,i}\right)y_{d-1,y,i} \quad \text{(A3)}$$

We modeled the freeze-over hazard ($h^{\text{fr}}_{d,y,i}$) with a logistic regression: the log of the odds ratio of the hazard for lake i on day of water year d in water year y was modeled as a linear function of lake- and day-varying covariates ($FDD_{d,y,i}$, $ASWR_{d,y,i}$), and a lake- and year-varying covariate ($antTDD_{y,i}$). Our model allowed for random lake-varying intercepts and regression coefficients (β_i).

$$ln\left(h^{\text{fr}}_{d,y,i} / \left(1 - h^{\text{fr}}_{d,y,i}\right)\right) = \beta_{1,i} + \beta_{2,i} \cdot FDD_{d,y,i} + \beta_{3,i} \cdot SWR_{d,y,i} + \beta_{4,i} \cdot antTDD_{y,i} \quad \text{(A4)}$$

Lake-varying intercept and slope random effects were modeled as draws from a multivariate normal distribution with a mean vector μ and covariance matrix $\sum$.

$$\begin{pmatrix} \beta_1 \\ \beta_2 \\ \beta_3 \\ \beta_4 \end{pmatrix}_i \sim \mathcal{MVN}\left(\begin{matrix} \mu^\beta_1 \\ \mu^\beta_2 \\ \mu^\beta_3 \\ \mu^\beta_4 \end{matrix}, \Sigma^\beta\right) \quad \text{(A5)}$$

We modeled the breakup hazard ($h^{\text{br}}_{d,y,i}$) analogously as a logit-linear function of lake- and day-varying covariates ($TDD_{d,y,i}$, $FDD_{d,y,i}$, $SWR_{d,y,i}$). For modelling the breakup hazard, d indexes days since 90% freeze-over, rather than days since start of water year.

$$ln\left(h^{\text{br}}_{d,y,i} / \left(1 - h^{\text{br}}_{d,y,i}\right)\right) = \gamma_{1,i} + \gamma_{2,i} \cdot TDD_{d,y,i} + \gamma_{3,i} \cdot SWR_{d,y,i} + \gamma_{4,i} \cdot FDD_{d,y,i} \quad \text{(A6)}$$

$$\begin{pmatrix} \gamma_1 \\ \gamma_2 \\ \gamma_3 \\ \gamma_4 \end{pmatrix}_i \sim \mathcal{MVN}\left(\begin{matrix} \mu^\gamma_1 \\ \mu^\gamma_2 \\ \mu^\gamma_3 \\ \mu^\gamma_4 \end{matrix}, \Sigma^\gamma\right) \quad \text{(A7)}$$

We used weakly informative priors for all model parameters (Table A1), including a scaled Wishart distribution [31] on the covariance matrices of multivariate distributions. Priors on μ^β and μ^γ were chosen to aid model convergence while permitting the data to dominate the posterior distribution [64]. To assess the influence of prior probability distributions on the regression parameters, we performed a sensitivity analysis, comparing regression estimates from models where the prior variance on μ^β and μ^γ were 1 and 2 orders of magnitude larger. We saw only minimal differences in parameter estimates (Appendix C).

Table A1. Prior distributions for parameters in lake-ice survival models.

Parameter	Prior Distribution	Characteristics
$\mu^\beta_1, \mu^\gamma_1$	$\mathcal{N}(0, 2.5)$	Weakly informative on logit scale
$\mu^\beta_{2:4}, \mu^\gamma_{2:4}$	$\mathcal{N}(0, 2.5)$	Weakly informative on logit scale
$\Sigma^\beta, \Sigma^\gamma$	scaled Wishart ($s_{1:4} = 10, df = 2$)	Uniform for correlation, half $- t_2(0, 10)$ for sd

From the modeled freeze-over and breakup hazards, we calculated the variables of interest in our study. "Survival", in the case of freeze-over being the probability that lake i remained unfrozen until day d in water year w, is calculated as the product of the complement of the freeze-over hazard over all days up to and including day d:

$$S^{\text{fr}}_{d,y,i} = \prod_{j=1}^{d} \left(1 - H^{\text{fr}}_{j,y,i}\right) \tag{A8}$$

The complement of survival is the probability that freeze-over had occurred by day d in water year y:

$$P^{\text{fr}}_{d,y,i} = 1 - S^{\text{fr}}_{d,y,i} \tag{A9}$$

Similarly, the probability that lake i remained frozen d days after the date of 90% freeze-over in water year w was calculated as the product of the complement of the breakup hazard over all days from the date of freeze-over until d days after the date of freeze-over:

$$S^{\text{br}}_{d,y,i} = \prod_{j=1}^{d} \left(1 - H^{\text{br}}_{j,y,i}\right) \tag{A10}$$

and the probability that breakup had occurred by day d in water year y:

$$P^{\text{br}}_{d,y,i} = 1 - S^{\text{br}}_{d,y,i} \tag{A11}$$

The date of freeze-over and the date of breakup were estimated as the day that the probability of freeze-over and probability of breakup exceeded 0.5, respectively. Years when the probability of freeze-over did not exceed 0.5 by day 180 of the water year were classified as no-freeze years.

Appendix B

Observed versus Predicted Figures for Freeze-Over and Breakup

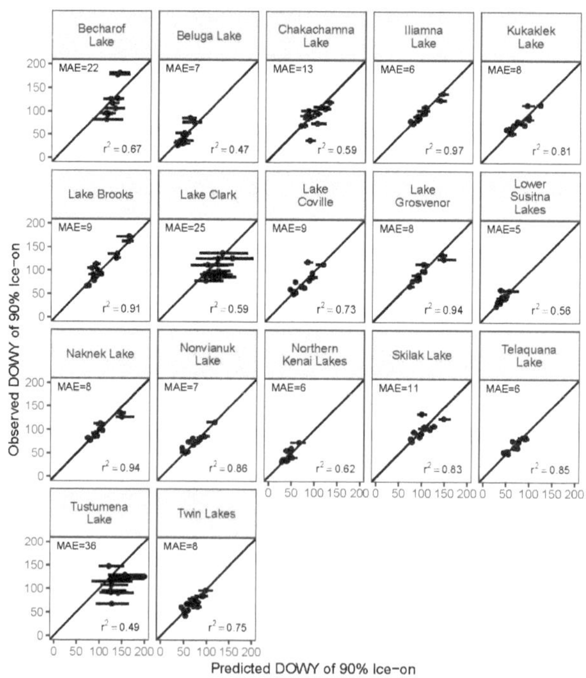

Figure A1. Model performance at each study lake. Observed versus predicted values (posterior median +/− 95% credible intervals) for freeze-over, 1:1 line in black. Mean absolute error and r^2 of model predictions are shown for each lake. Higher-volume lakes are at the top of figure.

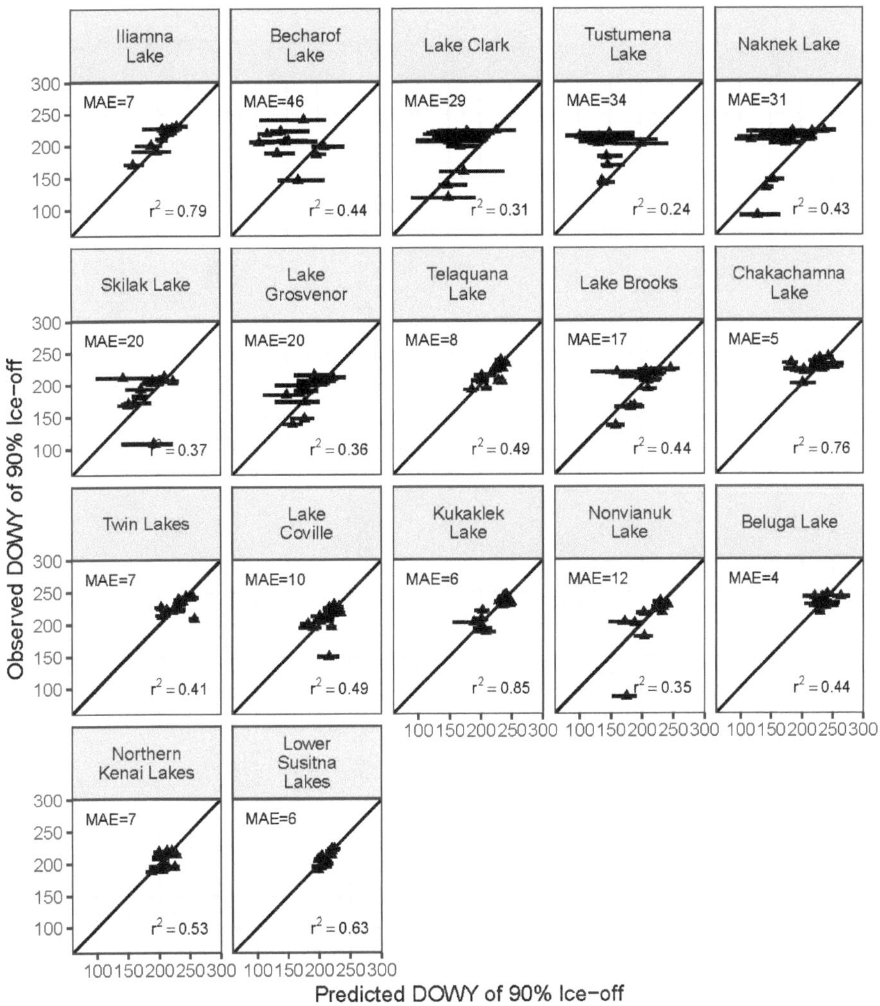

Figure A2. Model performance at each study lake. Observed versus predicted values (posterior median +/− 95% credible intervals) for breakup, 1:1 line in black. Mean absolute error and r^2 of model predictions are shown for each lake. Higher-volume lakes are at the top of figure.

Appendix C

Appendix C.1. Bayesian Prior Sensitivity Analysis

To understand the influence that model priors might have on our inferences, we tested our prior specification against two alternate prior specifications. For the mean climate effects ($\mu^{\beta}_{2:4}$ and $\mu^{\gamma}_{2:4}$), we fit models with our chosen prior, Normal (mean = 0, *sd* = 2.5), as well as more diffuse priors with variances covering a range of two orders of magnitude: Normal (mean = 0, *sd* = 10), Normal (mean = 0, *sd* = 31.7) (Figure A3).

A comparison of estimated parameter values from these three models revealed minimal sensitivity of the posterior to our prior specification (Figure A4). Generally, parameter estimates in our model were constrained to be slightly smaller than in models with more diffuse priors, but the effects were small, relative to the uncertainty.

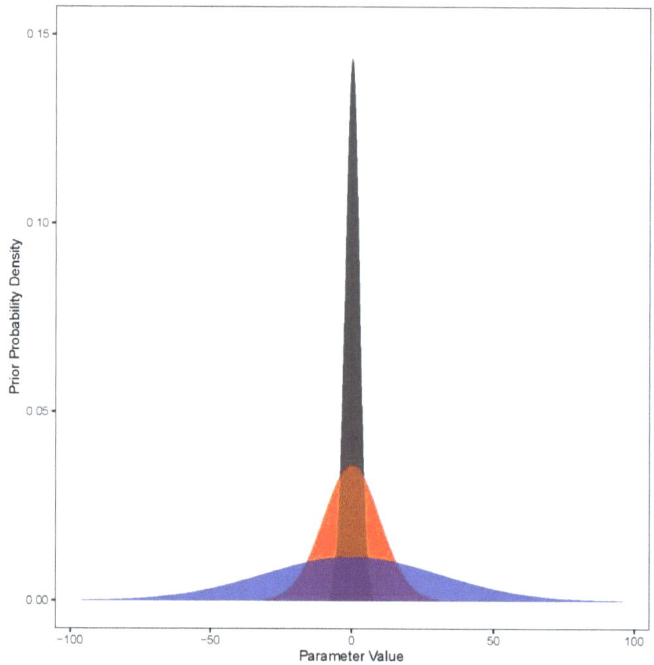

Figure A3. Prior probability distributions on regression coefficients, showing the prior used in analysis, N(0, 2.5) in grey, a N(0, 10) in red, and N(0, 31.7) in blue.

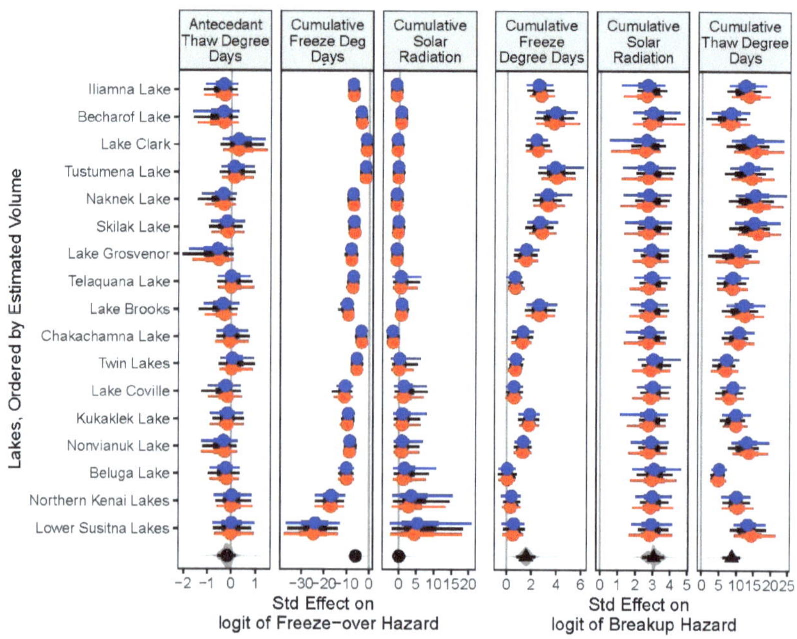

Figure A4. Model effects for model used in analysis (black) and models with alternate prior specifications, N(0, 10) in red, N(0, 31.7) in blue.

Appendix C.2. Climate Accumulation Periods

Our accumulation period for antecedent thaw-degree days (*antTDD*) was between June 1 and September 30. To confirm that this adequately represented the interannual variability in *antTDD*, we compared these values to *antTDD* accumulated from January 1 to September 30, predicting the latter with the former in a linear regression. This revealed that our chosen accumulation period accounted for 91% of the variability in the *antTDD* since the start of the calendar year (Figure A5). Furthermore, our chosen accumulation period largely avoids considering heat loading during times when the lakes may still be snow- and ice-covered, and insulated from air temperatures.

Our accumulation period for freeze-degree days (*FDD*) began on October 1, which accounts for most of the *FDD* at our study lakes. We calculated the *FDD* accumulated before October 1 for each lake and year of the satellite observation period. At 12 of the 17 lakes, no *FDD* accumulated before October 1 in any year of the satellite observation period. For the lakes that accumulated any *FDD* before October 1, we compared the amount accumulated by October 1 to the amount accumulated at the day of 90% freeze-over. *FDD* accumulated before October 1 accounted for 0–5.4% of the *FDD* accumulated on the day of 90% freeze-over.

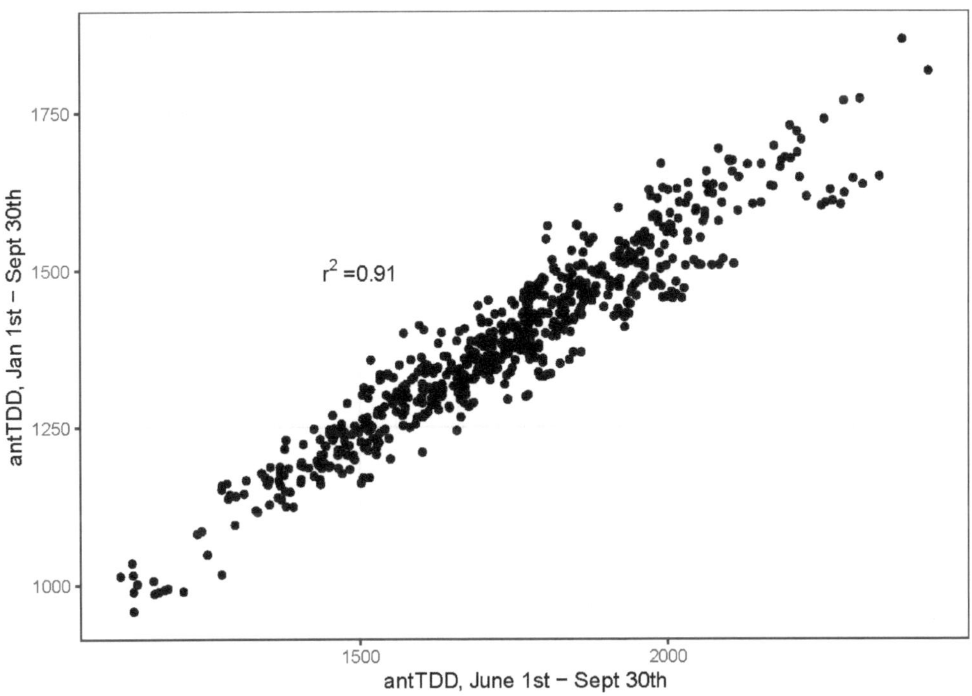

Figure A5. Regression showing strong agreement between antecedent thaw-degree days (*antTDD*) accumulated from 1 January to 30 September vs. *antTDD* accumulated from 1 June to 30 September for all study lakes and years, indicating minimal potential for model sensitivity to lengthening the period of accumulation.

References

1. Magnuson, J.J.; Robertson, D.M.; Benson, B.J.; Wynne, R.H.; Livingstone, D.M.; Arai, T.; Assel, R.A.; Barry, R.G.; Card, V.; Kuusisto, E.; et al. Historical Trends in Lake and River Ice Cover in the Northern Hemisphere. *Science* **2000**, *289*, 1743–1746. [CrossRef] [PubMed]
2. Newton, A.M.W.; Mullan, D.J. Climate Change and Northern Hemisphere Lake and River Ice Phenology from 1931–2005. *Cryosphere* **2021**, *15*, 2211–2234. [CrossRef]
3. Adrian, R.; Walz, N.; Hintze, T.; Hoeg, S.; Rusche, R. Effects of Ice Duration on Plankton Succession during Spring in a Shallow Polymictic Lake. *Freshw. Biol.* **1999**, *41*, 621–634. [CrossRef]
4. Cai, Y.; Duguay, C.R.; Ke, C.-Q. A 41-Year (1979–2019) Passive-Microwave-Derived Lake Ice Phenology Data Record of the Northern Hemisphere. *Earth Syst. Sci. Data* **2022**, *14*, 3329–3347. [CrossRef]
5. Serreze, M.C.; Francis, J.A. The Arctic Amplification Debate. *Clim. Chang.* **2006**, *76*, 241–264. [CrossRef]
6. Prowse, T.; Alfredsen, K.; Beltaos, S.; Bonsal, B.; Duguay, C.; Korhola, A.; McNamara, J.; Pienitz, R.; Vincent, W.F.; Vuglinsky, V.; et al. Past and Future Changes in Arctic Lake and River Ice. *Ambio* **2011**, *40*, 53–62. [CrossRef]
7. Cohen, J.; Screen, J.A.; Furtado, J.C.; Barlow, M.; Whittleston, D.; Coumou, D.; Francis, J.; Dethloff, K.; Entekhabi, D.; Overland, J.; et al. Recent Arctic Amplification and Extreme Mid-Latitude Weather. *Nat. Geosci.* **2014**, *7*, 627–637. [CrossRef]
8. Woolway, R.I.; Merchant, C.J. Worldwide Alteration of Lake Mixing Regimes in Response to Climate Change. *Nat. Geosci.* **2019**, *12*, 271–276. [CrossRef]
9. Sharma, S.; Blagrave, K.; Magnuson, J.J.; O'Reilly, C.M.; Oliver, S.; Batt, R.D.; Magee, M.R.; Straile, D.; Weyhenmeyer, G.A.; Winslow, L.; et al. Widespread Loss of Lake Ice around the Northern Hemisphere in a Warming World. *Nat. Clim. Chang.* **2019**, *9*, 227–231. [CrossRef]
10. Gaul, K. Nanutset ch'u Q'udi Gu, before Our Time and Now, An Ethnohistory of Lake Clark National Park and Preserve. USDOI, NPS, Lake Clark National Park and Preserve, 179. 2007. Available online: https://nps.gov/lacl/learn/historyculture/upload/Nanutset-Gaul-508.pdf (accessed on 20 October 2023).
11. Cold, H.S.; Brinkman, T.J.; Brown, C.L.; Hollingsworth, T.N.; Brown, D.R.N.; Heeringa, K.M. Assessing Vulnerability of Subsistence Travel to Effects of Environmental Change in Interior Alaska. *Ecol. Soc.* **2020**, *25*, 20. [CrossRef]
12. Bogdanova, E.; Lobanov, A.; Andronov, S.V.; Soromotin, A.; Popov, A.; Skalny, A.V.; Shaduyko, O.; Callaghan, T.V. Challenges of Changing Water Sources for Human Wellbeing in the Arctic Zone of Western Siberia. *Water* **2023**, *15*, 1577. [CrossRef]
13. Bengtsson, L.; Herschy, R.W.; Fairbridge, R.W. (Eds.) Ice Formation on Lakes and Ice Growth. In *Encyclopedia of Lakes and Reservoirs*; Springer: Dordrecht, The Netherlands, 2012.
14. Leppäranta, M. Modelling the Formation and Decay of Lake Ice. In *The Impact of Climate Change on European Lakes*; George, G., Ed.; Springer Science & Business Media: Dordrecht, The Netherlands, 2009; Volume 4.
15. Leppäranta, M. Ice Phenology and Thickness Modelling for Lake Ice Climatology. *Water* **2023**, *15*, 2951. [CrossRef]
16. Kirillin, G.; Leppäranta, M.; Terzhevik, A.; Granin, N.; Bernhardt, J.; Engelhardt, C.; Efremova, T.; Golosov, S.; Palshin, N.; Sherstyankin, P.; et al. Physics of Seasonally Ice-Covered Lakes: A Review. *Aquat. Sci.* **2012**, *74*, 659–682. [CrossRef]
17. Launiainen, J.; Cheng, B. Modelling of Ice Thermodynamics in Natural Water Bodies. *Cold Reg. Sci. Technol.* **1998**, *27*, 153–178. [CrossRef]
18. Brown, L.C.; Duguay, C.R. A Comparison of Simulated and Measured Lake Ice Thickness Using a Shallow Water Ice Profiler. *Hydrol. Process.* **2011**, *25*, 2932–2941. [CrossRef]
19. Rafat, A.; Pour, H.K.; Spence, C.; Palmer, M.J.; MacLean, A. An Analysis of Ice Growth and Temperature Dynamics in Two Canadian Subarctic Lakes. *Cold Reg. Sci. Technol.* **2023**, *210*, 103808. [CrossRef]
20. Golub, M.; Thiery, W.; Marcé, R.; Pierson, D.; Vanderkelen, I.; Mercado, D.; Woolway, R.I.; Grant, L.; Jennings, E.; Schewe, J.; et al. A framework for ensemble modelling of climate change impacts on lakes worldwide: The ISIMIP Lake Sector. *Geosci. Model Dev. Discuss.* **2022**, *15*, 4597–4623. [CrossRef]
21. Wynne, R.H. Statistical Modeling of Lake Ice Phenology: Issues and Implications. *SIL Proc.* **2000**, *27*, 2820–2825. [CrossRef]
22. Preston, D.L.; Caine, N.; McKnight, D.M.; Williams, M.W.; Hell, K.; Miller, M.P.; Hart, S.J.; Johnson, P.T.J. Climate Regulates Alpine Lake Ice Cover Phenology and Aquatic Ecosystem Structure. *Geophys. Res. Let.* **2016**, *43*, 5353–5360. [CrossRef]
23. Beyene, M.T.; Jain, S.; Gupta, R.C. Linear-Circular Statistical Modeling of Lake Ice-Out Dates. *Water Resour. Res.* **2018**, *54*, 7841–7858. [CrossRef]
24. Terres, M.A.; Gelfand, A.E.; Allen, J.M.; Silander, J.A. Analyzing First Flowering Event Data Using Survival Models with Space and Time-varying Covariates. *Environmetrics* **2013**, *24*, 317–331. [CrossRef]
25. Diez, J.M.; Pulliam, H.R. Hierarchical Analysis of Species Distributions and Abundance Across Environmental Gradients. *Ecology* **2007**, *88*, 3144–3152. [CrossRef] [PubMed]
26. Elmendorf, S.C.; Crimmins, T.M.; Gerst, K.L.; Weltzin, J.F. Time to Branch out? Application of Hierarchical Survival Models in Plant Phenology. *Agric. For. Meteorol.* **2019**, *279*, 107694. [CrossRef]
27. USGS Publications Warehouse, 1992, 1:2,000,000-Scale Digital Line Graph (DLG) Data. Available online: https://pubs.usgs.gov/publication/ds4 (accessed on 30 July 2009).
28. McCabe, S.; Bartz, K.; Young, D.; Gabriel, P. GIS bathymetry of selected southwest Alaska Lakes, from field measurements. In *GIS Data Layer, in Preparation*; National Park Service Southwest Alaska Inventory and Monitoring Network: Anchorage, AK, USA, 2024.

29. U.S. Bureau of Reclamation. *Chakachamna Project, Alaska Status Report*; Alaska District Office: Juneau, AK, USA, 1962; 21p.
30. Spafard, M.A.; Edmonson, J.A. *A Morphometric Atlas of Alaskan Lakes: Cook Inlet, Prince William Sound, and Bristol Bay Areas*; Alaska Department of Fish and Game, Commercial Fisheries Division: Anchorage, AK, USA, 2000. Available online: www.adfg.alaska.gov/fedaidpdfs/RIR.2A.2000.23.pdf (accessed on 20 June 2024).
31. Hollister, J.; Milstead, W.B. Using GIS to Estimate Lake Volume from Limited Data. *Lake Reserv. Manag.* **2010**, *26*, 194–199. [CrossRef]
32. Reed, B.; Budde, M.; Spencer, P.; Miller, A.E. Integration of MODIS-Derived Metrics to Assess Interannual Variability in Snowpack, Lake Ice, and NDVI in Southwest Alaska. *Remote Sens. Environ.* **2009**, *113*, 1443–1452. [CrossRef]
33. Verrier, M.; Shepherd, T.; Spencer, P.; Lindsay, C.; Kirchner, P. *Southwest Alaska Remote Sensing Lake Ice Phenology for 17 Lakes and Lake Clusters for 2002–2016 Water Years*; National Park Service—Southwest Alaska Inventory and Monitoring Network: Anchorage, AK, USA, 2024. [CrossRef]
34. Thorton, M.M.; Shrestha, R.; Wei, Y.; Thornton, P.E.; Kao, S.-C.; Wilson, B.E. *Daymet: Daily Surface Weather Data on a 1-km Grid for North America, Version 4R1*; ORNL DAAC: Oak Ridge, TN, USA, 2022. [CrossRef]
35. Plummer, M. JAGS Version 4.3.0 User Manual 2017, 73. Available online: https://sourceforge.net/projects/mcmc-jags/ (accessed on 20 September 2023).
36. Kellner, K. jagsUI: A Wrapper around "rjags" to Streamline "JAGS" Analyses 2018. Available online: https://CRAN.R-project.org/package=jagsUI (accessed on 1 April 2024).
37. Kay, M. Tidybayes: Tidy Data and Geoms for Bayesian Models 2022. Available online: https://mjskay.github.io/tidybayes/ (accessed on 1 April 2024).
38. R Core Team. R: A Language and Environment for Statistical Computing 2019. Available online: https://R-project.org/ (accessed on 20 September 2023).
39. Gelman, A.; Rubin, D.B. Inference from Iterative Simulation Using Multiple Sequences. *Stat. Sci.* **1992**, *7*, 457–511. [CrossRef]
40. Gelman, A.; Goodrich, B.; Gabry, J.; Vehtari, A. R-Squared for Bayesian Regression Models. *Am. Stat.* **2019**, *73*, 307–309. [CrossRef]
41. Nakagawa, S.; Schielzeth, H. A General and Simple Method for Obtaining R2 from Generalized Linear Mixed-effects Models. *Methods Ecol. Evol.* **2012**, *4*, 133–142. [CrossRef]
42. Eggleston, K. xmACIS Version 1.0.69a, NOAA Northeast Regional Climate Center Released 4-Dec-2023. Available online: https://xmacis.rcc-acis.org (accessed on 1 May 2023).
43. Zhang, S.; Pavelsky, T.M. Remote Sensing of Lake Ice Phenology across a Range of Lakes Sizes, ME, USA. *Remote Sens.* **2019**, *11*, 1718. [CrossRef]
44. Tuttle, S.E.; Roof, S.R.; Retelle, M.J.; Werner, A.; Gunn, G.E.; Bunting, E.L. Evaluation of Satellite-Derived Estimates of Lake Ice Cover Timing on Linnévatnet, Kapp Linné, Svalbard Using In-Situ Data. *Remote Sens.* **2022**, *14*, 1311. [CrossRef]
45. Bartz, K.; Gabriel, P. *Lake Temperature Monitoring in Southwest Alaska Parks: A Synthesis of Year-Round, Multi-Depth Data from 2006 through 2018*; National Park Service: Fort Collins, CO, USA, 2020. [CrossRef]
46. Ashton, G.D. Lake Ice Decay. *Cold Reg. Sci. Technol.* **1983**, *8*, 83–86. [CrossRef]
47. Ashton, G.D. River and Lake Ice Thickening, Thinning, and Snow Ice Formation. *Cold Reg. Sci. Technol.* **2011**, *68*, 3–19. [CrossRef]
48. Woolway, R.I.; Albergel, C.; Frölicher, T.L.; Perroud, M. Severe Lake Heatwaves Attributable to Human-Induced Global Warming. *Geophys. Res. Lett.* **2022**, *49*, e2021GL097031. [CrossRef]
49. Zhou, J.; Leavitt, P.R.; Rose, K.C.; Wang, X.; Zhang, Y.; Shi, K.; Qin, B. Controls of Thermal Response of Temperate Lakes to Atmospheric Warming. *Nat. Commun.* **2023**, *14*, 6503. [CrossRef] [PubMed]
50. Huang, B.; Thorne, P.; Banzon, V.; Boyer, T.; Chepurin, G.; Lawrimore, J.; Menne, J.; Smith, T.; Vose, S.; Zhang, H. NOAA Extended Reconstructed Sea Surface Temperature (ERSST), Version 5. NOAA National Centers for Environmental Information. 2017. Available online: https://ncei.noaa.gov/access/ (accessed on 29 December 2023).
51. Hartmann, B.; Wendler, G. The Significance of the 1976 Pacific Climate Shift in the Climatology of Alaska. *J. Clim.* **2005**, *18*, 4824–4839. [CrossRef]
52. Litzow, M.A.; Mueter, F.J. Assessing the Ecological Importance of Climate Regime Shifts: An Approach from the North Pacific Ocean. *Prog. Oceanogr.* **2014**, *120*, 110–119. [CrossRef]
53. Danielson, S.L.; Hennon, T.D.; Monson, D.H.; Suryan, R.M.; Campbell, R.W.; Baird, S.J.; Holderied, K.; Weingartner, T.J. Temperature Variations in the Northern Gulf of Alaska across Synoptic to Century-Long Time Scales. Deep-sea Research. Part 2. Topical Studies in Oceanography/Deep Sea Research. *Part II Top. Stud. Oceanogr.* **2022**, *203*, 105155. [CrossRef]
54. Benson, B.J.; Magnuson, J.J.; Jacob, R.L.; Fuenger, S.L. Response of Lake Ice Breakup in the Northern Hemisphere to the 1976 Interdecadal Shift in the North Pacific. *Verh. Int. Ver. Theor. Angew. Limnol.* **2000**, *27*, 2770–2774. [CrossRef]
55. Robertson, D.M.; Wynne, R.H.; Chang, W.Y.B. Influence of El Niño on Lake and River Ice Cover in the Northern Hemisphere from 1900 to 1995. *Verh. Int. Ver. Theor. Angew. Limnol.* **2000**, *27*, 2784–2788. [CrossRef]
56. Benson, B.J.; Magnuson, J.J.; Jensen, O.P.; Card, V.M.; Hodgkins, G.; Korhonen, J.; Livingstone, D.M.; Stewart, K.M.; Weyhenmeyer, G.A.; Granin, N.G. Extreme Events, Trends, and Variability in Northern Hemisphere Lake-Ice Phenology (1855–2005). *Clim. Chang.* **2011**, *112*, 299–323. [CrossRef]
57. Dauginis, A.A.; Brown, L.C. Recent Changes in Pan-Arctic Sea Ice, Lake Ice, and Snow-on/off Timing. *Cryosphere* **2021**, *15*, 4781–4805. [CrossRef]

58. Dibike, Y.; Prowse, T.; Saloranta, T.; Ahmed, R. Response of Northern Hemisphere Lake-ice Cover and Lake-water Thermal Structure Patterns to a Changing Climate. *Hydrol. Process.* **2011**, *25*, 2942–2953. [CrossRef]
59. Weyhenmeyer, G.A.; Livingstone, D.M.; Meili, M.; Jensen, O.; Benson, B.; Magnuson, J.J. Large Geographical Differences in the Sensitivity of Ice-Covered Lakes and Rivers in the Northern Hemisphere to Temperature Changes. *Global Chang. Biol.* **2010**, *17*, 268–275. [CrossRef]
60. Wang, X.; Feng, L.; Qi, W.; Cai, X.; Zheng, Y.; Gibson, L.; Tang, J.; Song, X.; Liu, J.; Zheng, C.; et al. Continuous Loss of Global Lake Ice Across Two Centuries Revealed by Satellite Observations and Numerical Modeling. *Geophys. Res. Lett.* **2022**, *49*, e2022GL099022. [CrossRef]
61. Warren, K.; Margaret, W. U.S. Fish & Wildlife Service, Region 7, Water Resources Branch, Egegik River/Becharof Lake Watershed Navigability Research Report, October 1997. Available online: https://dnr.alaska.gov/mlw/paad/nav/rdi/gegik_becharof/becharof_draft_report.pdf (accessed on 21 February 2024).
62. Lindsay, C.; Zhu, J.; Miller, A.E.; Kirchner, P.; Wilson, T.L. Deriving Snow Cover Metrics for Alaska from MODIS. *Remote Sens.* **2015**, *7*, 12961–12985. [CrossRef]
63. LaPerriere, J.D. Limnological Monitoring of Lake Becharof, Final Report to the U.S. Fish and Wildlife Service, Research Work Order 58, Alaska Cooperative Fish and Wildlife Research Unit April 1998. Available online: https://ecos.fws.gov/ServCat/DownloadFile/108087 (accessed on 15 February 2024).
64. Gelman, A.; Jakulin, A.; Pittau, M.G.; Su, Y.S. A Weakly Informative Default Prior Distribution for Logistic and Other Regression Models. *Ann. Appl. Stat.* **2008**, *2*, 1360–1383. [CrossRef]

water

MDPI

Article

Method for Producing Columnar Ice in Laboratory and Its Application

Yujia Zhang *, Zuoqin Qian and Weilong Huang

School of Naval Architecture, Ocean and Energy Power Engineering, Wuhan University of Technology, Wuhan 430063, China; qzq@whut.edu.com (Z.Q.); hwl220@whut.edu.cn (W.H.)
* Correspondence: zhangyujia@whut.edu.cn; Tel.: +86-027-8658-2035

Abstract: This study presents the design of a small open-circuit wind tunnel for laboratory use and a method for preparing columnar ice. The ice formation process was analyzed in terms of temperature and ice thickness variations under varying environmental temperatures and wind speeds. Observations revealed that as wind speed increased, the grain size of the columnar ice decreased. Key findings include the following: (1) the selection and validation of two cubic arcs for the wind tunnel contraction section, achieving an acceleration ratio of 6.7–6.8 and stable wind speeds of 1–10 m/s; (2) real-time temperature monitoring indicated rapid cooling before freezing and slower cooling post-freezing, with lower ambient temperatures and higher wind speeds accelerating the icing process; (3) the $-1/2$ power of grain size was found to be positively correlated with wind speed; and (4) the method's feasibility for studying mechanical properties of polar columnar ice was confirmed. This technique offers a controlled approach for producing columnar ice in the laboratory, facilitating comprehensive research on ice properties and providing a foundation for future studies on the mechanical behavior of ice under windy polar conditions.

Keywords: columnar ice; wind tunnel; ice formation; grain size

Citation: Zhang, Y.; Qian, Z.; Huang, W. Method for Producing Columnar Ice in Laboratory and Its Application. *Water* **2024**, *16*, 2558. https://doi.org/10.3390/w16182558

Academic Editors: Pentti Kujala, Zhijun Li, Fang Li, Weiping Li and Shifeng Ding

Received: 2 August 2024
Revised: 7 September 2024
Accepted: 9 September 2024
Published: 10 September 2024

1. Introduction

Ice plays a significant role in cold regions, affecting human activities and infrastructure. The internal structure of ice is quite complex and can be classified into granular ice and columnar ice based on crystal structure types [1]. Columnar ice is composed of vertically aligned ice crystal columns exhibiting distinct crystallographic orientation. In contrast, granular ice consists of randomly oriented small ice crystals lacking any regular crystallographic alignment. Granular ice forms in the early stages of freezing, influenced by wave agitation. When the water surface is calm or when a surface ice layer has already formed, ice crystals elongate vertically to form columnar ice. Granular ice is isotropic, while columnar ice exhibits significant anisotropy [2]. Ice formed in rivers, lakes, and oceans is generally predominantly columnar ice. In the Arctic and high-latitude inland waterways, the presence of ice is seasonal—water surfaces begin to freeze in winter and melt in summer [3,4]. The harsh winter environment results in complex ice conditions on the water surface. Issues such as ice-bound waterways and vessel damage not only affect shipping efficiency but also pose safety risks. When navigating during the ice season, ships need to select appropriate routes and icebreaking strategies [5–7]. Therefore, studying the mechanical properties and formation process of columnar ice is crucial for guiding the selection of navigation strategies for ships.

Ice strength is often a primary focus in the study of its mechanical properties, as it is a critical parameter for designing ships and offshore structures in icy regions. The failure modes of ice include compression, tension, and flexure, corresponding to compressive strength, tensile strength, and flexural strength, respectively [8–10]. Moslet [11] studied the uniaxial compressive strength of columnar sea ice in Svalbard, Norway, and found that sea ice exhibits brittle and ductile behavior under different loading rates. Tests on both horizontal and vertical samples concluded that strength mainly depends on temperature and also summarized the relationship between Young's modulus and porosity. Li [12], addressing the abnormal ice conditions in the Bohai Sea during the winter of 2009–2010, collected columnar sea ice samples from Liaodong Bay and conducted loading tests parallel to the ice surface. The uniaxial compressive strength of sea ice was measured, and the surface relationship between uniaxial compressive strength and porosity within the selected strain rate range was summarized, quantitatively describing the transition point of mechanical behavior with changes in porosity. Bonath [13] conducted uniaxial tensile tests on first-year ice ridges in the Svalbard region, finding that brine volume is a significant parameter affecting the tensile strength of columnar ice. Menge [14] conducted tensile tests on columnar ice perpendicular to the growth direction, showing that maximum tensile stress is most influenced by temperature, while failure strain and modulus are affected by loading rate. Han [15] studied the flexural properties of large columnar freshwater ice, revealing that flexural strength is not affected by loading direction but shows a clear correlation with temperature and strain rate. As test temperature decreases, ice strength increases; as strain rate increases, strength first increases and then decreases. Karulina [16] conducted full-scale flexural strength tests on sea ice and freshwater ice beams from the Svalbard region. The results indicated that sea ice has lower flexural strength than freshwater ice and that flexural strength is independent of the direction of the flexural force applied to the free end of the beam.

For studying the ice formation process, the most direct approach is to use specialized tools for real-time measurement of ice thickness. Common methods include drilling, echo sounding, electromagnetic (EM) sounding, visual ship-based observations, and video observations [17]. Worby [18] evaluated the applicability of using portable electromagnetic induction (EMI) devices for determining sea ice thickness under Antarctic winter and spring conditions. Uto [19] used ship-based electromagnetic induction devices to detect sea ice thickness in the southern Sea of Okhotsk, with results closely matching those obtained from drilling measurements. Upward-looking sonar is another classic method for monitoring sea ice thickness, typically mounted on submarines. This method was first used in Arctic sea ice surveys [20], providing technical support for Arctic exploration and yielding accurate data on Arctic sea ice thickness [21]. With technological advancements, radar has also been employed for ice thickness detection. Initially, it was primarily used in the polar regions [22,23], but it has since been used for ice thickness observations in high-latitude inland rivers [24].

During in situ measurements, the climate and environmental conditions at measurement locations are generally harsh, often leading to negative factors such as installation difficulties, instrument damage, and challenges in equipment retrieval due to external causes [25]. Additionally, the relatively high time and economic costs associated with field measurements result in certain limitations. Consequently, scholars are choosing to conduct studies on mechanical properties and the ice formation process in laboratory settings. Cole [26] prepared polycrystalline ice samples in the laboratory and reviewed several ice-making methods. Deng [27] conducted a series of uniaxial compression tests on laboratory-made ice, investigating the strain rate range during the transition from ductile to brittle behavior and the dispersion of compressive strength measurement data. Zhang [28] used a low-temperature laboratory to prepare distilled water ice at different temperatures, studying the relationship between uniaxial compressive strength, strain rate, and ice crystal grain size. Rosa [29] equipped a laboratory tank with wave conditions and thermal effects to observe the process of frazil ice crystals gradually accumulating to form a grease ice layer. Roscoe [30] studied the growth and composition of frost flowers in the laboratory, comparing laboratory results with field observations and finding consistency between the two.

Cultivating sea ice and freshwater ice in the laboratory provides a method for observing the physical properties of ice. Researchers can control the growth of ice through stringent variable control by adjusting environmental conditions [31]. Environmental temperature and wind speed are significant factors influencing the ice formation process. Generally, a temperature-controlled room can achieve the desired atmospheric temperature to control temperature effectively [32–34]. In some studies, fans are used to blow cold air over the water surface to increase the heat transfer rate and ensure the uniform distribution of cold air [35,36]. However, this method neither provides a stable wind speed nor allows precise wind speed adjustment. In the study of atmospheric ice, wind tunnel devices are used to provide stable and adjustable wind speeds for the icing process [37,38] and thus can be utilized in low-temperature laboratories to simulate cold region environments. This study proposed a method for preparing columnar ice in the laboratory, combining low-temperature laboratory and wind tunnel to simulate the cold and windy conditions of polar regions in a laboratory setting. This approach will facilitate future testing of columnar ice in the laboratory, thereby supplementing and improving the study of ice properties.

2. Methods

2.1. Design of Wind Tunnel

To ensure stable wind speed during the process of freezing, it is necessary to control the wind speed using a wind tunnel. In conventional air mediums, the maximum speed in a typical low-speed wind tunnel does not exceed 130 m/s [39]. In this research, the wind tunnel is housed within a low-temperature laboratory. Considering manufacturing costs and spatial constraints, the wind tunnel is designed as an open-circuit low-speed wind tunnel, with its structural schematic shown in Figure 1.

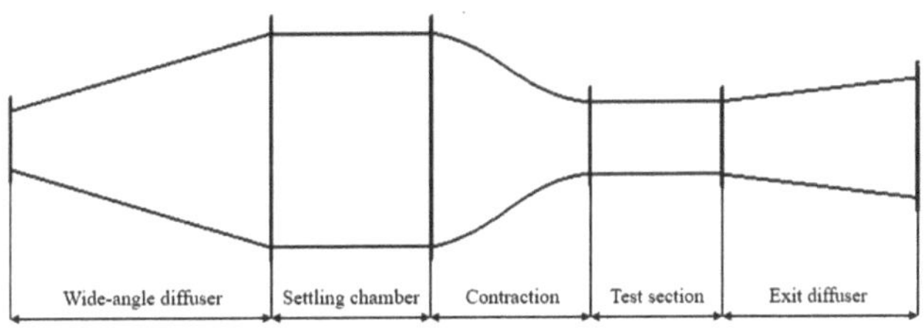

Figure 1. Structure of a typical open-circuit low-speed wind tunnel.

The wind tunnel should be designed as a whole for the purpose of sealing. As the settling chamber necessitates an internal flow rectification device while the contraction section does not, both sections are fabricated as a single unit. The remaining sections are individually processed and interconnected via flanges, supplemented with 2 mm thick rubber for damping and sealing purposes. For observational requirements, the test section is constructed using acrylic material with a thickness of 4 mm, while the remaining components are fabricated from stainless steel plates.

2.1.1. Design of Test Section and Settling Chamber

Based on the practical usage of the wind tunnel, the cross-sectional shape of the test section is designed to be square, and the dimensions are 150 mm × 150 mm. For low-speed wind tunnels, the length of the test section is typically 1.75 to 2.5 times the hydraulic diameter of the inlet cross-section [40]. Considering the size of the icing pool, the length of the test section is set at 285 mm. Thus, the final dimensions of the test section are determined to be 285 mm × 150 mm × 150 mm (length × width × height).

The size of the cross-sectional area of the settling chamber depends on the contraction ratio of the wind tunnel's contraction section. The contraction ratio is the ratio of the cross-sectional area between the settling chamber and the test section. For small-scale, low-speed open-circuit wind tunnels, the contraction ratio typically ranges from 6 to 9 [40]. In this study, the contraction ratio of the wind tunnel is set to 7. Calculations yield a settling chamber cross-sectional area of 1.6×10^5 mm^2, resulting in a square cross-section size of

400 mm × 400 mm. When the contraction ratio exceeds 5, the settling section's length is generally 0.5 to 1.0 times the diameter [40]. Therefore, the length of the settling chamber is determined to be 300 mm. To enhance the quality of the airflow field, a flow rectification device should be installed within the settling chamber to rectify and stabilize the flow field. The honeycomb structure was selected as the flow rectification device, as shown in Figure 2.

Figure 2. Flow rectification device.

2.1.2. Design of the Contraction

The contraction section is the core functional structure of the wind tunnel. The design of the contraction section mainly focuses on the axial length, contraction curve, and contraction ratio. Generally, the axial length of the contraction section is 0.5–1 times its inlet hydraulic diameter (the side length of the settling chamber) [40], so the length range of the contraction section is 200–400 mm. Selecting a longer length of the contraction section can make the contraction curve smooth, which is conducive to no separation of airflow. Therefore, the length of the contraction section was taken as 400 mm. The contraction curves commonly used for wind tunnel equipment with better performance are the Witosznski curve and the two cubic arcs [41]. Witosznski curve was defined by a 2nd-order polynomial as follows:

$$R = \frac{R_2}{\sqrt{\left\{1 - \left[1 - \left(\frac{R_2}{R_1}\right)^2\right]\frac{\left(1 - \frac{3x^2}{a^2}\right)^2}{\left(1 + \frac{x^2}{a^2}\right)^3}\right\}}} \tag{1}$$

where R_1, R_2 are the section radius at the inlet and outlet of the contraction section (mm), R is the section radius at the axial distance x (mm), and a is the length (mm) of $\sqrt{3}$ times the contraction section.

The formula of the two cubic arcs is given as follows:

$$\frac{R - R_2}{R_1 - R_2} = \begin{cases} 1 - \left(\frac{1}{X_m}\right)^2 \left(\frac{x}{L}\right)^3, & (x/L) \leq X_m \\ \frac{1}{(1 - X_m)^2}\left[1 - \left(\frac{x}{L}\right)\right]^3, & (x/L) > X_m \end{cases} \tag{2}$$

where x/L is a dimensionless local-contraction length, $X_m = x_m/L$ is the dimensionless contraction length where the matching of the two cubic arcs occurs, and R_1, R_2 are the section radius at the inlet and outlet of the contraction section (mm). The flow characteristics of the contraction section based on two kinds of curve modeling were analyzed by the computational fluid dynamics (CFD) method. It was determined that the contraction section should be modeled using the two cubic arcs. The computational results are elaborated in Section 3.

2.1.3. Design of Exit Diffuser and Wide-Angle Diffuser

The exit diffuser is located behind the test section and is directly connected to the atmosphere. The area ratio of the sections at both ends of the diffuser is generally designed to be about 2. The calculated outlet section size of the diffuser was 210 mm × 210 mm. The diffusion angle was taken as 10°, and the length of the diffuser was 140 mm. The wide-angle diffuser is located between the fan and the settling chamber. The diameter of the axial flow fan is 310 mm, so the inlet size of the diffusion section is set at 300 mm × 300 mm to ensure a good fit with the fan and reduce the airflow loss. The length of the wide-angle diffuser was determined to be 300 mm.

According to the above design of each part of the open-circuit low-speed wind tunnel, the wind tunnel device with a total length of 1425 mm was finally obtained. The structures of each part were connected by flanges and nuts, as shown in Figure 3. Wind tunnel performance testing has determined that the maximum wind speed achievable by the wind tunnel is 10 m/s. The results of the performance testing are discussed in Section 3.

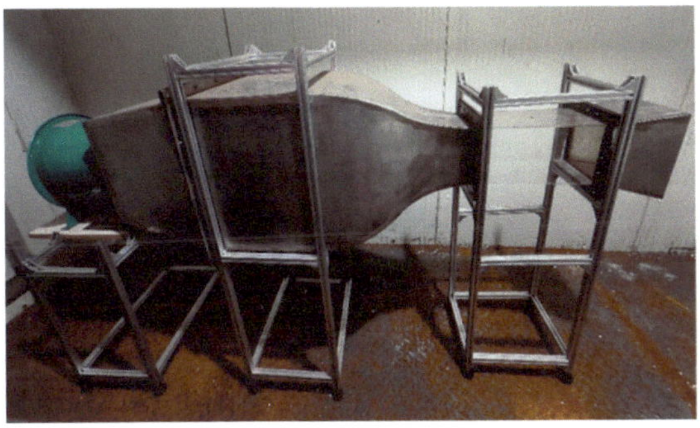

Figure 3. Open-circuit low-speed wind tunnel for this study.

2.2. *Ice Formation Test*

2.2.1. Test Devices

Regarding the selection of materials for the ice formation tank, while acrylic glass offers the advantage of transparency, facilitating direct observation of the ice formation process, its relatively high thermal conductivity causes heat exchange between the tank walls and the cold air. This exchange can affect the ice growth direction and observation accuracy. The objective of the experiment is to produce columnar ice, and to ensure that the ice grows naturally from top to bottom, a foam box is chosen as the ice formation tank. The reasons for this choice include the following: foam material has a low thermal conductivity, excellent sealing properties, resistance to water leakage, ease of repair in case of leaks, and stability in low-temperature environments. Additionally, the foam box's elasticity allows it to conform tightly to the test section and prevent air leakage. It is also lightweight, which facilitates handling during experiments and is cost-effective, thus reducing overall costs.

The ice formation tests were conducted in a low-temperature laboratory, where the environmental temperature could be lowered to a minimum of −40 °C, with a temperature control precision of 0.1 °C. To ensure that the initial water temperature in the ice formation tank of each test remains consistent, the foam ice pools were placed in a high–low temperature test chamber for temperature stabilization before the tests. The specific parameters of the low-temperature laboratory and the high–low temperature test chamber have been described in a previous study [42]. The required wind speed for the tests was provided by the wind tunnel. The test section of the wind tunnel was modified, with holes opened at the upper part of the test section to facilitate the insertion of temperature and wind measuring instruments; the lower part of the test section was not sealed, allowing the water to be in direct contact with the airflow, as shown in Figure 4.

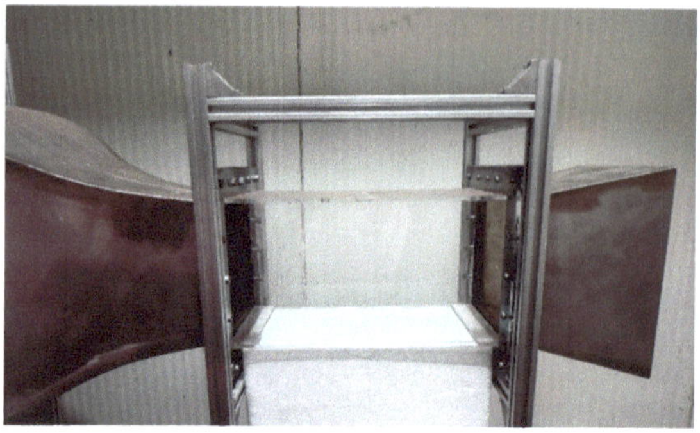

Figure 4. Ice formation tank and test section.

To measure temperature using a PT100 temperature sensor (Sigma-Aldrich, St. Louis, MA, USA), the sensor probe is welded onto a stainless-steel tube, forming a temperature chain. The high thermal conductivity of stainless steel may affect the heat transfer between air, ice, and water during the freezing process. To minimize this impact, the thickness of the stainless-steel tube is set at 1 mm. After welding the probe, the tube was vacuum-sealed. Temperature probes are placed at intervals of 2 cm along the temperature chain. The temperature sensor has a measurement range of −50 °C to 450 °C and an accuracy of 0.1 °C. The measurement data from the sensors are recorded using a paperless recorder, which has a reading accuracy of 0.1 °C. Wind speed is measured using the Testo 425 hot-wire anemometer (Testo SE, Titisee-Neustadt, Germany), with a resolution of 0.01 m/s and an accuracy of ±(0.03 m/s + 4.0% of the measured value). The arrangement of the temperature chain and the anemometer is illustrated in Figure 5.

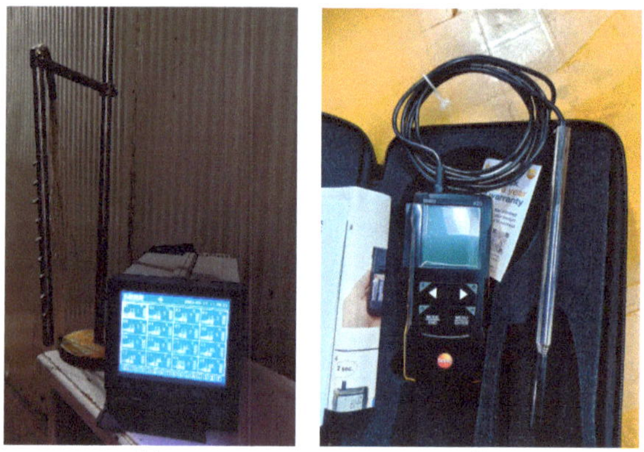

Figure 5. Temperature and wind speed measurement device. (**a**) Temperature chain and paperless recorder. (**b**) Hot-wire anemometer.

2.2.2. Test Procedure

Before the test, the ice formation tank filled with water is placed in the high–low temperature test chamber until thermal equilibrium is reached. Once the low-temperature laboratory reaches the desired test temperature and thermal equilibrium is maintained, the tank is positioned below the test section. The temperature chain is arranged vertically downward in the center of the foam ice pool, with five probes submerged in the water, the uppermost probe being at the water surface. Ambient temperatures in the low-temperature laboratory were set to −10 °C, −15 °C, −20 °C, −25 °C, and −30 °C, with wind speeds of 1 m/s, 2 m/s, 4 m/s, 6 m/s, and 8 m/s for each temperature. Ice thickness is measured hourly. At the same depth, three positions are selected to insert a fine iron pin horizontally into the ice pool's side. If the pin cannot be inserted, the depth is frozen; if it inserts smoothly, the depth contains water. The pin is then withdrawn, and waterproof tape is used to seal the hole. After the final measurement, the foam box is destroyed to drain the remaining water. The ice sample is photographed, its thickness measured, and its shape recorded.

3. Results and Discussion

3.1. Selection of the Wind Tunnel Contraction Section

3.1.1. The Distribution of Pressure and Velocity

In Section 2, it was mentioned that the contraction section is the core structure of the wind tunnel design, and two common contraction section curves were proposed. This section introduces the selection of the contraction section for the experimental wind tunnel. Finite element models of the two contraction sections were established, as shown in Figure 6. Computational fluid dynamics (CFD) methods were used to analyze the flow fields inside the two contraction sections and the test section, with inlet velocities set at 0.1 m/s, 1 m/s, 3 m/s, and 10 m/s, respectively. The CFD simulations were conducted using FLUENT 2020 R2.

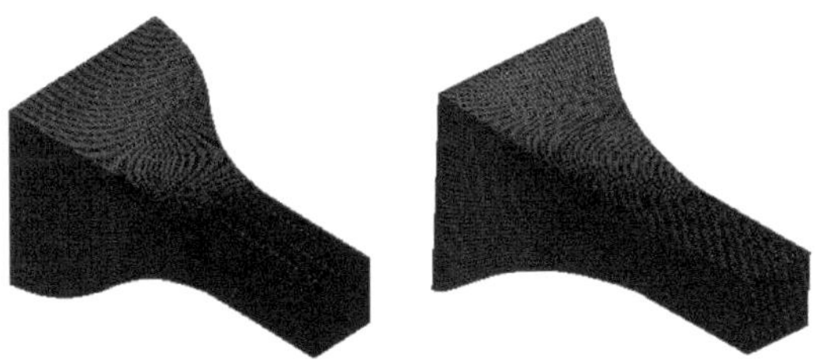

(a) (b)

Figure 6. The finite element models of the contraction sections and the test section with different contraction curve profiles. (**a**) The contraction section constructed using two cubic arcs. (**b**) The contraction section constructed using the Witosznski curve.

Under the same inlet velocity conditions, comparing the static pressure (gauge pressure) contour maps of the two types of contraction sections shown in Figure 7, taking 10 m/s as an example, it can be observed that the static pressure decreases along the flow direction from the entrance of the contraction section and tends to zero at the exit of the test section. In the contraction section constructed using two cubic arcs, the static pressure distribution is relatively uniform, with the maximum pressure evenly distributed at the entrance of the contraction section, and the rate of static pressure decrease is relatively gentle. In contrast, in the contraction section constructed using the Witosznski curve, it can be clearly observed that the static pressure distribution at the entrance is not uniform, with the maximum pressure concentrated at the corners of the entrance of the contraction section. This may lead to adverse pressure gradients causing gas recirculation, which could affect the overall quality of the flow field.

As shown in Figure 8, it can be found that at each inlet wind speed, the ratio of the velocity at the end of the contraction section to the inlet velocity is approximately 7, which matches the contraction ratio $C = 7$ designed for the wind tunnel model. Taking 10 m/s as an example, velocity streamlines in the test section in both types of models are relatively straight, indicating that both can provide a stable flow field. However, it was observed that the Witosznski curve contraction section exhibits more drastic changes in inlet velocity, while the two-cubic arcs contraction section experiences minimal variation in inlet velocity, primarily concentrated in the middle section for acceleration, and the velocity gradient at the exit returns to be gentle. A qualitative analysis of pressure and velocity distributions suggests that the flow characteristics of the two-cubic arcs contraction section are favorable.

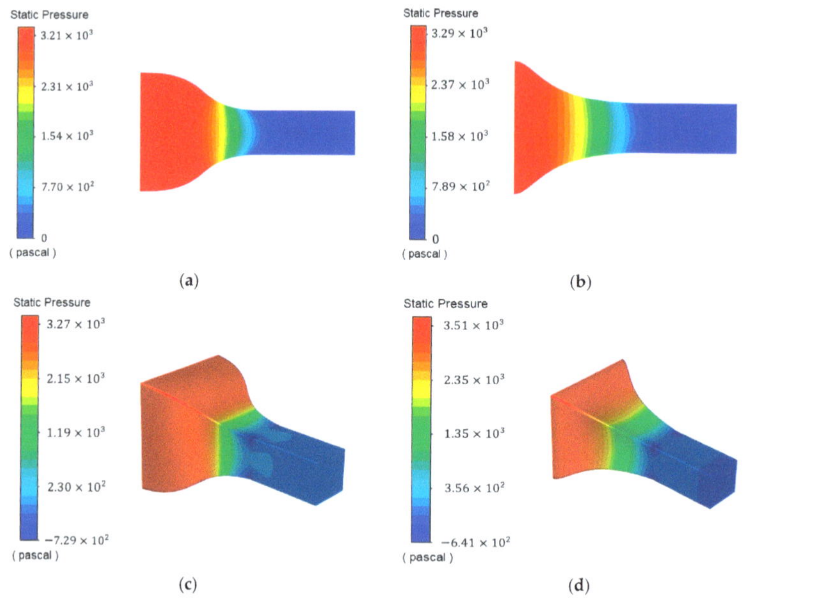

Figure 7. When the inlet wind speed is 10 m/s, the distribution of sectional static pressure and overall static pressure within the contraction and test section of the wind tunnel varies between the two types of contraction sections. (**a**,**c**) Contour maps of the two-cubic arcs model. (**b**,**d**) Contour maps of the Witosznski curve model.

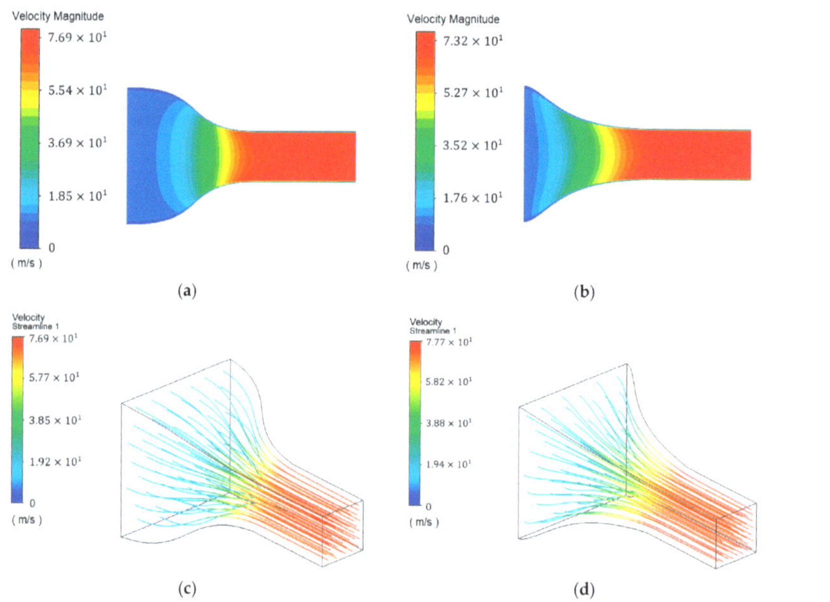

Figure 8. Contour maps and streamlines of the velocity distribution of two models when the inlet wind speed is 10 m/s. (**a**,**c**) Contour map and streamline of the two-cubic arcs model. (**b**,**d**) Contour map and streamline of the Witosznski curve model.

3.1.2. Comparison of Dynamic Pressure Coefficient and Velocity Non-Uniformity

The dynamic pressure coefficient and velocity non-uniformity are crucial parameters for evaluating the uniformity of airflow in the test section. By comparing their magnitudes, the performance of the contraction curves can be quantitatively assessed, with smaller values indicating better flow uniformity. The formulas for calculating the dynamic pressure coefficient α_i and velocity non-uniformity η are given as follows [43]:

$$\alpha_i = \left| \frac{q_i}{\overline{q}} - 1 \right| \tag{3}$$

$$\eta = \left| \frac{v_i}{\overline{v}} - 1 \right| \tag{4}$$

where q_i is the dynamic pressure at point i, $\overline{q}$ is the average dynamic pressure measured at different points, v_i is the velocity at point i, and $\overline{v}$ is the average velocity measured at different points.

The maximum dynamic pressure coefficient and maximum velocity non-uniformity in the central region of the test section for both wind tunnel models are calculated at different wind speeds (Tables 1 and 2). As the wind speed increases, both parameters decrease, indicating a more uniform flow field within the test section. When the wind speed is 0.1 m/s, the velocity non-uniformity of the Witosznski curve model is slightly better than that of the two-cubic arcs model. However, at inlet wind speeds of 1 m/s, 3 m/s, and 10 m/s, the maximum dynamic pressure coefficient and velocity non-uniformity for the model with the two-cubic arcs contraction section are smaller. Therefore, the model with the two-cubic arcs contraction section exhibits better flow characteristics than the one with the Witosznski curve contraction section and is chosen for use in this study.

Table 1. Maximum dynamic pressure coefficient in the middle of the test section of two wind tunnel models under different wind speeds.

Wind Speed (m/s)	0.1	1	3	10
Two cubic arcs	0.071	0.025	0.022	0.019
Witosznski curve	0.061	0.046	0.036	0.032

Table 2. Maximum velocity non-uniformity in the middle of the test section of two wind tunnel models under different wind speeds.

Wind Speed (m/s)	0.1	1	3	10
Two cubic arcs	0.035	0.012	0.011	0.009
Witosznski curve	0.030	0.023	0.018	0.016

3.2. Wind Tunnel Test Section Velocity Verification

After the completion of the wind tunnel construction, performance testing is required. In this study, the primary purpose of the wind tunnel is to provide stable wind speeds during the ice formation process. Therefore, the wind speed within the test section has been selected as the subject for performance verification. The contraction ratio was defined in Section 2. According to the law of mass conservation, the numerical ratio of the wind speed within the test section to the wind speed at the entrance of the contraction section should equal the contraction ratio. Therefore, measuring the wind speed can also verify whether the acceleration effect matches the design expectations. The wind speed within the wind tunnel is provided by an axial flow fan equipped with a variable transformer. The rotational speed of the fan blades can be continuously adjusted by varying the output voltage of the transformer. The transformer is adjusted to a specific output voltage until the fan speed remains constant. Then, the anemometer probe is positioned at the midpoint of the contraction section entrance cross-section, and the test section midpoint and the wind speeds are recorded, respectively. The output voltage is gradually adjusted from low to high, and the wind speed is measured at the midpoint of the test section. This process will determine the minimum and maximum stable wind speeds that the wind tunnel can provide under operational conditions. At each output voltage, the wind speed measurements are repeated three times, and the average is taken to minimize errors. The results of the wind speed measurements are shown in Table 3.

Table 3. Output voltages and results of wind speed measurements.

Output Voltages (V)	Average Wind Speed at the Midpoint of the Test Section (m/s)	Average Wind Speed at the Contraction Section Entrance (m/s)	Speed Ratio
90	1.08	0.16	6.75
105	1.97	0.29	6.79
120	3.04	0.45	6.76
130	4.12	0.61	6.75
150	6.23	0.92	6.77
180	8.13	1.21	6.72
220	10.56	1.57	6.73

The calculated ratio of the wind speed at the midpoint of the test section to the wind speed at the entrance of the contraction section ranges from 6.7 to 6.8, which is marginally less than the designed contraction ratio of 7. This deviation can be attributed to the fact that theoretical calculations typically assume idealized conditions, such as smooth wall surfaces and negligible flow losses. In contrast, practical implementations involve pressure gradients and energy dissipation within the wind tunnel, which reduce the kinetic energy of the airflow. Additionally, the formation of a boundary layer due to viscous effects between the airflow and the walls reduces the effective cross-sectional area, leading to a contraction ratio that is slightly lower than the design value. The results indicate that the wind tunnel in this study can provide a stable wind speed range from 1 m/s to 10 m/s under operational conditions. Integrating the midpoint wind speed of the test section with the output voltage into Figure 9 reveals a linear positive correlation between the output voltage and the midpoint wind speed of the test section. The relationship is expressed by the following equation:

$$\frac{v}{v_0} = 0.076\frac{U}{U_0} - 5.759 \quad (5)$$

where U is output voltage and v is wind speed of test section. v_0 and U_0 are used for non-dimensionalization.

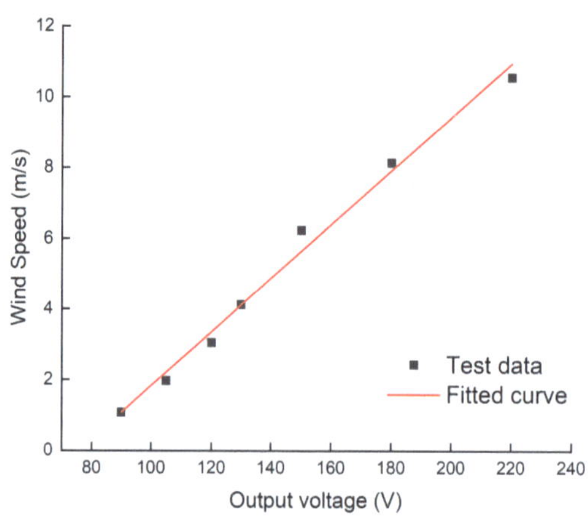

Figure 9. Wind speed at the midpoint of the test section at different output voltages.

3.3. Results and Discussion of Ice Formation Test

In the initial stages of the ice formation observed according to the experimental methods in Section 2, five temperature sensor probes were arranged every 2 cm along the water surface, with Probe 1 located at the water surface. The temperature data were collected at a frequency of once per minute, resulting in temperature variation curves during the experiment, as partially shown in Figures 10 and 11.

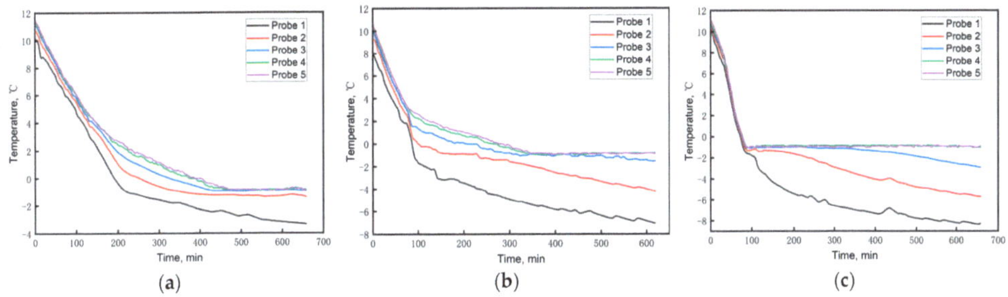

Figure 10. Temperature variation curves in the condensation tank at an ambient temperature of −10 °C at different wind speeds. The wind speeds for (**a**–**c**) are 1 m/s, 4 m/s, and 8 m/s, respectively.

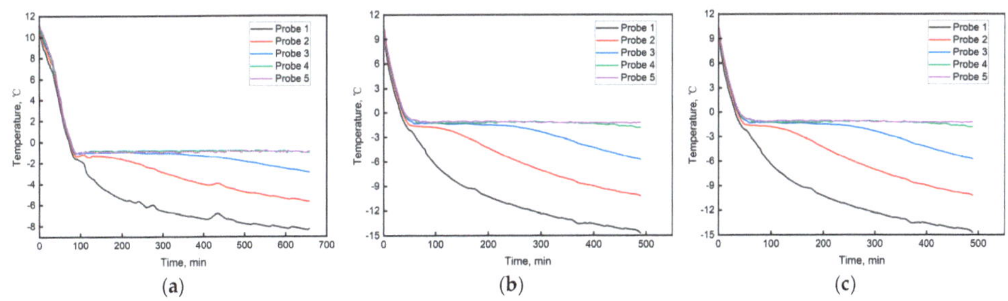

Figure 11. Temperature variation curves in the condensation tank at a wind speed of 8 m/s at different ambient temperatures. The temperatures for (**a**–**c**) are −10 °C, −20 °C, and −30 °C, respectively.

It can be observed that, under all conditions, the temperature variation in the condensation tank can be divided into two stages. The first stage is characterized by a rapid overall decline in water temperature, with the slopes of the temperature curves obtained by each probe being largely consistent. This slope can be used to calculate the cooling rate of the water before icing, with the calculated results shown in Table 4. By comparing the temperature curves at the same temperature but different wind speeds (Figure 10) and at the same wind speed but different temperatures (Figure 11), it is found that higher wind speeds and lower temperatures lead to faster cooling rates. Furthermore, it is evident that wind speed has a more significant impact on the cooling rate than ambient temperature at this stage. This is because the heat transfer modes in this study include both convective heat transfer and conduction. At this stage, the water surface has not yet frozen, and the convective heat transfer between water and air dominates. In the second stage, the cooling rate of the temperature at each measuring point in the condensation tank slows down, and the temperature differences between measuring points gradually increase. The closer to the air, the lower the temperature and the faster the cooling rate. The reason for the slowdown in cooling is that the water starts to freeze during this period, releasing heat during the phase transition. Additionally, the thermal conductivity of ice is lower than that of water. Once the water surface freezes, the ice layer acts as a thermal resistance, providing some insulation and affecting the heat exchange process between the underlying water and the cold air. This is reflected in the graphs, showing that the deeper the water, the slower the temperature drops.

Table 4. Calculation results of cooling rate at different wind speeds and temperatures (unit: °C/h).

Wind Speed/Temperature	−10 °C	−15 °C	−20 °C	−25 °C	−30 °C
1 m/s	3.15	3.56	3.81	4.37	4.71
2 m/s	3.94	5.13	6.06	7.72	8.05
4 m/s	5.56	7.20	7.91	8.98	9.35
6 m/s	6.97	8.20	9.73	10.23	11.40
8 m/s	9.13	13.09	14.71	16.88	19.63

After each test, the condensation tank was destroyed, and the remaining water was drained to reveal the final ice formation, with some results shown in Figure 12. It can be observed that despite the low thermal conductivity of the foam box, which minimizes boundary effects, there are still some protrusions around the edges of the ice. Additionally, the thickness in the middle of the ice layer is slightly greater due to the influence of the temperature chain. The thickness of the ice was measured after removing the protruding parts. The condensation tank is 280 mm long, and measurements were taken at 10 mm, 75 mm, 140 mm, 205 mm, and 270 mm from the end near the fan. The average value of these measurements was then calculated. Combined with the duration of test, the average icing rate was recorded in Table 5.

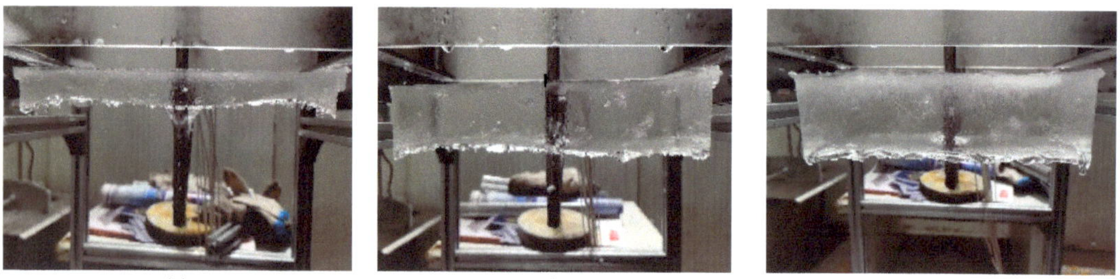

(a) (b) (c)

Figure 12. The results of ice shapes after the formation test at −10 °C. The wind speeds for (**a–c**) are 1 m/s, 4 m/s, and 8 m/s, respectively.

Table 5. Calculation results of icing rate at different wind speeds and temperatures (unit: mm/h).

Wind Speed/Temperature	−10 °C	−15 °C	−20 °C	−25 °C	−30 °C
1 m/s	3.31	4.19	4.54	5.39	5.54
2 m/s	3.88	5.41	5.49	5.57	6.36
4 m/s	4.67	6.00	6.01	6.80	7.53
6 m/s	4.92	6.29	6.63	7.08	7.99
8 m/s	5.09	6.57	7.03	7.57	8.26

The results indicate that both the cooling rate of the water body before the onset of freezing and the initial icing rate are influenced by the combined effects of ambient temperature and wind speed. By plotting the data in Figure 13 and applying surface fitting, the relationships between the cooling rate, the icing rate, and the temperature and wind speed can be expressed by Equations (6) and (7):

$$u_c = 2.051 - 0.183T - 0.581v - 0.003T^2 + 0.115v^2 - 0.046Tv \quad (6)$$

$$u_i = 0.092 - 0.022T + 0.061v - 2.95 \times 10^{-4}T^2 - 0.005v^2 - 5.94 \times 10^{-4}Tv \quad (7)$$

where u_c is the cooling rate of the water body, u_i is icing rate, T is ambient temperature, and v is wind speed.

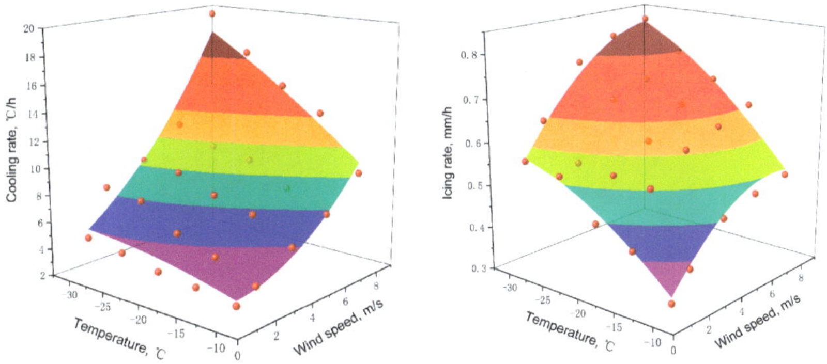

(**a**) (**b**)

Figure 13. Surface fitting results for the relationships between cooling rate (**a**), icing rate (**b**), temperature, and wind speed.

3.4. Results and Discussion of Ice Crystal Measurement

The objective of this study is to simulate the low-temperature, windy conditions of polar regions in a laboratory setting to produce columnar ice. Consequently, it is necessary to measure the crystal structure of the ice fabricated using the techniques in this study. To ensure the integrity of the ice and facilitate its smooth extraction, temperature chains were not inserted for temperature monitoring during the freezing process of the ice used for crystal structure observation. The methods for observing ice crystal structures are well established and have been utilized in previous studies [28,42]. Thin ice sections attached to glass slides are observed under polarized light, with sections categorized as either horizontal or vertical based on their orientation relative to the ice growth direction. The method for preparing sections for observing the ice crystal structure is as follows: after extracting the ice sample, sections of appropriate size and smooth cross-sections are cut from different positions, oriented either vertically (parallel to the ice growth direction) or horizontally (perpendicular to the ice growth direction), with a thickness of approximately 1–2 cm. A planer is used to gradually flatten one side of the section to ensure it can adhere seamlessly to a clean glass slide. The preheated glass slide, maintained in a water bath slightly above 0 °C, is then used to bond the flattened surface of the ice section. After a few minutes, the ice section is firmly frozen to the glass slide. The section is then thinned to a thickness of less than 1 mm using a planer to facilitate the distinction of crystal boundaries. Vertical sections are used to observe the crystal type, while the results of horizontal sections are analyzed using the equivalent circular diameter method to calculate the grain size. The equivalent circle diameter method involves counting the number of complete grains on a known cross-sectional area and treating each grain as a circle to calculate its diameter. The calculation formula is as follows:

$$D_g = 2\sqrt{\frac{S}{n\pi}} \tag{8}$$

where D_g is the average grain diameter (mm); S is the area of ice section (mm^2); and n is number of grains on the section.

The temperature of the low-temperature laboratory was set to −30 °C, and ice samples were prepared at different wind speeds. As shown in Figure 14, the ice samples grown under various wind speed conditions all exhibited columnar ice structures.

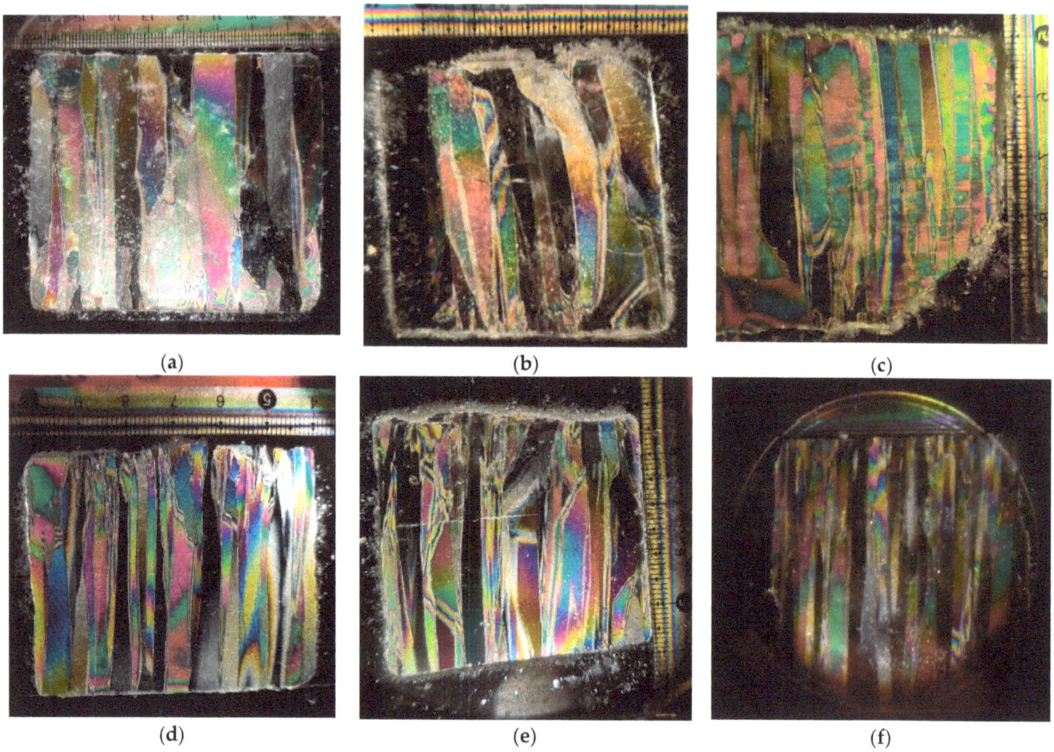

Figure 14. Ice crystal morphologies at varying wind speeds. Wind speeds of (a–f) are 0 m/s, 1 m/s, 2 m/s, 4 m/s, 6 m/s, and 8 m/s.

During the entire freezing process, ambient temperature remained almost constant, and wind speed only caused some disturbance to the water surface at the initial stage of freezing. Once the surface was frozen, the heat exchange between the stagnant ice and the underlying water and air diminished. This is reflected in the slower temperature fluctuations observed in Section 3.3. Consequently, the ice crystals had sufficient time and space to grow. Each crystal's growth was constrained by surrounding crystals, resulting in downward growth only, ultimately leading to the formation of columnar ice structures. The data for grain size calculations are presented in Figure 15. It is evident that the grain size increases with depth and decreases with increasing wind speed. The reason for this trend is that slower crystallization rates result in larger grain sizes, while higher wind speeds accelerate the icing rate, leading to smaller grain sizes. As the ice thickness gradually increases, the thermal resistance between the water and the cold air also increases, causing the icing rate to slow down and the grain size to increase.

From the results of ice grain size, it typically ranges from millimeters and does not easily achieve micron-scale sizes like other materials primarily due to its formation mechanisms, thermodynamic properties, and growth environment. The growth of ice grains is closely related to the cooling rate; in slower cooling conditions, ice crystals have more time to grow, leading to the formation of larger grains. Consequently, both experimental and natural environments with slow cooling rates typically result in millimeter-scale ice grains. In contrast, the formation of micron-scale grains theoretically requires extremely rapid cooling rates, which are often not achievable in ice formation processes. Additionally, the low thermal conductivity of ice leads to a slower heat dissipation during freezing, further limiting the reduction in grain size. The hexagonal crystal structure of ice, along with its unique intermolecular hydrogen bonding, also contributes to the tendency for ice to form larger crystal clusters during growth, making it challenging to achieve micron-scale grains under natural or conventional experimental conditions.

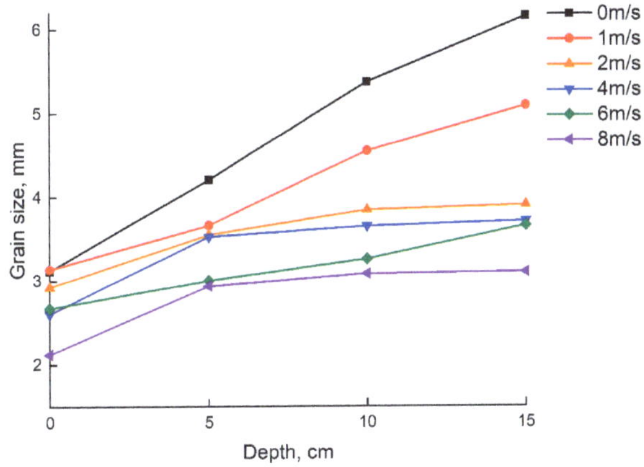

Figure 15. Grain size vs. depth curves of distilled water ice grown at different wind speeds.

Previous studies have indicated that the strength of ice is linearly related to −1/2 power of grain size [44–46]. The analysis above reveals that wind speed is one of the factors influencing grain size. As summarized, there is a linear relationship between wind speed and −1/2 power of average grain size, as shown in Figure 16. This implies that, under constant temperature conditions, the desired grain size can be achieved by controlling the wind speed, thereby facilitating further research into the physical properties of ice. Future research could further investigate the combined effects of temperature and wind speed on grain size.

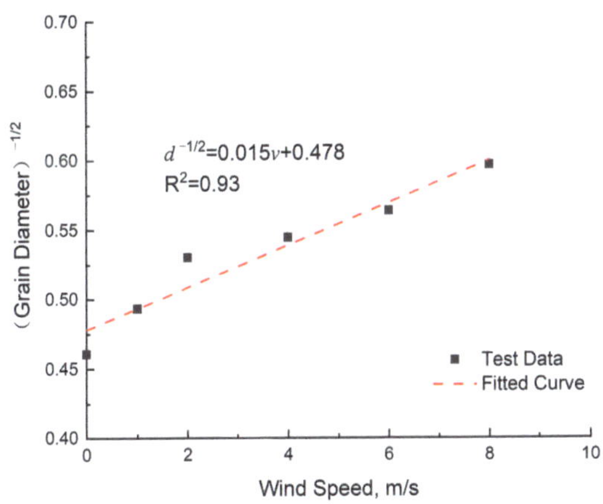

Figure 16. Grain size versus wind speed and fitted curve.

3.5. Application Discussion

This study integrates a low-temperature laboratory with a custom-made wind tunnel apparatus to develop a technique for producing columnar ice under laboratory conditions. This method allows for the simulation of the microstructure of ice formed under cold and windy conditions. In previous research [42], it was applied to study the uniaxial compressive strength of ice, investigating the strain rate sensitivity of columnar ice's uniaxial compressive strength. The results elucidated the relationship between grain size and compressive strength, explaining the physical mechanism behind the higher ice strength observed in polar regions under low temperature and high wind speed conditions. This technique provides the means for comprehensive research into the mechanical properties of polar ice formed under windy conditions. It allows for better control of environmental variables during the freezing process, laying the foundation for studying the effects of various factors on the physical properties of columnar ice.

In future work, this ice production method can be used to provide test samples for mechanical property testing of columnar ice formed under windy polar conditions. Zhang [47] utilized confined single-sided shear tests and Brazilian disc splitting tests to study the shear and tensile strengths of freshwater and seawater, examining the impact of temperature on both strengths. The ice samples were formed by freezing water collected from different bodies of water in molds within the laboratory. Wang [48] collected sea ice samples from Prydz Bay and, after transporting them back to the laboratory, investigated the bending and compressive strengths of the sea ice using three-point bending tests and uniaxial compression tests, respectively, studying the effects of strain rate and porosity on strength. Ji [49], based on field observations, studied the impact of loading direction on the uniaxial compressive strength of sea ice. The columnar sea ice was subjected to uniaxial compression tests at different loading rates in both horizontal (parallel to the grain columns) and vertical (across the grain columns) directions, exploring the anisotropy of the compressive properties of columnar sea ice. The columnar ice production method developed in this study can also be applied to the aforementioned research to supply test samples. This method allows for more precise control of variables and can be completed under laboratory conditions, addressing the challenges and costs associated with field sample collection. This technique facilitates supplementary and comprehensive studies on ice properties and can be utilized in numerous experimental investigations.

4. Conclusions

This study designed a small open-circuit wind tunnel for laboratory use and proposed a method for the preparation of columnar ice. Using this method, columnar ice was produced, and the freezing process was observed under different environmental temperatures and wind speeds. Temperature changes and ice thickness variations during the freezing process were analyzed. The shape and grain size of the columnar ice were observed, and the relationship between grain size and wind speed was investigated. The application prospects of this method were discussed, leading to the following conclusions:

1. The contraction section of the wind tunnel was selected. Through qualitative and quantitative analyses, it was found that using two cubic arcs for the contraction section was more appropriate based on the flow field characteristics of models established with different curves. Wind speed verification tests were conducted on the processed wind tunnel, showing that the actual acceleration ratio of the wind tunnel contraction section was 6.7–6.8, providing a stable wind speed range of 1 m/s to 10 m/s.

2. The temperature chain was used for real-time monitoring of the temperature at different depths in the condensation tank during the freezing process. It was observed that the water temperature experienced two stages: a rapid overall decline before freezing and a slow decline after freezing began. The relationships between water cooling rate, icing rate, ambient temperature, and wind speed were analyzed. Lower ambient temperatures and higher wind speeds resulted in faster cooling and icing rates. These relationships were summarized through surface fitting.

3. The crystal structure of ice samples produced using this method was observed, revealing that all samples were columnar ice. The grain size of the ice crystals was calculated, showing that under the same temperature, higher wind speeds resulted in smaller grain sizes. It was concluded that $-1/2$ power of the grain size was positively linearly correlated with wind speed.

4. The feasibility of this method was discussed in conjunction with previous research on the mechanical properties of ice. This method can be used in the future to test various mechanical properties of columnar ice formed under windy polar conditions, providing supplementary and refined methods for ice property research.

These conclusions highlight the effectiveness and potential applications of the columnar ice production method developed in this study, offering a controlled laboratory technique to study the physical and mechanical properties of ice.

Author Contributions: Conceptualization, Y.Z. and Z.Q.; methodology, Y.Z.; software, Y.Z. and W.H.; validation, Y.Z. and W.H.; formal analysis, Y.Z.; investigation, Y.Z.; resources, Y.Z.; data curation, Y.Z.; writing—original draft preparation, Y.Z.; writing—review and editing, Y.Z. and W.H.; visualization, Y.Z.; supervision, Z.Q.; project administration, Z.Q. and Y.Z.; funding acquisition, Z.Q. All authors have read and agreed to the published version of the manuscript.

Funding: This research was funded by the National Natural Science Foundation of China, grant number 51676144.

Data Availability Statement: The original contributions presented in the study are included in the article, and further inquiries can be directed to the corresponding author.

Acknowledgments: The authors would like to thank the Wuhan University of Technology (WUT).

Conflicts of Interest: The authors declare no conflicts of interest.

References

1. Cole, D.M. The microstructure of ice and its influence on mechanical properties. *Eng. Fract. Mech.* **2001**, *68*, 1797–1822. [CrossRef]
2. Timco, G.W.; Weeks, W.F. A review of the engineering properties of sea ice. *Cold Reg. Sci. Technol.* **2010**, *60*, 107–129. [CrossRef]
3. Moline, M.A.; Karnovsky, N.J.; Brown, Z.; Divoky, G.J.; Frazer, T.K.; Jacoby, C.A.; Torres, J.J.; Fraser, W.R. High latitude changes in ice dynamics and their impact on polar marine ecosystems. *Ann. N. Y. Acad. Sci.* **2008**, *1134*, 267–319. [CrossRef]
4. Xu, H.; Yin, Z.; Jia, D.; Jin, F.; Ouyang, H. The potential seasonal alternative of Asia–Europe container service via Northern sea route under the Arctic sea ice retreat. *Marit. Pol. Mgmt.* **2011**, *38*, 541–560. [CrossRef]
5. Goerlandt, F.; Goite, H.; Banda, O.A.V.; Höglund, A.; Rainio, P.A.; Lensu, M. An analysis of wintertime navigational accidents in the Northern Baltic Sea. *Saf. Sci.* **2017**, *92*, 66–84. [CrossRef]
6. Lee, H.W.; Roh, M.; Kim, K.S. Ship route planning in Arctic ocean based on POLARIS. *Ocean Eng.* **2021**, *234*, 109297. [CrossRef]
7. Shu, Y.; Zhu, Y.; Xu, F.; Gan, L.; Lee, P.T.; Yin, J.; Chen, J. Path planning for ships assisted by the icebreaker in ice-covered waters in the Northern Sea Route based on optimal control. *Ocean Eng.* **2023**, *267*, 113182. [CrossRef]
8. Petrovic, J.J. Review mechanical properties of ice and snow. *J. Mater. Sci.* **2003**, *38*, 1–6. [CrossRef]
9. Dempsey, J.P. Research trends in ice mechanics. *Int. J. Solids Struct.* **2000**, *37*, 131–153. [CrossRef]
10. Timco, G.W.; Brien, S.O. Flexural strength equation for sea ice. *Cold Reg. Sci. Technol.* **1994**, *22*, 285–298. [CrossRef]
11. Moslet, P.O. Field testing of uniaxial compression strength of columnar sea ice. *Cold Reg. Sci. Technol.* **2007**, *48*, 1–14. [CrossRef]
12. Li, Z.; Zhang, L.; Lu, P.; Leppäranta, M.; Li, G. Experimental study on the effect of porosity on the uniaxial compressive strength of sea ice in Bohai Sea. *Sci. China Technol. Sci.* **2011**, *54*, 2429–2436. [CrossRef]
13. Bonath, V.; Edeskär, T.; Lintzén, N.; Fransson, L.; Cwirzen, A. Properties of ice from first-year ridges in the Barents Sea and Fram Strait. *Cold Reg. Sci. Technol.* **2019**, *168*, 102890. [CrossRef]
14. Menge, J.A.R.; Jones, K.F. The tensile strength of first-year sea ice. *J. Glaciol.* **1993**, *39*, 609–618. [CrossRef]
15. Han, H.; Jia, Q.; Huang, W.; Li, Z. Flexural strength and effective modulus of large columnar-grained freshwater ice. *J. Cold Reg. Eng.* **2016**, *30*, 04015005. [CrossRef]
16. Karulina, M.; Marchenko, A.; Karulin, E.; Sodhi, D.; Sakharov, A.; Chistyakov, P. Full-scale flexural strength of sea ice and freshwater ice in Spitsbergen Fjords and North-West Barents Sea. *Appl. Ocean Res.* **2019**, *90*, 101853. [CrossRef]
17. Alekseeva, T.A.; Frolov, S.V.; Serovetnikov, S.S. Review of methods and main results of sea ice thickness measurements in the Arctic. *Russ. Arctic.* **2021**, *1*, 33–49. (In Russian) [CrossRef]
18. Worby, A.P.; Griffin, P.W.; Lytle, V.I.; Massom, R.A. On the use of electromagnetic induction sounding to determine winter and spring sea ice thickness in the Antarctic. *Cold Reg. Sci. Technol.* **1999**, *29*, 49–58. [CrossRef]
19. Uto, S.; Toyota, T.; Shimoda, H.; Takeyama, K.; Shirasawa, K. Ship-borne electromagnetic induction Sounding of Sea-ice thickness in the Southern Sea of Okhotsk. *Ann. Glaciol.* **2006**, *44*, 253–260. [CrossRef]
20. McLaren, A.S. The under-ice thickness distribution of the Arctic Basin as recorded in 1958 and 1970. *J. Geophys. Res.-Ocean.* **1989**, *94*, 4971–4983. [CrossRef]
21. Vinje, T.; Nordlund, N.; Kvambekk, Å. Monitoring ice thickness in Fram Strait. *J. Geophys. Res.-Ocean.* **1998**, *103*, 10437–10449. [CrossRef]

22. Giles, K.A.; Laxon, S.W.; Ridout, A.L. Circumpolar thinning of Arctic sea ice following the 2007 record ice extent minimum. *Geophys. Res. Lett.* **2008**, *35*, L22502. [CrossRef]
23. Griggs, J.A.; Bamber, J.L. Antarctic ice-shelf thickness from satellite radar altimetry. *J. Glaciol.* **2011**, *57*, 485–498. [CrossRef]
24. Fu, H.; Liu, Z.; Guo, X.; Cui, H. Double-frequency ground penetrating radar for measurement of ice thickness and water depth in rivers and canals: Development, verification and application. *Cold Reg. Sci. Technol.* **2018**, *154*, 85–94. [CrossRef]
25. Kwok, R.; Cunningham, G.F. ICESat over Arctic sea ice: Estimation of snow depth and ice thickness. *J. Geophys. Res.-Ocean.* **2008**, *113*, 179–185. [CrossRef]
26. Cole, D.M. Preparation of polycrystalline ice specimens for laboratory experiments. *Cold Reg. Sci. Technol.* **1979**, *1*, 153–159. [CrossRef]
27. Deng, K.; Feng, X.; Tian, X.; Hu, Y. Experimental research on compressive mechanical properties of ice under low strain rates. *Mater. Today Commun.* **2020**, *24*, 101029. [CrossRef]
28. Zhang, Y.; Qian, Z.; Lv, S.; Huang, W.; Ren, J.; Fang, Z.; Chen, X. Experimental investigation of uniaxial compressive strength of distilled water ice at different growth temperatures. *Water* **2022**, *14*, 4079. [CrossRef]
29. Ghobrial, T.R.; Loewen, M.R.; Hicks, F. Laboratory calibration of upward looking sonars for measuring suspended frazil ice concentration. *Cold Reg. Sci. Technol.* **2012**, *70*, 19–31. [CrossRef]
30. Roscoe, H.K.; Brooks, B.; Jackson, A.V.; Smith, M.H.; Walker, S.J.; Obbard, R.W.; Wolff, E.W. Frost flowers in the laboratory: Growth, characteristics, aerosol, and the underlying sea ice. *J. Geophys. Res. Atmos.* **2011**, *116*, D12301. [CrossRef]
31. Haas, C. Ice tank investigations of the microstructure of artificial sea ice grown under different boundary conditions during INTERICE II. In Proceedings of the HYDRALAB Workshop, Hannover, Germany, 19 February 1999.
32. Weeks, W.F. Tensile strength of NaCl ice. *J. Glaciol.* **1962**, *4*, 25–52. [CrossRef]
33. Cottier, F.; Eicken, H.; Wadhams, P. Linkages between salinity and brine channel distribution in young sea ice. *J. Geophys. Res.-Ocean.* **1999**, *104*, 15859–15871. [CrossRef]
34. Loose, B.; McGillis, W.R.; Schlosser, P.; Perovich, D.; Takahashi, T. Effects of freezing, growth, and ice cover on gas transport processes in laboratory seawater experiments. *Geophys. Res. Lett.* **2009**, *36*, 1–5. [CrossRef]
35. Perovich, D.K.; Grenfell, T.C. Laboratory studies of the optical properties of young sea ice. *J. Glaciol.* **1981**, *27*, 331–346. [CrossRef]
36. Marks, A.A.; Lamare, M.L.; King, M.D. Optical properties of laboratory grown sea ice doped with light absorbing impurities (black carbon). *Cryosphere* **2017**, *11*, 2867–2881. [CrossRef]
37. Liu, R.; Liu, Y.; Wang, Q.; Yi, X. Investigation of microstructure and density of atmospheric ice formed by high-wind-speed in-cloud icing. *Crystals* **2023**, *13*, 1015. [CrossRef]
38. Bansmer, S.E.; Baumert, A.; Sattler, S.; Knop, I.; Leroy, D.; Schwarzenboeck, A.; Jurkat-Witschas, T.; Voigt, C.; Pervier, H.; Esposito, B. Design, construction and commissioning of the Braunschweig Icing Wind Tunnel. *Atmos. Meas. Tech.* **2018**, *11*, 3221–3249. [CrossRef]
39. Blocken, B.; Stathopoulos, T.; Van Beeck, J.P.A.J. Pedestrian-level wind conditions around buildings: Review of wind-tunnel and CFD techniques and their accuracy for wind comfort assessment. *Build. Environ.* **2016**, *100*, 50–81. [CrossRef]
40. Mehta, R.D.; Bradshaw, P. Design rules for small low speed wind tunnels. *Aeronaut. J.* **1979**, *83*, 443–453. [CrossRef]
41. Zanoun, E.S. Flow characteristics in low-speed wind tunnel contractions: Simulation and testing. *Alex. Eng. J.* **2018**, *57*, 2265–2277. [CrossRef]
42. Zhang, Y.; Qian, Z.; Huang, W.; Chen, X.; Zhang, Z.; Ren, J. Effect of grain size on the uniaxial compressive strength of ice forming with different wind speeds in a cold laboratory. *Water* **2024**, *16*, 2049. [CrossRef]
43. Morel, T. Design of Two-Dimensional Wind Tunnel Contractions. *J. Fluids Eng.* **1977**, *99*, 371–377. [CrossRef]
44. Schulson, E.M. The brittle compressive fracture of ice. *Acta Metall. Mater.* **1990**, *38*, 1963–1976. [CrossRef]
45. Cole, D.M. Strain-rate and Grain-size Effects in Ice. *J. Glaciol.* **1987**, *33*, 274–280. [CrossRef]
46. Nixon, W.A. Wing crack models of the brittle compressive failure of ice. *Cold Reg. Sci. Technol.* **1996**, *24*, 41–55. [CrossRef]
47. Zhang, Y.; Zhang, Y.; Guo, R.; Cui, B. Method for testing shear and tensile strengths of freshwater/seawater ice. *Water* **2022**, *14*, 1363. [CrossRef]
48. Wang, Q.; Li, Z.; Lu, P.; Xu, Y.; Li, Z. Flexural and compressive strength of the landfast sea ice in the Prydz Bay, East Antarctic. *Cryosphere* **2022**, *16*, 1941–1961. [CrossRef]
49. Ji, S.; Chen, X.; Wang, A. Influence of the loading direction on the uniaxial compressive strength of sea ice based on field measurements. *Ann. Glaciol.* **2020**, *61*, 86–96. [CrossRef]

water

MDPI

Article

Threshold Ranges of Multiphase Components from Natural Ice CT Images Based on Watershed Algorithm

Shengbo Hu [1], Qingkai Wang [1], Chunjiang Li [2] and Zhijun Li [1,*]

[1] State Key Laboratory of Coastal and Offshore Engineering, Dalian University of Technology, Dalian 116024, China; hushengbo@mail.dlut.edu.cn (S.H.); wangqingkai@dlut.edu.cn (Q.W.)

[2] School of Energy and Environment, Inner Mongolia University of Science and Technology, Baotou 014010, China; lichunjiang0405@imust.edu.cn

* Correspondence: lizhijun@dlut.edu.cn

Abstract: The multiphase components of natural ice contain gas, ice, unfrozen water, sediment and brine. X-ray computed tomography (CT) analysis of ice multiphase components has the advantage of high precision, non-destructiveness and visualization; however, it is limited by the segmentation thresholds. Due to the proximity of the CT value ranges of gas, ice, unfrozen water, sediment and brine within the samples, there is uncertainty in the artificial determination of the CT image segmentation thresholds, as well as unsuitability of the global threshold segmentation methods. In order to improve the accuracy of multi-threshold segmentation in CT images, a CT system was used to scan the Yellow River ice, the Wuliangsuhai lake ice and the Arctic sea ice. The threshold ranges of multiphase components within the ice were determined by watershed algorithm to construct a high-precision three-dimensional ice model. The results indicated that CT combined with watershed algorithm was an efficient and non-destructive method for obtaining microscopic information within ice, which accurately segmented the ice into multiphase components such as gas, ice, unfrozen water, sediment, and brine. The gas CT values of the Yellow River ice, the Wuliangsuhai lake ice and the Arctic sea ice ranged from −1024 Hu~−107 Hu, −1024 Hu~−103 Hu, and −1024 Hu~−160 Hu, respectively. The ice CT values of the Yellow River ice, the Wuliangsuhai lake ice and the Arctic sea ice ranged from −103 Hu~−50 Hu, −100 Hu~−38 Hu, −153 Hu~−51 Hu. The unfrozen water CT values of the Yellow River ice and the Wuliangsuhai lake ice ranged from −8 Hu~18 Hu, −8 Hu~13 Hu. The sediment CT values of the Yellow River ice and the Wuliangsuhai lake ice ranged from 20 Hu~3071 Hu, 20 Hu~3071 Hu, and the brine CT values of the Arctic sea ice ranged from −6 Hu~3071 Hu. The errors between the three-dimensional ice model divided by threshold ranges and measured sediment content were less than 0.003 g/cm^3, which verified the high accuracy of the established microscopic model. It provided a scientific basis for ice engineering, ice remote sensing, and ice disaster prevention.

Keywords: threshold ranges; natural ice; multiphase components; CT images; watershed algorithm

Citation: Hu, S.; Wang, Q.; Li, C.; Li, Z. Threshold Ranges of Multiphase Components from Natural Ice CT Images Based on Watershed Algorithm. *Water* **2024**, *16*, 3330. https://doi.org/10.3390/w16223330

Academic Editor: Chang Huang

Received: 8 October 2024
Revised: 15 November 2024
Accepted: 18 November 2024
Published: 19 November 2024

1. Introduction

Ice is a prevalent molecular crystal in nature, which inevitably contains gas, unfrozen water and sediment due to complex natural conditions [1]. The microstructure of ice reflects its internal "skeleton" characteristics. The non-uniform distribution of components within ice alters its internal structure, which directly impacts its physical properties [2]. For instance, the interaction between ice and structures forms the foundation of ice engineering research, while the mechanical properties of ice are determined by its microstructure [3]. The microstructure of ice also leads to variability in the thermal, optical and electrical properties, which further influences subglacial water ecosystems and forms the basis for ice thermodynamics and ice remote sensing research [4]. Information on the ice growth process is stored in the ice microstructure. Analyzing the microstructure can facilitate the study of ice growth and melting, thereby improving the accuracy of predictions related to ice sheet breakup, ice jamming, and ice damming [5]. Consequently, obtaining information on the ice microstructure is the key to research on ice engineering, ice remote sensing and ice disaster prevention.

The microstructure of ice has its fundamental physical properties, which mainly includes stratification, crystal structure, and impurities. Currently, common instruments for obtaining ice microstructural information are a Federov rotary stage [6], a scanning electron microscope [7], a nuclear magnetic resonance instrument [8], and an X-ray scanning system [9]. Among the various microanalysis tools, X-ray scanning systems are widely used for analyzing the internal microstructures of samples due to the unique advantages of non-destructiveness and visualization [9]. Researchers have conducted numerous studies to characterize the ice pore structure. Michel et al. [10] qualitatively classified the ice structure by in situ observations and examinations of ice crystals. Shokr et al. [11] conducted field experiments in Resolute Bay to characterize the microstructure characteristics of one-year and multi-year ice based on crystal observations. Cole et al. [12] analyzed the main microstructural types of ice and their origins and discussed the microstructural changes that occurred during deformation. Li et al. [13] performed uniaxial compression experiments on 117 columnar-grained sea ice specimens along the direction parallel to ice surface under different test temperatures and strain rates. The results supported the curved-surface relationship between the uniaxial compressive strength and porosity within a wide range of strain rate. Sammonds et al. [14] considered that ice occurred as polycrystalline aggregates in which the bulk behavior was the result of the behavior of the ensemble of individual grains; therefore, it was dependent on the microstructure, that was to say, the whole arrangement of grains, their internal substructure, impurities, and second phases. Hammonds et al. [15] conducted uniaxial compression experiments on polycrystalline ice samples at different strain rates and temperatures. The extent of cracking from each test is characterized via micro-CT imaging and is quantified via a newly proposed variant of the crack density tensor. Salomon et al. [16] analyzed the three-dimensional microstructure of sea ice by means of X-ray computed tomography. Microscopic (brine and air pore sizes, numbers and connectivity) and macroscopic (salinity, density, porosity) properties of young Arctic sea ice were analyzed. The current research mainly focuses on the relationship between microstructural information obtained through CT systems and macroscopic physical properties. Thus, accurately determining the threshold range of ice components is essential for advancing the study of ice mechanics, thermodynamics and optics, which is also crucial for improving the accuracy of ice numerical modeling.

Recently, X-ray computed tomography (CT) has been increasingly applied to micro-scale studies of materials, including non-destructive testing [17], pore structure analysis [18], compositional delineation [19], and numerical simulation based on CT images [20]. This rise in application is attributed to its advantages of non-destructiveness, dynamic imaging, and continuity [21]. However, compared to common rock microstructures, the CT value ranges of ice, gas, unfrozen water, and sediment within ice microstructures are close to each other, complicating threshold segmentation. Additionally, recent research indicates that there is significant uncertainty in the manual determination of CT image segmentation thresholds, which can lead to errors in analysis [22].

The major objectives of the study included (a) the CT value ranges of different components in the Yellow River ice, Wuliangsuhai Lake ice and Arctic sea ice were summarized, by which the pore morphology was explored. Moreover, the study compared the two-dimensional CT image threshold segmentation results with the field observation; (b) based on the reconstructed three-dimensional ice model, the morphology, distribution of intra-ice sediment and intra-ice pore were extracted, and the reasons for the formation of different morphological bubbles in different types of ice were discussed; (c) the three-dimensional model of the Yellow River ice was reconstructed using CT and digital image processing techniques. The feasibility of the watershed algorithm for ice image threshold segmentation was validated through actual measurements of sediment content at different depths. The study improved the segmentation accuracy of CT images, which provided an essential foundation for accurate extraction of ice microstructure information.

2. Materials and Methods

2.1. Material

The ice samples for CT analysis were collected from the Yellow River ice, Wuliangsuhai lake ice and Arctic sea ice. The Yellow River ice samples were obtained from Baotou to Toketo County. River ice typically contains not only air bubbles but also impurities such as sediment, which is introduced during the flow of river. Seven ice samples were collected in the Yellow River, which were named in order, such as the Yellow River No. 1 ice sample. Wuliangsuhai, situated in Ulatqian Banner, Bayannur City, Inner Mongolia Autonomous Region, experiences a freeze period from November to March each year. During frozen period, water flows slowly, and the microstructure of lake ice is primarily influenced by localized air bubbles [23]. Two ice samples were collected according to ice surface in the Wuliangsuhai, named the Wuliangsuhai No. 1 ice sample and the Wuliangsuhai No. 2 ice sample. Arctic sea ice samples were collected from field stations during China's ninth scientific expedition, located at 79°13′ N, 168°49′ W and 84°24′ N, 156°08′ E, respectively. The growth of sea ice is influenced by seawater salinity, which leads to widely distributed brine bubbles and channels. Two ice samples were collected in the Arctic, named the Arctic No. 1 ice sample and the Arctic No. 2 ice sample. Field-collected ice samples were placed in ice core bags, which were stored in a cooler covered with crushed ice and snow (temperature −5 °C). A total of 2987 ice CT images were obtained by scanning the ice samples layer by layer, which were used to statistically analyze threshold range of multiphase components of natural ice. The flow of experimental processing is shown in Figure 1.

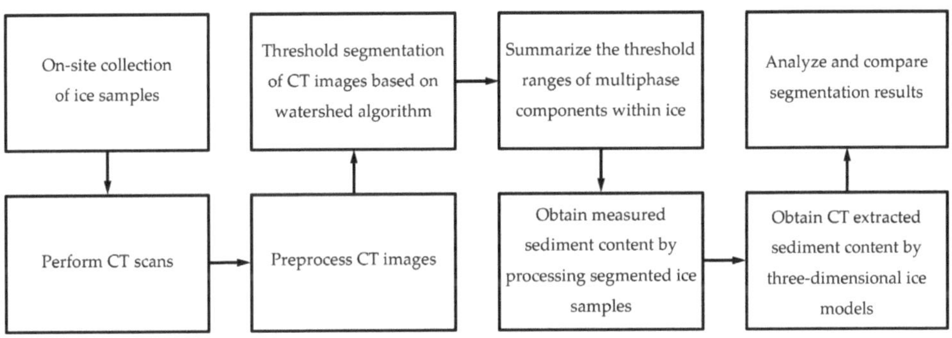

Figure 1. Flow chart of the experimental processing.

2.2. Experimental Equipment and Scanning Principle

The CT scanning test was performed by a Philips Brilliance 16 CT scanning experimental machine in Cold and Arid Regions Environmental and Engineering Research Institute, Chinese Academy of Sciences, Lanzhou. The instrument's main technical parameters are shown in Table 1.

Table 1. Main technical parameters of CT instrument.

	Scan Cross Section (mm^2)	Scan Layer Thickness (mm)	Scan Voltage (kV)	Scan Current (mA)	Reconstruction Matrix
Technical parameters	200 × 200	3	120	313	1024 × 1024

The CT scan system typically comprises an X-ray source, a sample platform, a detector, and a computer system for data analysis. As X-ray penetrates the sample, the X-ray energy decreases due to photoelectric absorption, Compton effect, and electron pair effect [24]. The intensity decay law of X-ray is shown in Equation (1) [25]:

$$I = I_0 \exp[-\mu x] \quad (1)$$

where I is the X-ray transmission intensity, I_0 is the X-ray incident intensity, μ is the linear attenuation coefficient of the material, x is the path length of the X-rays through the material.

The CT system operates by determining the attenuation coefficient of X-rays within the scanned object, and the attenuation coefficient distribution matrix on the scanning cross section is established by computer system. In practical applications, the differences in attenuation coefficients between materials are minimal, so the CT values (Hounsfield unit, Hu) are introduced to amplify this difference, and the CT value of water is set to 0 Hu [26], as shown in Equation (2):

$$CT_{number} = \frac{\mu - \mu_w}{\mu_w} \times 1000 \quad (2)$$

where CT_{number} is the CT value, μ_w is the attenuation coefficient of water.

The CT value increases with the material's density, which can provide information about the density of the scanning section. In the experiment, seven samples of the Yellow River ice, two samples of Wuliangsuhai lake ice and two samples of Arctic sea ice were scanned by CT, yielding microstructural images of these three different types of ice. The scanning section was parallel to the ice core profile, then, the ice samples were scanned layer by layer along the depth direction with a scanning section of 200 mm × 200 mm, a layer spacing of 3 mm, and a CT image reconstruction matrix size of 1024 × 1024 pixels.

2.3. CT Image Preprocessing

During the acquisition of CT images, inherent electronic device perturbations and environmental influences result in the production of noise and distortion, which negatively impact the segmentation and extraction of components [27]. To mitigate image noise, the Median filtering algorithm was employed. Additionally, manual cutting of the ice samples resulted in surface imperfections, such as breakages and protrusions. Moreover, the outer air portion of the CT image was not the test subject. To eliminate the potential interference from these parts, the images were cropped, as shown in Figure 2, with the research area highlighted in the red frame.

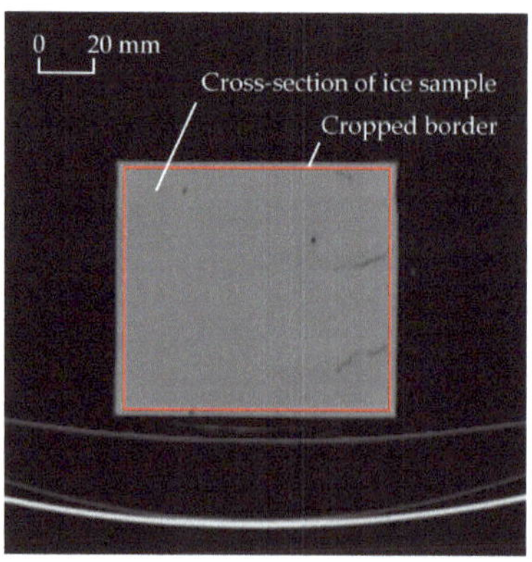

Figure 2. CT original image and research area frame.

2.4. Watershed Algorithm

Natural ice contains a significant number of impurities that vary in composition, leading to complex and heterogeneous structures. The CT value denotes the attenuation coefficient of X-rays reaching the detection point, which is related to the density of each component within the ice sample. The range of machine detection is from −1024 Hu to 3071 Hu. However, the CT values displayed in the images correspond only to a grayscale range of 0 to 255. As shown in Figure 3, the CT value ranges of gas, ice, unfrozen water, and sediment within the samples are similar without significant intervals. This proximity leads to uncertainty in artificially determining the segmentation thresholds and unsuitability of the global threshold segmentation methods. The boundary between the multiphase components within ice is indistinct. To extract the distribution, morphology, and spatial location of the multiphase components within ice, the generation of closed region boundaries is necessary.

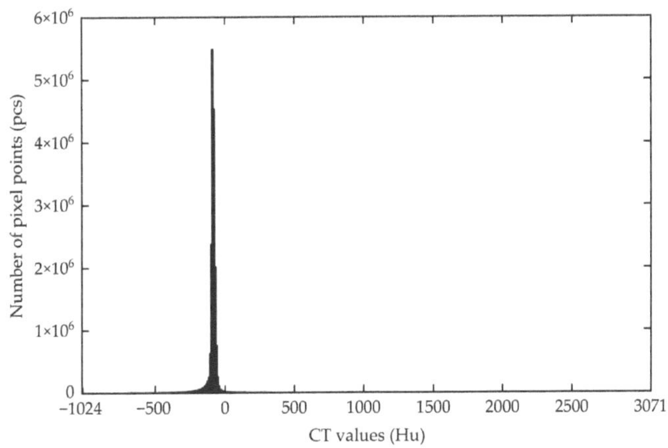

Figure 3. Histogram of ice sample CT values. There are no peaks and valleys in the CT value histograms, which proves that the CT values range of gas, unfrozen water, ice, and sediment within the samples are similar without significant intervals.

The watershed algorithm simulates the process of water immersion, which is suitable for processing multiple objects and complex edge structures in images to form closed boundaries. The algorithm treats the image as a topographical surface, with the gray value of each pixel representing the terrain elevation [28]. Local minima in the image serve as the points from which water continuously immerses, and the water gradually floods the corresponding basin of image. As the water levels in two different regions rise and converge, a dam is formed at their junction. Upon completion of the overflow process, each local minimum is encircled by the dam corresponding to its water accumulation basin, and each dam serves as the watershed. Consequently, the boundary of multiphase components within ice is clearly extracted [29], as shown in Figure 4. Different water accumulation basins represent different partitions of the image, thereby achieving image segmentation.

The CT values of gas, ice and unfrozen water are all less than or equal to zero. In contrast, high CT values in freshwater ice are indicative of sediment, while high CT values in sea ice are indicative of brine. The ice sample is considered to be composed of gas, ice, water, and sediment (brine). The CT value ranges of gas, ice, water, and sediment (brine) within the ice samples are determined by combining with the watershed algorithm, which offers a novel technical method for the segmentation of ice CT images.

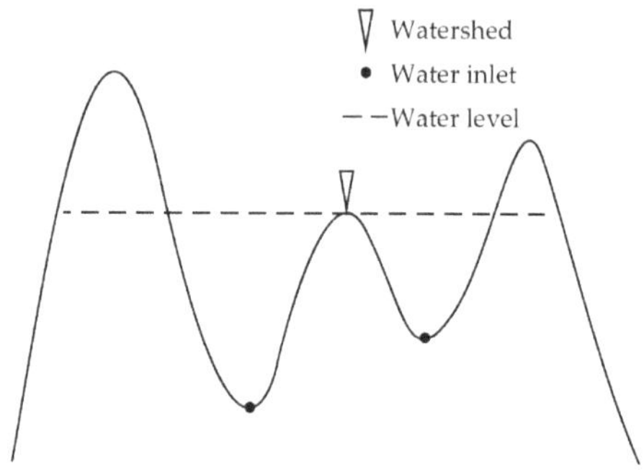

Figure 4. Schematic diagram of watershed algorithm model.

The watershed algorithm first defines the two-dimensional image I as a grayscale image set to discrete values of [0, N], where N is the positive integer, D is the set of positive integers, and p is the pixel of the image. The set of pixels T_h with height less than h is filtered in the two-dimensional image I, where h is the threshold value [30], as shown in Equation (3).

$$T_h = \{p \in D|\, I(p) \leq h\} \tag{3}$$

For the set A and B, d_A (a, b) denotes the shortest distance between two points a, b in the set A. If $B \subseteq A$ is satisfied, then B is randomly divided into k mutually interconnected parts, which is denoted as B_i ($i = 1, 2, \ldots, k$), and B_i corresponds to the geodesic influence zone solution formula is defined as Equation (4) [30].

$$iz_A(B_i) = \{p \in A|d_A(p, B) < d_A(p, B/B_i)\} \tag{4}$$

The set $IZ_A(B)$ is the union of B's geodetic influence zone defined as Equation (5) [30].

$$IZ_A(B) = \bigcup_{i=1}^{k} iz_A(B_i) \tag{5}$$

In A, the complementary set of $IZ_A(B)$ is called $SKIZ_A(B)$ (Skeleton of Geodetic Influence Zone) as in Equation (6) [30].

$$SKIZ_A(B) = A/IZ_A(B) \tag{6}$$

The pixels in the foreground area are aggregated to obtain the $SKIZ$, as shown in iterative Equation (7) [30].

$$\begin{cases} X_{h_{\min}} = \{p \in D|I(p) \leq h_{\min}\} = T_{h_{\min}} \\ X_{h+1} = MIN_{h+1} \cup IZ_{h+1}(X_h), h \in [h_{\min}, h_{\max}] \end{cases} \tag{7}$$

where h is the range of gray value, X_{hmin} is the pixel point with the smallest gray value in image I, h_{min} is the smallest gray value, h_{max} is the largest gray value, T_{hmin} is the set of pixels with minimum value points in each basin, X_{h+1} is all pixels with gray value less than $h + 1$, MIN_{h+1} is the set of pixel points with minimum gray value regenerated at the $h + 1$, $IZ_{h+1}(X_h)$ denotes the set of individual regions divided on the basis of the shortest distance within all T_{h+1} connected regions.

In D, the complementary set of X_{hmin} is watershed region X_{Wshed}, as shown in Equation (8) [30].

$$X_{Wshed} = D/X_{h_{\min}} \tag{8}$$

2.5. Sediment Content

Due to the proximity of the CT value ranges of gas, ice, unfrozen water and sediment within the samples, which makes it impossible to validate the accuracy of segmentation results through subjective awareness. Furthermore, there is no precise and effective method to obtain the content of each component in the ice samples except sediment. In order to quantitatively validate the accuracy of the segmentation results, this study measured the sediment content per unit volume of ice sample by melting a certain volume of the Yellow River ice samples into a bottle, followed by filtering, drying, and weighing the sediment in the ice sample, as shown in Equation (9).

$$S_m = m/V_i \tag{9}$$

where S_m is the measured sediment content, m is the dried sediment mass in the ice sample, V_i is the volume of corresponding ice sample.

Moreover, the sediment content was extracted from the reconstructed three-dimensional model. This study compared the error between the measured sediment content and the CT extracted sediment content, as shown in Equation (10).

$$S_{CT} = (\rho_s \times V_s)/V_i \tag{10}$$

where S_{CT} is the sediment content extracted from the CT three-dimensional model, ρ_s is the density of dried sand, V_s is the volume of sediment extracted from the CT three-dimensional model.

3. Results

3.1. Threshold Range and Two-Dimensional Image Segmentation Results

According to the multi-threshold segmentation results, the CT values of various components within the ice were statistically summarized. The lower limit of gas CT values fluctuated widely. For example, the Yellow River No.2 ice sample was dominated by trapped bubbles (The CT values of No.2 ice sample's gas ranged from −212 Hu~−107 Hu). Bubbles were small and hermetic, which contained high water vapor. Strip shaped pores were present in the Yellow River No.3 ice sample (The CT values of No.3 ice sample's gas ranged from −269 Hu~−115 Hu), but the pore structure was elongated and hermetic, restricting air circulation. However, open and flaky cracks were present in the ice samples of the Yellow River No. 1, 4, 5, 6 and 7, resulting in the lower limit values being close to air's CT value (−1000 Hu). To ensure the universal applicability of the CT value ranges, the minimum and maximum values of various components in multiple ice samples were, respectively, used as the lower and upper limits for CT values, as shown in Table 2. The CT values of gas ranged from −1024 Hu~−107 Hu. The CT values of ice ranged from −103 Hu to −50 Hu. The CT values of unfrozen water ranged from −8 Hu to 18 Hu. Due to the differing mineral compositions in the sediment, the upper limit of sediment CT values fluctuated widely, and this value did not have an effect on the image segmentation. In this paper, the CT maximum value of the instrument's technical parameters was taken as the upper limit value of sediment, the CT values of sediment ranged from 20 Hu~3071 Hu.

Table 2. Statistical results of various components' CT values in the Yellow River ice.

	The Yellow River Ice Lower Limit Value (Hu)	Upper Limit Value (Hu)
gas	-	−107
ice	−103	−50
unfrozen water	−8	18
sediment	20	-

In the case of the Yellow River ice, the field sample exhibited a white and transparent appearance, as shown in Figure 5a. The sediment content was particularly high in the middle portion of the ice sample (17 cm~30 cm from the bottom of the ice sample), with a greater number of spherical bubbles in the upper portion of the sediment layer. Deng et al. found that the bubbles in the Yellow River ice during the freezing period were mainly spherical using the Federov rotary stage. Moreover, there is no significant correlation between the equivalent diameter of bubbles in ice and depth [31]. The CT image was segmented into four regions based on the CT value ranges, where gray represented gas, dark blue represented unfrozen water, orange represented sediment, and light blue represented ice, as shown in Figure 5b–j. Figure 5h showed the top of the ice sample, which exhibited an abundance of gasses and lacked both sediment and unfrozen water. Figure 5i showed the layer at 25 cm from the bottom of ice sample, containing sediment, unfrozen water, gas, and ice. The sediment within the ice sample was randomly distributed on the plane, with some sediment adhering to the edges of irregularly shaped bubbles. Unfrozen water occurred with the sediment as well as in small, isolated bubbles. The shapes of bubbles were predominantly single spheres or combinations of multiple spheres. Figure 5j showed the bottom of the ice sample, which contained a little sediment and unfrozen water. It matched the visual observation of samples collected in the field.

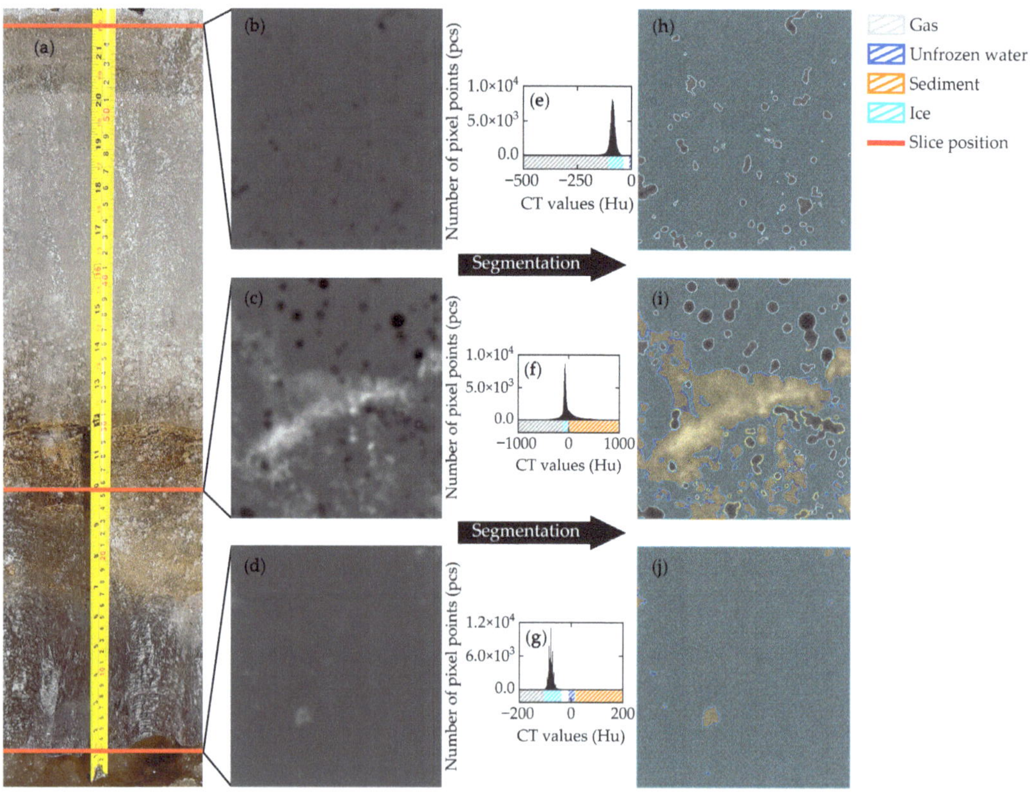

Figure 5. Schematic of sample collection and two-dimensional image threshold segmentation. (**a**) The Yellow River ice sample. Original two-dimensional CT images of (**b**) top layer, (**c**) 25 cm from the bottom layer, (**d**) bottom layer. Histograms of CT values for (**e**) top layer, (**f**) 25 cm from the bottom layer, (**g**) bottom layer. Two-dimensional image multi-threshold segmentation results of (**h**) top layer, (**i**) 25 cm from the bottom layer, (**j**) bottom layer.

The bubbles in the Wuliangsuhai lake ice were small and sealed, as shown in Table 3. The CT values of Wuliangsuhai No. 1 ice sample's gas ranged from −186 Hu~−103 Hu, which were comparable to the CT values of the Yellow River No. 2 ice sample's gas (−212 Hu~−107 Hu). Notably, the CT lower limit value of Wuliangsuhai lake ice's gas was slightly larger than that of the Yellow River ice's gas. It could be inferred that the bubbles in the Wuliangsuhai lake ice may contain higher concentrations of gasses, such as oxygen and carbon dioxide produced by biological communities, as well as water vapor. Consequently, the density inside the Wuliangsuhai lake ice's bubbles was slightly higher than the density inside the Yellow River ice's bubbles. The CT values of Wuliangsuhai lake ice's gas ranged from −1024 Hu~−103 Hu. The CT values of Wuliangsuhai lake ice ranged from −100 Hu~−38 Hu. The lower limit was close to that of the Yellow River ice, while the upper limit was slightly higher. This difference can be attributed to the reduced hydrodynamic influence during the freezing process in Wuliangsuhai lake, leading to a more compact and stable ice structure. The CT values of unfrozen water ranged from −8 Hu~13 Hu. Additionally, the stable hydrodynamic environment also contributed to low sediment content in the Wuliangsuhai lake ice, with only tiny amounts of sediment found in the Wuliangsuhai No. 1 ice sample.

Table 3. Statistical results of various components' CT values in the Wuliangsuhai lake ice.

	Wuliangsuhai Lake Ice Lower Limit Value (Hu)	Upper Limit Value (Hu)
Gas	-	−103
Ice	−100	−38
Unfrozen Water	−8	13
Sediment	20	-

The pore structures in the Arctic sea ice were dominated by strip-shaped brine channels, egg-shaped trapped bubbles, and irregular-shaped extruded bubbles, which were large and interconnected. The CT lower limit value of gas in sea ice was close to that of air, indicating that the pores were connected to the outside. As shown in Table 4, the CT values of gas ranged from −1024 Hu~−160 Hu. The CT values of Arctic sea ice ranged from −153 Hu~−51 Hu. Ji et al. reviewed sea ice density measurement data during 2000–2015 and found Arctic sea ice density ranged from 675 kg/m^3 to 954 kg/m^3 [32]. Zhang et al. measured the densities of the Yellow River ice and Wuliangsuhai lake ice during 2017–2020. The Yellow River ice density ranged from 703 kg/m^3 to 965 kg/m^3 and Wuliangsuhai lake ice density ranged from 883 kg/m^3 to 907 kg/m^3 [2]. Since the density of sea ice was lower than the density of freshwater ice, the CT lower limit value of Arctic sea ice was also lower than that of the Yellow River ice and the Wuliangsuhai lake ice. The CT values of brine ranged from −6 Hu~3071 Hu, with the upper limit being determined by salinity, which increases with increasing salinity.

Table 4. Statistical results of various components' CT values in the Arctic sea ice.

	Arctic Sea Ice Lower Limit Value (Hu)	Upper Limit Value (Hu)
Gas	-	−160
Ice	−153	−51
Brine	−6	-

3.2. Three-Dimensional Reconstructed Images of Ice

Common bubbles in ice can be categorized into three types: trapped bubbles, closed bubbles, and extruded bubbles. These bubbles differ significantly in their formation processes and physical properties [33]. Trapped bubbles form due to the lower solubility of air in ice compared to its solubility in water. During the freezing process, air is expelled from the ice to the freezing front. When the gas concentration at the freezing front reaches a supersaturation level, bubble nucleation occurs. As the freezing front moves, the high gas concentration surrounding the bubble diffuses into it, facilitating its growth. If the gas escape rate is slower than the freezing rate, the freezing front gradually covers the air bubbles, ultimately resulting in the formation of trapped bubbles, which typically exhibit egg-like or needle-like shapes. Closed bubbles are commonly found in high-latitude alpine lakes, and the formation process involves the freezing of methane and other gasses produced by microbial fermentation on the lakebed, resulting in primarily disk-shaped bubbles. Extruded bubbles are found in the deep ice cores of polar ice sheets. The formation process involves the gradual transformation of fluffy polar snow into ice through gravitational extrusion, resulting in the entrapment of irregularly shaped bubbles within the ice. Two-dimensional CT images can only provide localized information. To comprehensively represent the segmentation effect of ice samples, a three-dimensional model of the ice is reconstructed intuitively and stereoscopically by stacking multiple layers of two-dimensional images. According to model data, the surface of the Yellow River ice had begun to melt and formed vertical ice, which was dominated by strip-shaped pores, as shown in Figure 6a,b, and trapped bubbles, as shown in Figure 6c,d. The high sediment content (20 to 400 g/L) of the Yellow River significantly impacted the generation and elimination process of river ice, with sediment randomly frozen within the ice [34], as shown in Figure 6c.

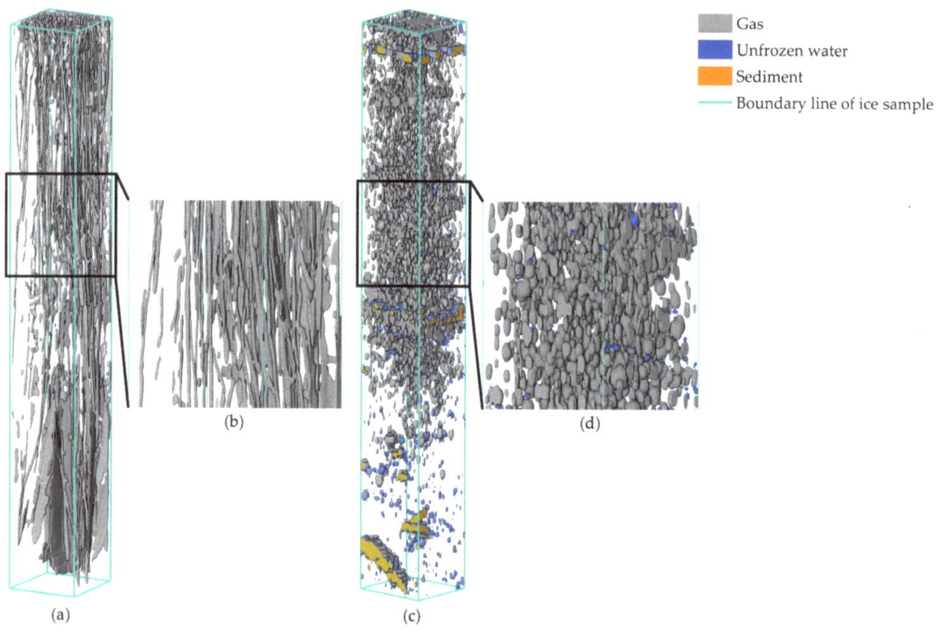

Figure 6. Three-dimensional reconstructed images of the Yellow River ice samples. Global three-dimensional image of (**a**) the Yellow River No. 3 ice sample, and (**c**) the Yellow River No. 4 ice sample. Local three-dimensional images of (**b**) the Yellow River No. 3 ice sample, and (**d**) the Yellow River No. 4 ice sample.

In contrast to the complex hydrodynamic conditions of the Yellow River, the Wuliangsuhai lake ice experienced stable hydrodynamic conditions during its freezing process. As a result, the lake ice had a complete structure, devoid of open cracks or flaky bubbles, whose porosity was significantly lower than that of the Yellow River. The collection zone of the Wuliangsuhai No. 1 ice sample appeared white with distinct boundary lines, as shown in Figure 7a. The air bubbles within the Wuliangsuhai No. 1 ice sample were predominantly closed bubbles, which were disk-shaped and in bunches, as shown in Figure 7b,c. These closed bubbles formed when gasses produced by aquatic organisms' respiration could not be discharged in time and were frozen within the ice, suggesting the presence of substantial, stable plant or microbial communities beneath the area where the Wuliangsuhai No. 1 ice sample was collected. Zhang et al. simulated and analyzed changes in dissolved oxygen during growth and stability period of ice. The Wuliangsuhai's maximum daily average oxygen production rate was 7.19 mg/(L·d) [35]. The low porosity and small-volume bubbles of Wuliangsuhai No. 2 ice sample, as shown in Figure 7d,e, proved that Wuliangsuhai No. 2 ice sample was less affected by the external environment during the process of generation.

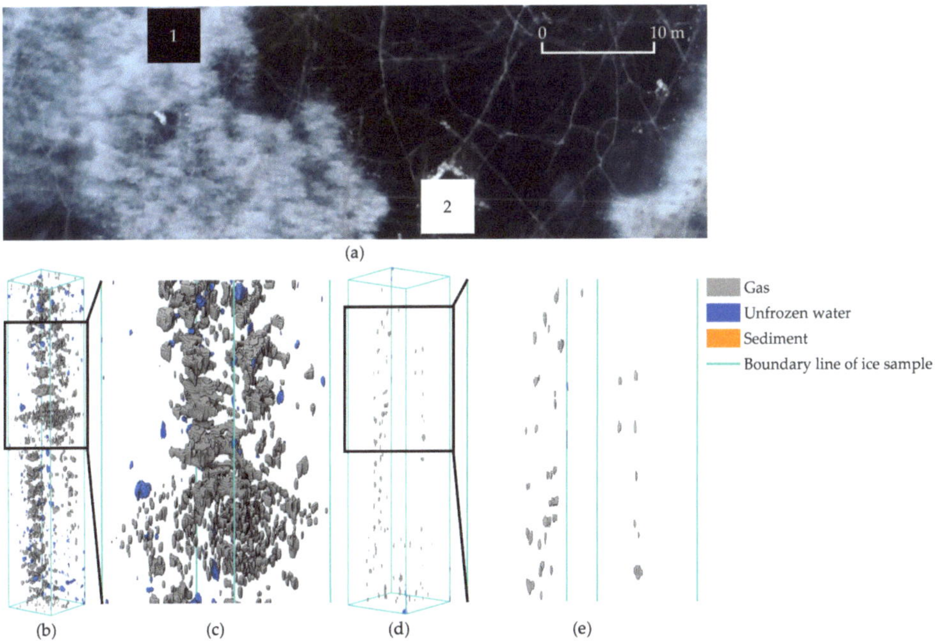

Figure 7. Study area and local three-dimensional reconstructed images of the Wuliangsuhai lake ice. (**a**) the Wuliangsuhai No. 1 and No. 2 ice samples collection areas. Global three-dimensional images of (**b**) the Wuliangsuhai No. 1 ice sample, and (**d**) the Wuliangsuhai No. 2 ice sample. Local three-dimensional images of (**c**) the Wuliangsuhai No. 1 ice sample, and (**e**) the Wuliangsuhai No. 2 ice sample.

In contrast to the Yellow River ice and Wuliangsuhai lake ice, Arctic sea ice was not affected by sediment during its generation and elimination process, but rather by seawater salinity. Although, more than 80% of the salt was expelled during freezing, but salt still remained in the form of brine cells within the sea ice. Under constant low-temperature conditions (0 °C~−30 °C), the high concentration brine within the sea ice always remained in the liquid state. Due to the higher specific gravity of brine compared to that of ice crystals, it was influenced by gravity to move down along the ice crystal gaps, resulting in the formation of brine channels. As shown in Figure 8a,b, the spherical or strip-shaped brine cells were mainly attached to the pore structures, while the pores were mainly downward strip-shaped brine channels.

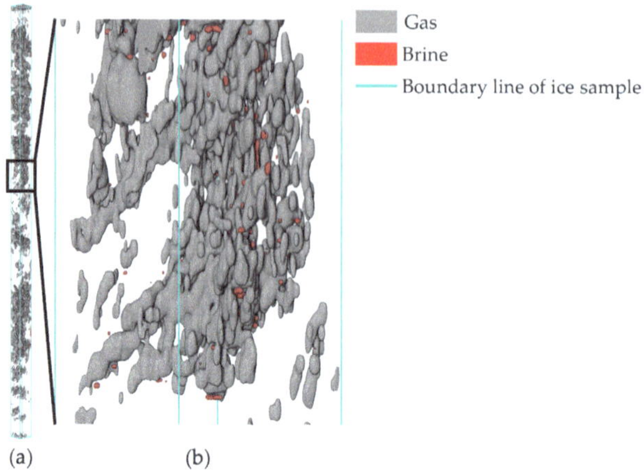

Figure 8. Global three-dimensional image of (**a**) the Arctic No. 1 ice sample. Local three-dimensional image of (**b**) the Arctic No. 1 ice sample.

3.3. Comparison of CT Extracted Sediment Content with Measured Sediment Content

The CT value range is −1024 Hu~3071 Hu, which corresponds to 0~255 levels of gray value on the CT image, with each gray level representing 16 Hu. Furthermore, the proximity of the CT value ranges of gas, ice, and unfrozen water within the samples, as well as the large amount of image data, makes it impossible to validate the accuracy of segmentation results through subjective awareness. In order to quantitatively validate the accuracy of the segmentation results, this study measured the sediment content per unit volume of ice sample by melting a certain volume of the Yellow River ice samples into a bottle, followed by filtering, drying, and weighing the ice sample. Meanwhile, the volume of sediment within the ice sample was obtained based on the CT three-dimensional model. The dry density of sediment was set to 1.4 g/cm^3 to calculate the sediment content in the CT data [36], as shown in Figure 9. The errors between the measured data and the CT extracted data were less than 0.003 g/cm^3. The results showed that the ice three-dimensional model based on the watershed algorithm aligned well with the measured data, demonstrating consistent trends in both sets of results.

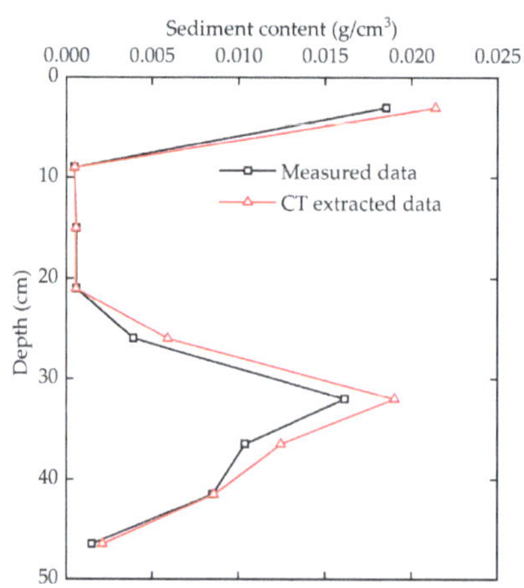

Figure 9. Distribution of sediment content in the Yellow River ice along the depth direction.

4. Conclusions

Ice multiphase components not only reflected the ice growth process, but also significantly influenced the mechanical, thermal, optical, and electrical properties of ice. However, the CT value ranges of gas, ice, unfrozen water, and sediment were close to each other, making it challenging to determine segmentation thresholds artificially. This study determined the threshold ranges for multiphase components of the Yellow River ice, Wuliangsuhai lake ice and Arctic sea ice based on X-ray computed tomography and watershed algorithm, which provided a reference for the threshold ranges of ice experiments. The main conclusions were as follows.

1. X-ray computed tomography combined with the watershed algorithm was an efficient and reliable method for obtaining internal microstructural information of ice without destruction. This approach could determine the multi-threshold value of CT image to segment the ice samples into various components, such as gas, ice, unfrozen water, and sediment (brine). According to the multi-threshold segmentation results, the three-dimensional ice model was constructed to obtain the morphology and spatial distribution of various components within the ice samples, which provided a scientific basis for ice engineering, ice remote sensing, and ice disaster prevention.
2. The gas CT values of the Yellow River ice, the Wuliangsuhai lake ice, and the Arctic sea ice ranged from −1024 Hu~−107 Hu, −1024 Hu~−103 Hu, −1024 Hu~−160 Hu, respectively. The Yellow River ice and the Wuliangsuhai lake ice were dominated by egg-shaped trapped bubbles and disk-shaped closed bubbles. The Arctic sea ice was dominated by strip-shaped brine channels and irregular-shaped extruded bubbles. The ice CT values of the Yellow River ice, the Wuliangsuhai lake ice and the Arctic sea ice ranged from −103 Hu~−50 Hu, −100 Hu~−38 Hu, −153 Hu~−51 Hu. In contrast to the Yellow River ice and the Arctic sea ice, the Wuliangsuhai lake ice had a more compact structure. The unfrozen water CT values of the Yellow River ice and the Wuliangsuhai lake ice ranged from −8 Hu~18 Hu, −8 Hu~13 Hu. The sediment CT values of the Yellow River ice and the Wuliangsuhai lake ice ranged from 20 Hu~3071 Hu, 20 Hu~3071 Hu, and the brine CT values of the Arctic sea ice ranged from −6 Hu~3071 Hu.
3. High sediment content of the Yellow River significantly impacted the generation and elimination process of river ice, with sediment randomly frozen within the ice. At the same time, the sediment was surrounded by a lot of unfrozen water, which was significantly higher than the unfrozen water content in the sediment-free zone.
4. The three-dimensional ice model based on X-ray computed tomography and watershed algorithm was in good agreement with the measured data, exhibiting errors of less than 0.003 g/cm^3. It could provide a new idea for quantitative study of ice microstructure information. This study only verified the accuracy of sediment content. In further research, we need to use nuclear magnetic resonance instruments to verify the accuracy of gas and liquid phases in ice samples.

Author Contributions: Conceptualization, Z.L.; methodology, S.H. and Z.L.; investigation, S.H. and Q.W.; data curation, S.H., C.L. and Z.L.; writing—original draft preparation, S.H.; writing—review and editing, S.H., Q.W. and C.L.; project administration, Q.W. and Z.L.; funding acquisition, Q.W. and Z.L. All authors have read and agreed to the published version of the manuscript.

Funding: This research was supported by Joint Funds of the National Natural Science Foundation of China (U23A2012) and the National Natural Science Foundation of China (42276242, 52192692).

Data Availability Statement: The data are available in the case that it is required.

Conflicts of Interest: The authors declare no conflicts of interest.

References

1. Müller-Stoffels, M.; Langhorne, P.J.; Petrich, C.; Kempema, E.W. Preferred crystal orientation in fresh water ice. *Cold Reg. Sci. Technol.* **2009**, *56*, 1–9. [CrossRef]
2. Zhang, Y.D.; Li, Z.J.; Li, C.J.; Zhang, B.S.; Deng, Y. Microstructure characteristics of river ice in Inner Mongolia section of the Yellow River and its influencing factors. *J. Hydraulic Eng.* **2021**, *52*, 1418–1429. [CrossRef]
3. Timco, G.; Weeks, W. A review of the engineering properties of sea ice. *Cold Reg. Sci. Technol.* **2010**, *60*, 107–129. [CrossRef]
4. Huang, W.F.; Han, H.W.; Shi, L.Q.; Niu, F.J.; Deng, Y.S.; Li, Z.J. Effective thermal conductivity of thermokarst lake ice in Beiluhe Basin, Qinghai-Tibet Plateau. *Cold Reg. Sci. Technol.* **2013**, *85*, 34–41. [CrossRef]
5. Rødtang, E.; John, J.; Alfredsen, K.; Høyland, K. In-situ ice strength distribution of anchor ice dams. *Cold Reg. Sci. Technol.* **2023**, *215*, 103982. [CrossRef]
6. Deng, Y.; Li, Z.K.; Wang, J.; Xu, L.K. The microstructure of Yellow River ice in the freezing period. *Crystals* **2019**, *9*, 484. [CrossRef]
7. Misra, R.; Chung, T.F.; Lin, S.C. Understanding microstructural evolution during three–axial thermos–mechanical processing involving severe plastic deformation of magnesium alloys. *Mater. Technol.* **2024**, *39*, 2350220. [CrossRef]
8. Mel'nichenko, N.; Tyveev, A.; Lazaryuk, A.Y.; Savchenko, V.; Kustova, E. Vertical distribution of brine and volume structure of thin annual ice in Amursky Bay based on the methods of nuclear magnetic resonance and magnetic resonance imaging. *Oceanology* **2019**, *59*, 777–786. [CrossRef]
9. Kawamura, T. Observations of the internal structure of sea ice by X-ray computed tomography. *J. Geophys. Res. Ocean.* **1988**, *93*, 2343–2350. [CrossRef]
10. Michel, B.; Ramseier, R.O. Classification of river and lake ice. *Can. Geotech. J.* **1971**, *8*, 36–45. [CrossRef]
11. Shokr, M.E.; Sinha, N.K. Arctic Sea ice microstructure observations relevant to microwave scattering. *Arctic* **1994**, *47*, 265–279. [CrossRef]
12. Cole, D.M. The microstructure of ice and its influence on mechanical properties. *Eng. Fract. Mech.* **2001**, *68*, 1797–1822. [CrossRef]
13. Li, Z.J.; Zhang, L.M.; Lu, P.; Leppäranta, M.; Li, G.W. Experimental study on the effect of porosity on the uniaxial compressive strength of sea ice in Bohai Sea. *Sci. China Technol. Sci.* **2011**, *54*, 2429–2436. [CrossRef]
14. Sammonds, P.; Montagnat, M.; Bons, P.; Schneebeli, M. Ice microstructures and microdynamics. *Phil. Trans. R. Soc. A.* **2017**, *375*, 20160438. [CrossRef] [PubMed]
15. Hammonds, K.; Baker, I. Quantifying damage in polycrystalline ice via X-ray computed micro-tomography. *Acta Mater.* **2017**, *127*, 463–470. [CrossRef]
16. Salomon, M.L.; Maus, S.; Petrich, C. Microstructure evolution of young sea ice from a Svalbard fjord using micro-CT analysis. *J. Glaciol.* **2022**, *68*, 571–590. [CrossRef]
17. Du Plessis, A.; Le Roux, S.G.; Guelpa, A. Comparison of medical and industrial X-ray computed tomography for non-destructive testing. *Case Stud. Nondestruct. Test. Eval.* **2016**, *6*, 17–25. [CrossRef]
18. Zeng, Q.; Chen, S.; Yang, P.C.; Peng, Y.; Wang, J.Y.; Zhou, C.S.; Wang, Z.D.; Yan, D.M. Reassessment of mercury intrusion porosimetry for characterizing the pore structure of cement-based porous materials by monitoring the mercury entrapments with X-ray computed tomography. *Cem. Concr. Compos.* **2020**, *113*, 103726. [CrossRef]
19. Wang, H.N.; Ni, W.K. Quantitative analysis of loess microstructure based on CT and SEM images. *Rock Soil Mech.* **2012**, *33*, 243–247. (In Chinese) [CrossRef]
20. Zhang, X.N.; Wan, C.; Wang, D.; He, L.F. Numerical simulation of asphalt mixture based on three-dimensional heterogeneous specimen. *J. Cent. South Univ. Technol.* **2011**, *18*, 2201–2206. [CrossRef]
21. Ge, Z.D.; Chen, L.X.; Luo, R.; Wang, Y.W.; Zhou, Y.C. The detection of structure in wood by X-ray CT imaging technique. *BioResources* **2018**, *13*, 3674–3685. [CrossRef]
22. Zhang, Z.B.; Zou, Y.N.; Huang, Y.L.; Li, Q. CT image crack segmentation method based on linear feature enhancement. *J. X-Ray Sci. Technol.* **2022**, *30*, 903–917. [CrossRef] [PubMed]
23. Cui, Z.M.; Zhao, S.N.; Shi, X.H.; Lu, J.P.; Liu, Y.; Liu, Y.H.; Zhao, Y.X. Vertical distribution characteristics and ecological risk assessment of mercury and arsenic in ice, water, and sediment at a cold-arid lake. *Toxics* **2024**, *12*, 540. [CrossRef]
24. Ketcham, R.A.; Carlson, W.D. Acquisition, optimization and interpretation of X-ray computed tomographic imagery: Applications to the geosciences. *Comput. Geosci.* **2001**, *27*, 381–400. [CrossRef]
25. Zhang, S.J.; Lai, Y.M.; Zhang, X.F.; Pu, Y.B.; Yu, W.B. Study on the damage propagation of surrounding rock from a cold-region tunnel under freeze–thaw cycle condition. *Tunn. Undergr. Space Technol.* **2004**, *19*, 295–302. [CrossRef]
26. Rabba, J.A.; Jaafar, H.A.; Suhaimi, F.M.; Jafri, M.Z.M.; Osman, N.D. A simplified low-cost phantom for image quality assessment of dental cone beam computed tomography unit. *J. Med. Radiat. Sci.* **2024**, *71*, 78–84. [CrossRef]
27. Jing, D.J.; Meng, X.X.; Ge, S.C.; Zhang, T.; Ma, M.X.; Tong, L.Q. Reconstruction and seepage simulation of a coal pore–fracture network based on CT technology. *PLoS ONE* **2021**, *16*, e0252277. [CrossRef]
28. Rambabu, C.; Chakrabarti, I. An efficient immersion-based watershed transform method and its prototype architecture. *J. Syst. Architect.* **2007**, *53*, 210–226. [CrossRef]
29. Qin, Y.B.; Wang, W.; Liu, W.; Yuan, N. Extended-maxima transform watershed segmentation algorithm for touching corn kernels. *Adv. Mech. Eng.* **2013**, *5*, 268046. [CrossRef]
30. Shen, X.J.; Wu, X.Y.; Han, D.J. Survey of research on watershed segmentation algorithms. *Comput. Eng.* **2015**, *41*, 26–30. (In Chinese) [CrossRef]

31. Deng, Y.; Wang, J.; Zhou, J.; Zhang, P. Quantitative analysis of the geometrically representative volume element of the Yellow River's granular ice microstructure during the freezing period. *Crystals* **2023**, *13*, 1021. [CrossRef]
32. Ji, Q.; Li, B.J.; Pang, X.P.; Zhao, X.; Lei, R.B. Arctic Sea ice density observation and its impact on sea ice thickness retrieval from CryoSat–2. *Cold Reg. Sci. Technol.* **2021**, *181*, 103177. [CrossRef]
33. Yoshimura, K.; Inada, T.; Koyama, S. Growth of spherical and cylindrical oxygen bubbles at an ice-water interface. *Cryst. Growth Des.* **2008**, *8*, 2108–2115. [CrossRef]
34. Ni, J.R.; Sun, L.Y.; Sun, W.L. Modification of chemical oxygen demand monitoring in the Yellow River, China, with a high content of sediments. *Water Environ. Res.* **2007**, *79*, 2336–2342. [CrossRef]
35. Zhang, F.; Shi, X.H.; Zhao, S.N.; Hao, R.N.; Zhai, J.L. Equilibrium analysis of dissolved oxygen in Lake Wuliangsuhai during ice-covered period. *J. Lake Sci.* **2022**, *34*, 1570–1583. (In Chinese) [CrossRef]
36. Qi, P.; Ji, M.X.; Sun, Z.Y. Formation mechanism of hyper-concentrated sediment flow caused by scouring of reservoir emptying. *J. Hydraul. Eng.* **2006**, *8*, 906–912. (In Chinese) [CrossRef]

water

MDPI

Article

An Experimental Investigation of the Flexural Strength and Fracture Toughness of Granular Snow Ice Under a Three-Point Bending Test

Hongwei Han [1,2], Wanyun Li [1], Yu Li [1,*], Zhi Liu [1] and Xingchao Liu [1,*]

[1] School of Water Conservancy and Civil Engineering, Northeast Agricultural University, Harbin 150030, China; hanhongwei@neau.edu.cn (H.H.); 18845147160@163.com (W.L.); 17853819179@163.com (Z.L.)

[2] Heilongjiang Provincial Key Laboratory of Water Resources and Water Conservancy Engineering in Cold Region, Northeast Agricultural University, Harbin 150030, China

* Correspondence: 18846771809@163.com (Y.L.); dnlxc@neau.edu.cn (X.L.)

Abstract: Ice is a common natural phenomenon in cold areas, which plays an important role in the construction of cold areas and the design of artificial ice rinks. To supplement our knowledge of ice mechanics, this paper investigates the mechanical properties of granular snow ice. The factors influencing the flexural strength of granular snow ice are analyzed through a three-point bending test. It is found that flexural strength is affected by strain rate. At low strain rates, flexural strength increases with increasing strain rate, whereas at high strain rates, flexural strength decreases with increasing strain rate. As temperature decreases, the flexural strength value of ice increases, but its brittleness becomes more pronounced, indicating that the strain rate corresponding to the maximum flexural strength is lower. Within the test temperature range, the tough-brittle transition range is from 6.67×10^{-5} s^{-1} to 3.11×10^{-4} s^{-1}. At −5 °C, the strain rate corresponding to the maximum bending strength is 3.11×10^{-4} s^{-1}, while at −10 °C, it is only 6.67×10^{-5} s^{-1}. Flexural strength is influenced by crystal structure. At −20 °C, the average flexural strength of granular snow ice is 2.85 MPa, compared to 1.93 MPa for columnar ice at the same temperature. Through observation, we found that there are straight cracks and oblique cracks. The fracture toughness of granular snow ice was investigated by cutting prefabricated cracks at the bottom of the ice beam and employing a three-point bending device. It is found that fracture toughness decreases with increasing strain rate. Temperature also affects granular snow ice. At −15 °C, fracture toughness is 181.60 kPa·m$^{1/2}$, but at −6 °C, it decreases to 147.28 kPa·m$^{1/2}$. However, at varying temperatures and strain rates, there is no significant difference in the fracture patterns of ice samples, which predominantly develop upward along the prefabricated cracks.

Keywords: granular snow ice; ice temperature; strain rate; flexural strength; fracture toughness

Citation: Han, H.; Li, W.; Li, Y.; Liu, Z.; Liu, X. An Experimental Investigation of the Flexural Strength and Fracture Toughness of Granular Snow Ice Under a Three-Point Bending Test. *Water* **2024**, *16*, 3358. https://doi.org/10.3390/w16233358

Academic Editor: Zhuotong Nan

Received: 10 October 2024
Revised: 18 November 2024
Accepted: 20 November 2024
Published: 22 November 2024

1. Introduction

Ice is a non-homogeneous composite material consisting of grains, grain boundaries, and initial defects [1,2], with types including sea ice, river ice, lake ice, reservoir ice, icebergs, permafrost, atmospheric ice, and polar glaciers in cold regions [3]. Ice, as a widespread natural phenomenon in cold regions, has beneficial aspects for human life, such as enabling the development of oil resources in the polar seas [4]. Moreover, ice itself serves as a load-bearing platform, capable of supporting and transporting both mobile and stationary heavy loads [5]. Ice on rivers and lakes has long been utilized for ice tourism, recreation, and ice transportation [6]. However, ice-related disasters, including river ice floods, reservoir ice impacts on gates, slope protection, and winter operations of power plants, as well as the effects of sea ice on offshore structures and navigation, pose significant threats to life and property [7]. In the design, construction, and operation of hydroelectric equipment, as well as in the design of structures such as icebreakers, special consideration must be given to ice loads to ensure operational safety in cold regions [8,9]. When ice interacts with inclined or conical structures, bending failure typically occurs [10]. It is not only an important mechanical property for assessing the climbing and impact strength of ice on inclined hydraulic structures [11], but also a key parameter for calculating ice loads [8]. Ice fracture destruction is also common. It is an important parameter in the design of hydraulic structures and the analysis of river ice dynamics. Following the formation of the ice layer, the force generated by its heating and expansion is the primary load that leads to the failure of reservoir slope protection [12]. Extreme ice pressure is related to the deformation and fracture behavior of the ice sheet under compression. To develop an extreme ice pressure model based on fracture mechanics theory, it is particularly necessary to understand the fracture toughness of ice [13].

In 1968, Weeks and Assur demonstrated that sea ice exhibits viscoelastic mechanical properties [14]. Sinha [15] argued that the viscoelastic–plastic constitutive relationship of sea ice allows it to exhibit a wide range of mechanical behaviors at different loading rates. Previous research has found that ice shows brittleness at higher strain rates, while lower rates lead to ductile behavior [16–19]. Specifically, the low-strain-rate region exhibits ductile failure, the high-strain-rate region is characterized by brittle failure, and there is a ductile-to-brittle transition stage in between. Zhang et al. [20] discovered that the transition range for ice from ductile to brittle is between 1.46×10^{-6} s^{-1} and 3.54×10^{-5} s^{-1} on freshwater ice. Schulson [21,22] first proposed the critical grain size of ice in analyzing the transition from brittle to ductile behavior, concluding that this transition in the properties of ice materials is related to grain size. Additionally, many scholars use theoretical or experimental methods to describe the transition of ice from exhibiting ductile behavior to brittle behavior [23,24].

Currently, experimental studies on ice bending mainly include three-point bending, four-point bending, and cantilever beam tests. Of these, three-point bending tests involve retrieving ice samples and preparing them indoors, while cantilever beam tests are predominantly conducted in the field [25]. In 1943, the Brazilian engineer Carneiro [26] proposed the famous Brazilian disk indirect tensile test method, which is now widely used to test the tensile strength of rocks, concrete, and other brittle materials [27]. There are a number of methods to test brittle materials for mixed-mode (I + II) or pure-mode II fracture toughness, such as using straight-notched Brazilian disks, V-notched Brazilian disks, and compression short-core tests [28–30]. Xiao et al. [31] concluded that under the premise of ensuring that ice is a brittle material, it can be tested as a rock and the fracture toughness and tensile strength can be measured by applying Brazilian disk splitting.

The mechanical properties of ice may depend on a combination of factors, such as crystal structure, temperature, porosity, grain size, and strain rate [32,33]. In general, the fracture toughness of ice is in the range of 50–150 kPa·m$^{1/2}$ [34], increasing slightly with decreasing temperature [35] and decreasing with increasing grain size [36] and porosity [37]. Wang [38] conducted a three-point bending test on artificial columnar ice and found that the flexural strength increased and then decreased with an increase in the strain rate, and a fitting relationship between flexural strength, elastic modulus, and strain rate was obtained. Gagnon [39] and Ji et al. [40] conducted bending tests on glacier ice and sea ice, respectively, and found that the flexural strength increased with an increase in strain rate and a decrease in temperature. Zhang et al. [41] studied freshwater granular ice and found that the flexural strength initially decreases, then increases, and then decreases again with increasing stress rate. Zhang [42] found that the variation in flexural strength with strain rate is similar to an inverted "W" shape. Xu et al. [43] conducted three-point bending tests with notches on pure polycrystalline ice at different temperatures (−20 °C, −30 °C, and −40 °C) and loading rates (1 to 100 mm/min). They found that when the strain rate at the crack tip was less than the critical value of 6×10^{-3} s^{-1}, the fracture toughness decreased with the increasing crack tip strain rate. Beyond this critical value, the fracture toughness remained constant. Litwin et al. [44] studied the tensile strength and fracture toughness within the temperature range of 260 K to 110 K and found that fracture toughness is not sensitive to temperature. Mulmule et al. [45] and Dempsey et al. [46] conducted fracture toughness tests on sea ice of different sizes and found a significant size effect. Dempsey et al. [46] suggested that the results of fracture tests can be characterized by tensile strength, which decreases with increasing specimen size.

Ice plays a role in transportation, the military, and other fields, and has also led to the development of numerous ice sports [47]. Some countries use refrigeration technology to manufacture artificial ice rinks, in which the artificial ice surfaces must meet the requirements of stiffness and bearing capacity, as well as ensure the normal requirements of ice sports. However, when moving on the ice surface, it may crack or be locally damaged, causing harm. Therefore, investigating the mechanical properties of artificial ice is essential. Due to variations in growth environments, different types of ice crystals form, including snow ice, granular ice, and columnar ice [48]. However, previous studies have provided relatively little research on the mechanics of granular ice. In field experiments, the relatively low content of granular ice and uneven thickness in the ice layer sometimes hinder the extraction of a sufficient number of samples, making it difficult to stably test its mechanical properties. Studies [8] also indicate that the bending strength of granular ice is higher than that of columnar ice, necessitating a serious consideration of its mechanical properties. Schwarz et al. [49] provided recommendations for the dimensions of ice specimens in bending tests. The focus of this study is to prepare artificial granular ice in the laboratory, cut a small ice sample measuring 35 mm × 35 mm × 180 mm, and investigate the influence of temperature and strain rate on the bending strength and fracture toughness of granular ice through three-point bending and fracture toughness tests, aiming to further supplement the mechanical properties of ice.

2. Method

2.1. Ice Sample Preparation

As a result of changes in climate and temperature, the three states of water are constantly shifting. As temperatures continue to drop to freezing point, bodies of water develop ice. During crystallization, water forms different crystal structures due to various external conditions such as temperature and pressure, and crystals with the same structure are affected by the environment, resulting in different grain sizes [32]. In rivers, for example, granular snow ice crystals may form during the early stages of freezing when the ice grows too fast or when snow falls before cooling, after which the vertical growth rate dominates and columnar ice crystals are formed [50].

The granular ice and columnar ice samples used in this study were prepared at Northeast Agricultural University in Harbin, Heilongjiang Province, China. The specific steps for the preparation of granular ice are as follows: add about one-third of tap water into a container wrapped with foam board, place it in a sub-zero environment and cool it down to 0 °C, then add slush and stir it to form an ice–water mixture, and finally invert the container to remove the ice after it is completely frozen. Columnar ice was prepared in the low-temperature laboratory, with a temperature control range of room temperature to −40 °C and a temperature fluctuation value of ±0.5 °C. Due to the addition of snow mud during the preparation process to form an ice–water mixture, we will use "granular snow ice" to describe the experimental object in the following description.

Cracks and large fractures in the middle of the ice body significantly impact test results, so any specimen exhibiting these conditions should be promptly discarded [2]. To prevent weathering and adhesion, the specimens were wrapped in cling film and transported in a foam box. The entire process of ice mechanics testing requires temperature control. Temperature control includes sample storage, keeping a constant temperature, and loading processes. After sample preparation is complete, the ice sample should be stored in a −15 °C freezer to ensure the long-term preservation and stability of the ice crystal structure. Before starting the experiment, the time required for the ice sample to reach thermal equilibrium should be calculated using the heat conduction equation, based on the sample size and temperature difference. Then, place the ice sample in the freezer for a duration exceeding the calculated time to ensure thermal equilibrium. After 48 h, the ice sample will achieve full thermal equilibrium [50].

2.2. Ice Crystal Structure Measurement

In general, the internal organization of ice reflects its growth history and determines its fundamental physical properties; thus, observing the structural characteristics of ice's internal organization is an essential task. Based on prior research [32], ice flakes were prepared in a low-temperature laboratory and observed using a Rigsby universal stage. The Rigsby universal stage employs a polarizing microscope to measure the spatial orientation of linear and planar elements within the flakes, enabling the direct visualization of the particle size and shape of the observed samples. The specific steps include the following: Select a vertical and intact ice sample, and use a planer to flatten the protruding defects on the observed surface. Secondly, place the ice sample in contact with a glass sheet at a temperature slightly above 0 °C, moving the sample left and right on the glass to expel air bubbles. And then, freeze the glass sheet with the ice sample at a low temperature, then use a planer to thin the ice sample to about a 1 mm thickness after it is solidly frozen, and mark it clearly. To avoid weathering, place the completed ice slices in a sealed plastic bag and store them at a low temperature and then observe the ice slices in a dark room using a Rigsby universal stage.

2.3. Three-Point Bending Test

2.3.1. Test Devices

This test uses the WDW-100 electronic universal testing machine, with a maximum test force of 100 kN, displacement measurement resolution of 0.01 mm, and a loading rate range of 0.005 to 1000 mm/min. It comes from Changchun Kexin Testing Instrument Co., Ltd. located in Changchun city, China. The test machine can perform tensile, bending, shear, and other tests in conventional and low-temperature environments [49]. In order to maintain a low temperature environment during the test, a low-temperature test chamber with a temperature accuracy of ±1 °C is used [38]. Additionally, to adjust the distance between the specimen and the indenter and effectively observe the fracture pattern of the specimen during the loading process, etc., the lighting switch can be turned on. Figure 1 shows the WDW-100 electronic universal testing machine and the testing zone.

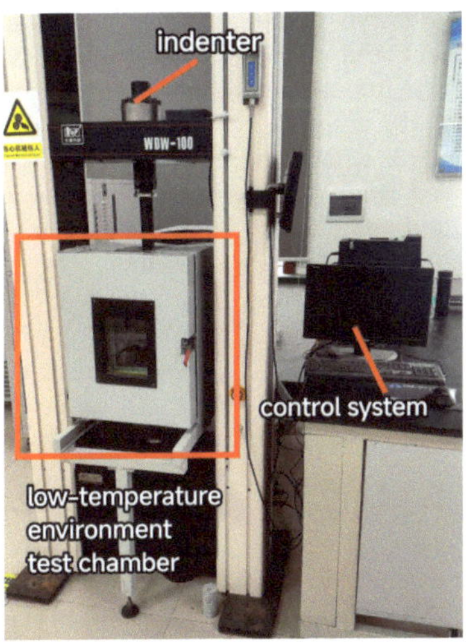

(a)

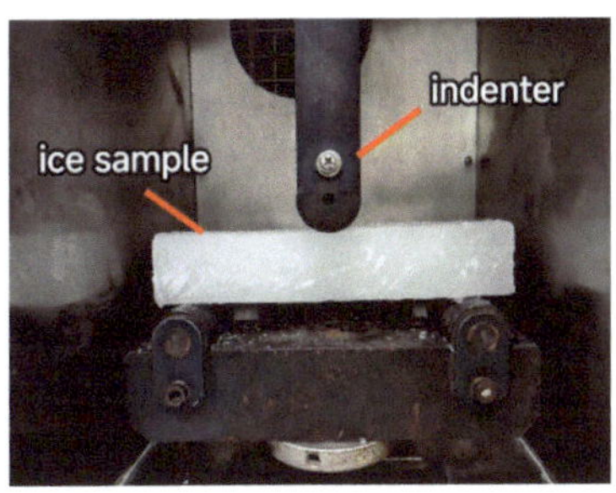

(b)

Figure 1. Test devices. (**a**) WDW–100 electronic universal testing machine. (**b**) Testing zone.

2.3.2. Test Principles and Procedures

Ice exhibits viscoelastic properties and is not entirely elastic. Gold demonstrated that between −40 °C and −3 °C, ice approximates to a purely elastic material, permitting the application of elastic theory to analyze its failure and determine effective flexural strength values [51].

Ice is a viscoelastic plastic material. Han et al. [32] utilized linear elasticity theory to estimate the bending strength of columnar granular freshwater ice subjected to three-point bending beam tests. This study also fulfills the application conditions of linear elasticity theory, that is, takes place under the action of midspan loads, whereby the bending strength formula for a rectangular cross-section of a linear elastic, uniformly simply supported beam is as follows:

$$\sigma_f = \frac{3PL}{2bh^2} \tag{1}$$

where P is the load at which a three-point simply supported beam fails, b is the width of the beam, h is the height of the beam, and L is the span of the ice beam which is 150 mm.

Due to the existence of dimensional differences between the specimens, the specimen size data need to be measured again before each test, and the displacement loading rate is converted into the strain rate to unify the independent variable. According to the method proposed by Han et al. [32], the strain and strain rate at the bottom of the span of an ice beam are estimated according to the relationship between the strain and the deflection of a three-point simply supported beam:

$$\varepsilon = \frac{6h\delta}{L^2} \tag{2}$$

$$\dot{\varepsilon} = \frac{6h\dot{\delta}}{L^2} \tag{3}$$

where ε is the strain at the load application point, $\dot{\varepsilon}$ is the strain rate at the bottom of the load application point, δ is the deflection, which refers to the displacement of the point of action, and $\dot{\delta}$ is the displacement loading rate. When the ice sample fractures during the bending test, the test is immediately terminated and data are recorded.

Before the test, adjust the temperature of the environmental test chamber. and pre-cool for 30 min. Before the experiment, observe whether there are bubbles, impurities, and cracks in the sample and record them, and measure the size of the sample [38]. After everything is ready, place the specimen into the environmental test chamber on the fixed three-point bending fixture, aligning the indenter with the center of the specimen. And then, set the loading rate within the test machine program, number the specimen, and start loading. Before the sample is destroyed, the testing machine program continuously collects data and automatically saves the data information. After the experiment is completed, record the form of sample damage and clean the testing machine.

In this study, three-point bending tests of simply supported beams are performed on granular snow ice at −5 °C, −8 °C, −10 °C, −15 °C, −18 °C, −20 °C, −25 °C, −30 °C, and −35 °C, for a total of nine working conditions, and on columnar ice at −20 °C.

2.4. Fracture Toughness Test

2.4.1. Test Principles

According to fracture mechanics, crack growth occurs when the incremental energy available from the release of stored potential energy is equal to, or exceeds, that required to create new fracture surfaces. Resistance to crack growth is defined by fracture toughness, which is the amount of work required to propagate a crack by unit area [43]. Wei et al. [52] believe that fracture toughness K_{IC} is the stress intensity factor corresponding to unstable fracture of materials. In a three-point bending test, the specimen does not undergo simple stretching, and, thus, K_{IC} lacks an analytical solution. Concerning the measurement of fracture toughness, there is no uniform standard, and various research results have a high degree of dispersion. Huang [53] calculated the fracture toughness of granular and columnar Yellow River ice. Based on the method given by Huang [53], the fracture toughness of granular ice was calculated in this paper:

$$K_{IC} = \frac{PL}{BW^{3/2}} P\left(\frac{a}{W}\right) \quad (4)$$

$$P\left(\frac{a}{W}\right) = 3(a/W)^{1/2} \times \frac{1.99 - (a/W)(1-a/W)\left[2.15 - 3.93(a/W) + 2.70(a/W)^2\right]}{2(1+2a/W)(1-a/W)^{3/2}} \quad (5)$$

where K_{IC} is the fracture toughness of the specimen, kPa·m$^{1/2}$.

2.4.2. Test Procedures

Unlike the bending test, the three-point bending fracture test requires pre-fabricated cracks to be machined in the center of the specimen to ensure that failure occurs at the loading point. The ratio of the length of the pre-fabricated crack to the height W of the specimen should be between 0.2 and 1 [2], while in this study, it is between 0.28 and 0.36. When using a saw bone machine for ice cutting, it is essential to ensure that the saw blade is perpendicular to the ice surface during processing and to apply even force throughout the process [2]. The loading method and structural dimensions of the specimens are shown in Figure 2.

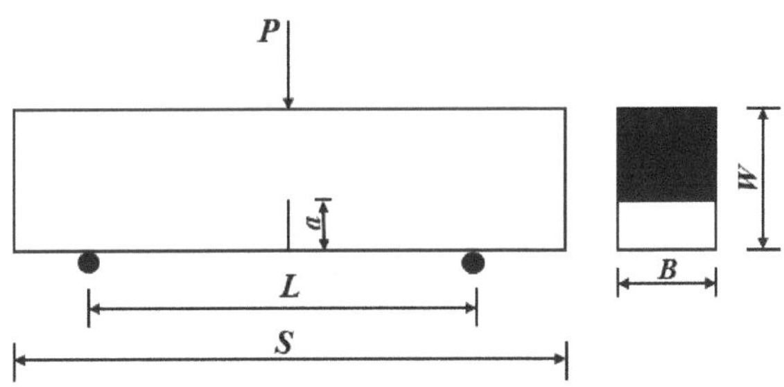

Figure 2. A schematic diagram of ice-sample loading (here, *L*, *S*, *W*, and *B* represent the ice beam's span, length, height, and thickness, respectively; a is the initial crack length, and *P* is the applied load).

The effects of temperature and loading rate on the ice beam were considered. Four temperatures were tested: −6 °C, −8 °C, −10 °C, and −15 °C. The displacement loading rates included eight types: 0.05 mm/min, 0.1 mm/min, 0.3 mm/min, 0.5 mm/min, 1 mm/min, 3 mm/min, 5 mm/min, and 10 mm/min.

For the preparation work before the experiment, refer to Section 2.3.2. After precooling is completed, remove the ice sample from the freezer. Prefabricated cracks may remain filled with foam or frost and should be cleaned with a planer to ensure they run through from top to bottom. Measure the width, height, and crack length of the ice sample and input these dimensions into the tester program.

3. Results and Analysis

3.1. Crystal Structure of Granular Snow Ice

Figure 3 shows a horizontal sheet of granular snow ice. As can be seen from the figure, the artificial ice prepared in the laboratory is a typical granular ice structure [32]. Slush is added in the icing process to form a mixture of ice water, and the growth rate of ice crystals is relatively high, resulting in the formation of granular ice crystals.

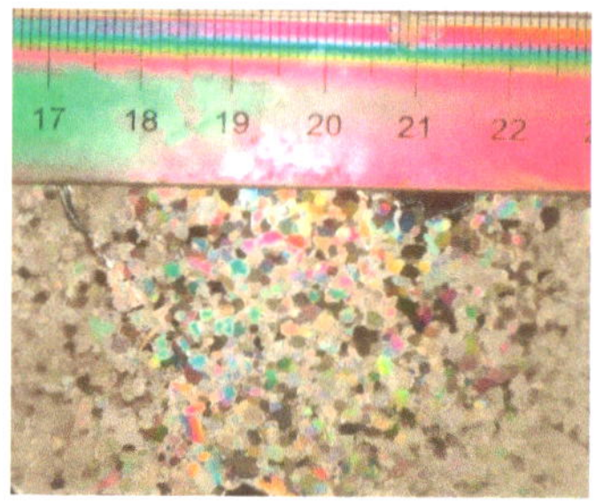

Figure 3. Horizontal slice of granular snow ice.

3.2. Three-Point Bending Test

3.2.1. Fracture Process Curve

Figure 4 presents a typical curve of flexural stress over time in a three-point bending test, which was conducted at −5 °C with a loading rate of 5 mm/min and a duration of 7.64 s. The ice beam undergoes three stages from loading to failure. At the beginning of loading, the load is small, the stress on the cross-section of the ice beam is also minimal, and microcracks produced at the edges of bubbles and impurities are negligible. The ice beam undergoes elastic deformation, with stress and strain maintaining a linear relationship; this is the first stage. As the load increases, because the tensile strength of ice is lower than its compressive strength, the tensile zone on the lower surface of the ice beam reaches its tensile strength, resulting in plastic deformation and crack formation, while the compressive zone remains in elastic deformation. This is the second stage. The load continues to increase until the strain at the lower surface of the ice beam exceeds the tensile limit strain, causing the original microcracks to develop into macrocracks. At this point, the ice near the neutral axis remains uncracked. With the increasing load, the cracks rapidly extend upward until the ice beam completely fractures. At this stage, the applied load exceeds the bending capacity of the ice, marking the third stage of ice beam failure [41].

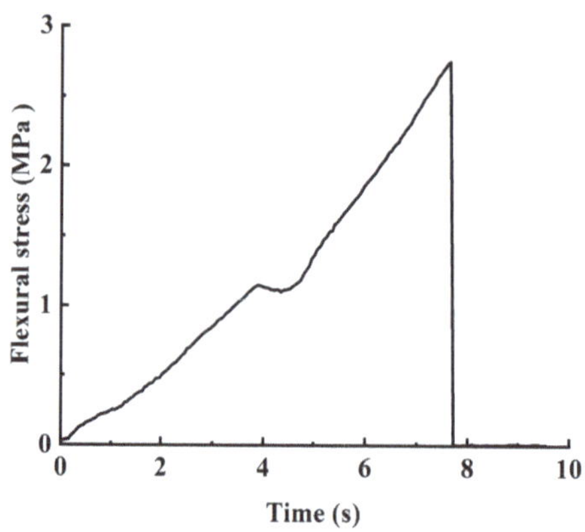

Figure 4. The curve of flexural stress over time in the three-point bending test.

Taking into account the uncontrollability of errors, four to six repeated tests were conducted for each loading rate and temperature, followed by the calculation of the average bending strength. The bending strength of granular snow ice ranged from 1.68 to 3.65 MPa, with an average of 2.89 ± 0.27 MPa; for columnar ice, it ranged from 1.50 to 2.36 MPa, averaging 1.69 MPa.

3.2.2. The Relationship Between Flexural Strength and Strain Rate

As shown in Figure 5, the relationship between the flexural strength of granular snow ice at different temperatures and loading rates is presented. The flexural properties of ice are influenced by the strain rate, showing a trend where flexural strength initially increases and then decreases with strain rate at each temperature. Previous research results have suggested that ice demonstrates toughness at low strain rates and brittleness at high strain rates [16–19]. Our results show that the ultimate flexural strength of ice occurs within the ductile–brittle transition interval, which ranges from 6.67×10^{-5} s^{-1} to 3.11×10^{-4} s^{-1} at the tested temperatures.

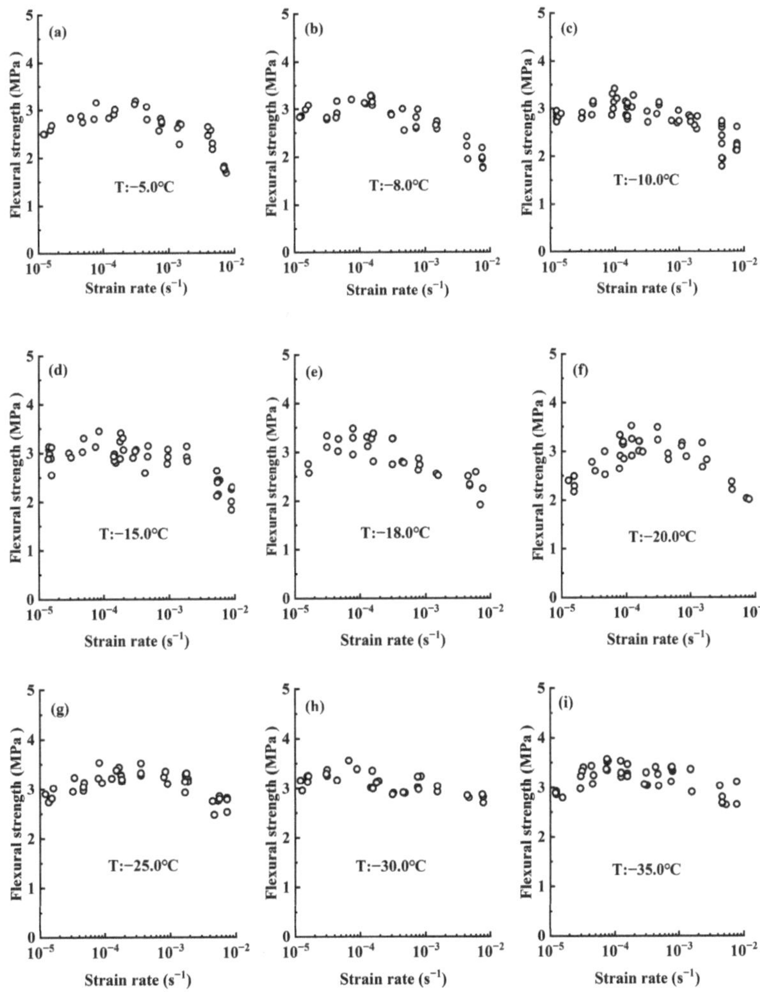

Figure 5. The flexural strength of granular snow ice at different temperatures and strain rates. (**a**–**i**) is the bending strength variation trend with strain rate at −5 °C, −8 °C, −10 °C, −15 °C, −18 °C, −20 °C, −25 °C and −30 °C and −35 °C, respectively.

Gagnon et al. [39] conducted bending tests on glacier ice at strain rates ranging from 10^{-5} s^{-1} to 10^{-3} s^{-1} and temperatures from −16 °C to −1 °C. They found that at −11 °C, the bending strength at a strain rate of 10^{-3} s^{-1} was approximately 26% higher than at 10^{-5} s^{-1}. The results of this paper align well with Gagnon's findings. For granular snow ice at −5 °C, the average flexural strength at a strain rate of 10^{-3} s^{-1} is 17% higher than at 10^{-5} s^{-1}. And when the strain rate is less than 3.11×10^{-4} s^{-1}, the flexural strength of ice increases with the increase in strain rate, and the maximum value of flexural strength is 3.19 MPa. Above this strain rate, the flexural strength decreases with the increase in strain rate, and the minimum value of flexural strength is 1.68 MPa when the strain rate is 6.89×10^{-3} s^{-1}.

3.2.3. The Relationship Between Flexural Strength and Temperature

At 0 °C, the flexural strength of ice is close to 0 MPa, and the linear model cannot accurately predict the mechanical properties of ice at this temperature. Wang et al. [38] used logarithmic fitting to obtain the relationship between the bending strength and temperature of artificial ice:

$$\sigma_f = A + B\, ln(|T/T_0|) \quad (6)$$

in the formula, to coordinate units, the independent variable is adjusted to T/T_0, where T_0 is the temperature of 1 °C.

Temperature affects the mechanical properties of ice. In this experiment, three-point bending tests were conducted on granular snow ice from −35 °C to −5 °C. Figure 6 shows the logarithmic fitting effect between the average flexural strength and ice temperature. It was found that the flexural strength increased as the temperature decreased. This occurs because lower ice temperatures increase the intermolecular linkage force, requiring more energy to produce cracks, thereby increasing ice strength and flexural strength.

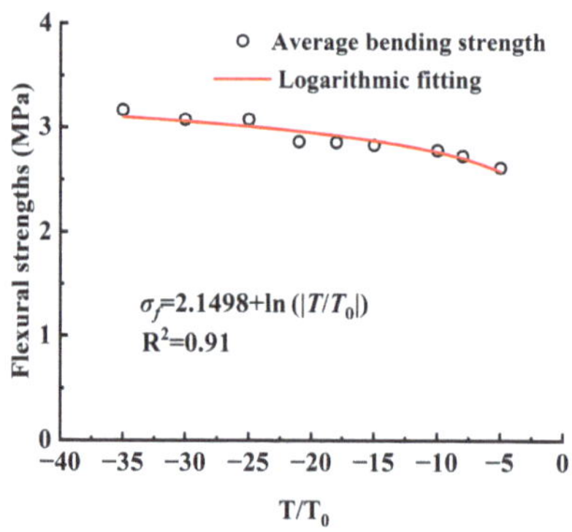

Figure 6. Logarithmic simulation of average flexural strength and ice temperature of granular snow ice.

Han et al. [54] concluded that the lower the temperature of the ice beam, the more pronounced the brittleness characteristics, which is manifested by the lower strain rate corresponding to the maximum ultimate flexural strength. The ductile–brittle transition range of granular snow ice at the test temperature is 6.67×10^{-5} s^{-1} to 3.11×10^{-4} s^{-1}. The maximum ultimate flexural strength and the corresponding strain rate at each test temperature are summarized in Table 1, which shows that the strain rate corresponding to the maximum ultimate flexural strength value tends to decrease as the temperature decreases. For example, at −5 °C, −8 °C, and −10 °C, the maximum ultimate flexural strengths are 3.19 MPa, 3.29 MPa, and 3.41 MPa, with corresponding strain rates of 3.11×10^{-4} s^{-1}, 1.51×10^{-4} s^{-1}, and 9.78×10^{-5} s^{-1}, respectively, showing a clear reduction with decreasing temperature.

Table 1. Maximum ultimate flexural strength and corresponding strain rate of granular snow ice at different temperatures.

Temperature (°C)	Maximum Ultimate Flexural Strength (MPa)	Strain Rate (s^{-1})
−5	3.19	3.11×10^{-4}
−8	3.29	1.51×10^{-4}
−10	3.41	9.78×10^{-5}
−15	3.45	8.44×10^{-5}
−18	3.48	7.78×10^{-5}
−20	3.50	7.89×10^{-5}
−25	3.53	8.22×10^{-5}
−30	3.56	6.67×10^{-5}
−35	3.57	7.50×10^{-5}

3.2.4. The Relationship Between Flexural Strength and Ice Structure

In this study, three-point bending tests on columnar ice were conducted at −20 °C, with loading rates ranging from 0.1 mm/min to 30 mm/min. Figure 7 shows the variation in the flexural strength of columnar ice with strain rate at −20 °C. It reveals that the flexural strength of columnar ice first increases and then decreases with strain rate, but it is lower than that of granular snow ice. The average flexural strength of columnar ice at this temperature is 1.93 MPa, compared to 2.85 MPa for granular snow ice. This is consistent with the results of Timco et al. [37], and Blanchet et al. [55] attributed this phenomenon to the fact that granular ice has a smaller grain size than columnar ice. Cole et al. [56] concluded that stress of ice decreases with increasing grain size. For the same material, a smaller grain diameter results in larger grain boundaries, greater barriers to dislocation motion, higher resistance to deformation and macroscopic strength.

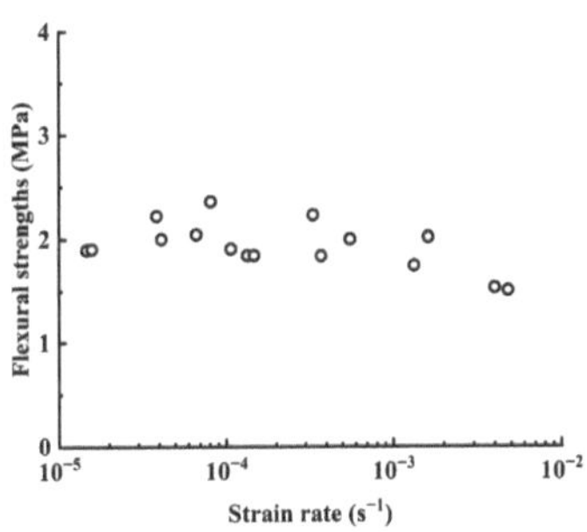

Figure 7. Relationship between flexural strength and strain rate of columnar ice at −20 °C.

Wang [38] conducted three-point bending tests on ice and obtained a fitting equation between flexural strength and temperature. By substituting it, the bending strength at −20 °C was found to be 1.91 MPa. In contrast, this study measures the flexural strength of columnar ice under the same conditions to be 1.70 MPa.

3.2.5. Flexural Failure Mode

Ice cracks can be roughly categorized into two types: one is straight cracks (Figure 8a), where, starting at the lower surface of the ice beam, because of the tension effect of cracks, with the increase in load, the cracks continue to develop upward until they exist throughout the entire specimen, and the destruction of the cross-section is relatively flat. These are type-I tension cracks. The second type is the oblique crack (Figure 8b), where the ice beam is mainly subjected to shear force, and the crack and the direction of tensile stress presents an angle of 30°~45°. This is a type-II shear crack [38].

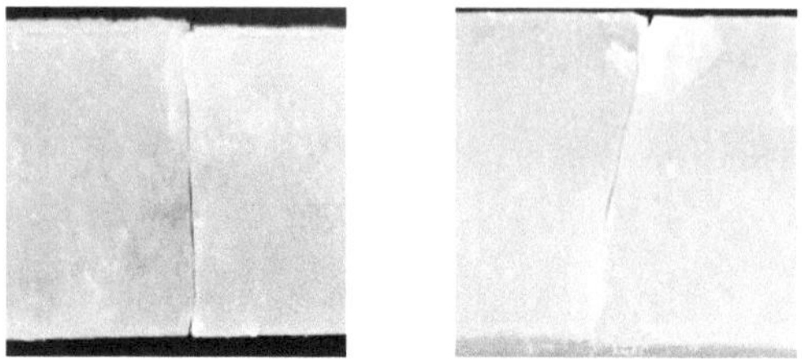

(a) Straight crack (b) Oblique crack

Figure 8. The forms of ice failure under different conditions.

3.3. Fracture Toughness Test

3.3.1. Fracture Process Curve

The fracture toughness was calculated using Formulas (4) and (5), obtaining the fracture toughness values of ice beams at different temperatures and strain rates. A typical granular snow ice-fracture curve is shown in Figure 9. The ice specimen exhibits brittle failure, and after the load reaches its peak, the test shows direct fracturing, with a decrease in bearing capacity to zero. The temperature of the ice specimen is −6 °C, and the loading rate is 0.1 mm/min.

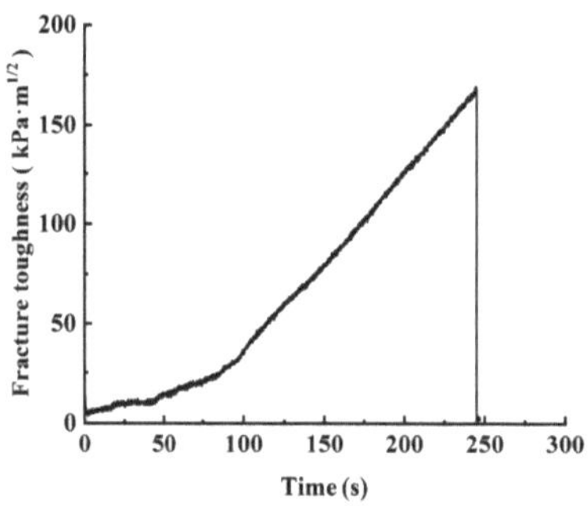

Figure 9. Time variation curve of load at −6 °C and loading rate of 0.1 mm/min.

3.3.2. The Relationship Between Fracture Toughness and Strain Rate

The relationship between ice mechanical properties and loading rate has always been a research focus of engineering ice. Fracture toughness decreases with increasing strain rate due to stress relaxation at the crack tip and material creep [33]. Figure 10 shows the fracture toughness versus strain rate for granular snow ice in the temperature range of −15 °C to −6 °C. At all four temperatures, the fracture toughness tends to decrease with increasing strain rate.

Observation of Figure 10 shows that the fracture toughness has a relatively obvious linear relationship with multiples of strain rate. According to Huang's research [53], this article performs logarithmic fitting on strain rate and fracture toughness:

$$K_{IC} = A \ln\left(\dot{\varepsilon}/\dot{\varepsilon}_0\right) + B \tag{7}$$

In the formula, K_{IC} represents the fracture toughness of granular snow ice, while A and B are parameters that are temperature-dependent parameters.

As shown in Figure 11, at −10 °C, the fracture toughness peaks at 269.93 kPa·m$^{1/2}$ when the strain rate is 8.0×10^{-6} s^{-1}, and decreases to 98.99 kPa·m$^{1/2}$ when the strain rate increases to 1.53×10^{-3} s^{-1}, representing a 63.33% decrease in fracture toughness.

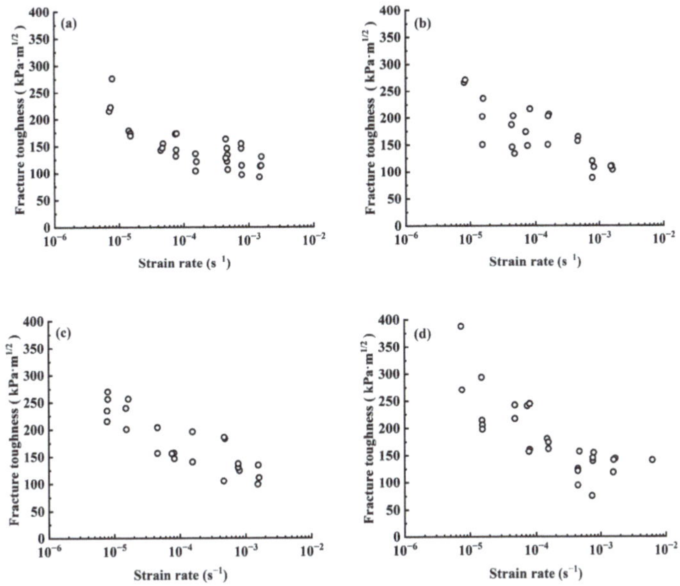

Figure 10. (**a–d**) is the relationship between fracture toughness and strain rate (10^{-6} s^{-1}~10^{-2} s^{-1}) of granular snow ice at −6 °C, −8 °C, −10 °C, −15 °C, respectively.

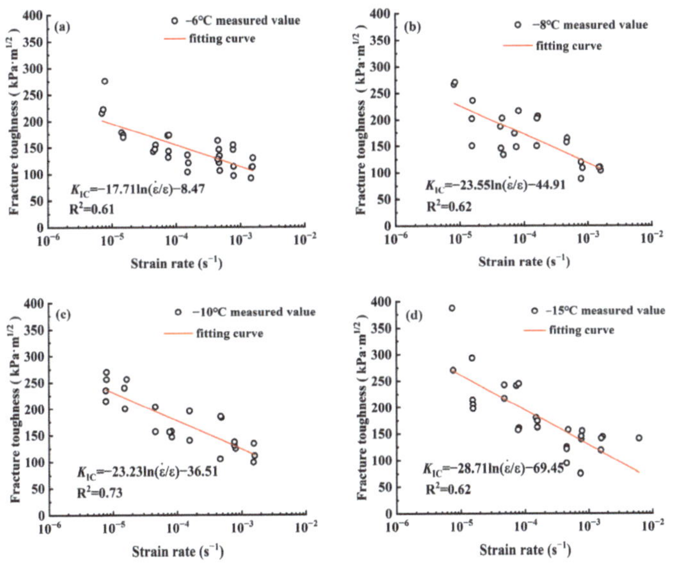

Figure 11. (**a–d**) Fit the fracture toughness and strain rate of granular snow ice at −6 °C, −8 °C, −10 °C, and −15 °C, respectively.

Xu et al. [43] conducted three-point bending tests on pure polycrystalline ice with notch at under high loading rates (1 mm/min to 100 mm/min) from −40 °C to −20 °C. They concluded that the fracture toughness of pure polycrystalline ice decreases with increasing strain rate, showing a strong power law relationship between the two. Ji et al. [40] studied the fracture toughness of sea ice and found that the loading rate significantly affects fracture toughness, with the K_{IC} value increasing as the loading rate decreases.

3.3.3. The Relationship Between Fracture Toughness and Temperature

Temperature affects the fracture toughness of ice. By analyzing and organizing data from the three-point flexural fracture test on granular snow ice, changes in fracture toughness at various temperatures were plotted. The average fracture toughness values were 181.60 kPa·m$^{1/2}$, 175.53 kPa·m$^{1/2}$, 167.50 kPa·m$^{1/2}$, and 147.28 kPa·m$^{1/2}$ at temperatures of −15 °C, −10 °C, −8 °C, and −6 °C, respectively. It can be observed that fracture toughness does not differ significantly from −15 °C to −8 °C but decreases at −6 °C.

Ji et al. [40] conducted fracture tests on sea ice in the Bohai Sea at temperatures ranging from −18 °C to −3 °C and analyzed the relationship between sea ice temperature and fracture toughness. Liu et al. [35] studied artificial columnar ice and found that fracture toughness decreased with increasing temperature in the range of −30 °C to −1 °C under a loading rate of 10 mm·s^{-1}. Table 2 presents the average fracture toughness values for the three tests at different temperatures. The fracture toughness values obtained by Liu et al. [35] are slightly lower than those reported in this study due to the lower loading rate. In contrast, the results of Ji et al. [40] are more consistent with the present study. For example, the average fracture toughness in this study at −10 °C is 175.53 kPa·m$^{1/2}$, compared to 169.0 kPa·m$^{1/2}$ reported by Ji et al. [40]. Although the test conditions resulted in differing fracture toughness values, all three tests concluded that the higher the temperature, the lower the fracture toughness.

Table 2. Average values of fracture toughness in different temperature ranges.

Temperatures (°C)	Ji [40] Fracture Toughness (kPa·m$^{1/2}$)	Liu [35] Fracture Toughness (kPa·m$^{1/2}$)	Fracture Toughness of This Test (kPa·m$^{1/2}$)
−6	131.94	109.74	147.28
−10	169.00	115.18	175.53
−15	215.31	121.97	181.60

Fracture toughness approaches 0 kPa·m$^{1/2}$ as temperature approaches 0 °C. According to Huang's research [53], logarithmic fitting was performed on the relationship between statistical data and fracture toughness with temperature:

$$K_{IC} = C + D\ln(|T/T_0|) \quad (8)$$

In the formula, to coordinate units, adjust the independent variable to T/T_0, where T_0 is a temperature of 1 °C, and C and D are parameters that are temperature-dependent.

Fit the average fracture toughness to the ice temperature. Figure 12 shows the fitting relationship between fracture toughness values and average values of granular snow ice at different temperatures. It is evident that fracture toughness decreases with increasing temperature.

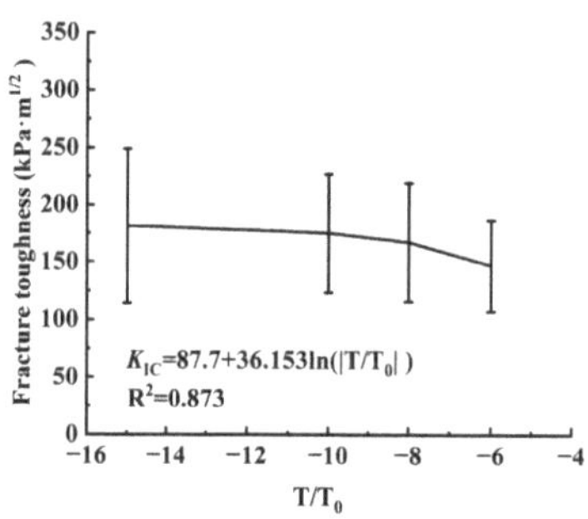

Figure 12. Fitting curves of fracture toughness values and average values of granular snow ice at different temperatures (−6 °C, −8 °C, −10 °C, −15 °C).

3.3.4. Fracture Toughness Failure Mode

In this test, the ice temperature is at −15 to −6 °C, and the displacement loading rate is between 0.05 mm/min and 10 mm/min. In accordance with Formula (3), the strain rate can be calculated. The granular snow ice exhibited brittle failure even at the lowest strain rate. When granular snow ice breaks, a slight sound is produced, and the specimen breaks into two halves. During the test, peeling may sometimes occur near the indenter and at the edge of the granular snow ice. The fracture patterns of the specimens were recorded in this test. Figure 13 shows the fracture morphology of granular snow ice at different temperatures. In the granular snow ice fracture toughness test, 102 ice samples were examined, and we found that 96 samples exhibited straight cracks that progressed upwards from the tip of the prefabricated cracks, with a relatively flat cross-section. Huang [53] conducted three-point flexural fracture tests on Yellow River ice to investigate its fracture failure modes. The results indicated that cracks predominantly developed upward along the tips of pre-existing fractures, and the crack traces were not clearly defined, which aligns with the findings of this study.

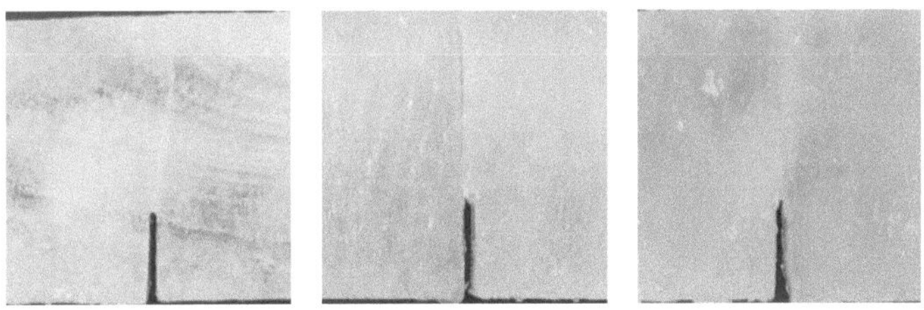

(**a**) −6 °C (**b**) −10 °C (**c**) −15 °C

Figure 13. Failure modes of granular snow ice under different temperatures.

4. Conclusions and Future Prospective

To investigate the relationship between bending strength, the fracture toughness of granular ice, temperature, and strain rate, a three-point bending device was employed for both the three-point bending test and the fracture toughness test. Since the fracture toughness cannot be determined directly from the bending test, this study cut prefabricated cracks and then determined the fracture toughness by calculating the critical stress intensity factor. This study supplemented the mechanical properties of granular ice and provided valuable insights for engineering design, construction, and ice sports in cold regions. Specifically, the key findings of this study are as follows:

1. The flexural strength of granular snow ice is influenced by the strain rate and temperature. The bending performance of ice is affected by the strain rate; within the temperature range of −35 to −5 °C, the flexural strength exhibits a trend of first increasing and then decreasing with increasing strain rate. Ice exhibits ductility at low strain rates and brittleness at high strain rates. There exists a ductile–brittle transition interval for ice, ranging from 6.67×10^{-5} s^{-1} to 3.11×10^{-4} s^{-1}. The lower the temperature, the higher the flexural strength of ice, but the more pronounced its brittle characteristics, manifesting as a lower strain rate corresponding to the maximum flexural strength. For example, at −5 °C, the strain rate corresponding to the maximum flexural value is 3.11×10^{-4} s^{-1}, while at −10 °C, the strain rate corresponding to the maximum flexural strength is only 6.67×10^{-5} s^{-1}. The flexural strength is influenced by the crystal structure: at −20 °C, the average flexural strength of granular snow ice is 2.85 MPa, while under the same temperature, the average flexural strength of columnar ice is 1.93 MPa.
2. The fracture toughness of granular snow ice is influenced by strain rate and temperature. Within the range of −15 to −6 °C, fracture toughness decreases as the strain rate increases. Temperature similarly affects fracture toughness, with higher temperatures

resulting in lower values. At −15 °C, the fracture toughness is 181.60 kPa·m$^{1/2}$, but it decreases to 147.28 kPa·m$^{1/2}$ at −6 °C.

3. In the three-point bending test, the crystal structure, temperature, and strain rate of ice do not significantly influence the fracture mode. Ice cracks can be categorized into two types: straight cracks that develop along grain boundaries and oblique cracks that form at a specific angle to the direction of tensile stress. In the three-point bending test of granular snow ice with a notch, 102 ice samples were analyzed, 96 of which exhibited vertical fractures.

In addition, there are still some limitations in this study that need to be addressed and explored in future research:

1. In practical applications, the temperature of the ice layer varies with depth. The experiment controls the sample temperature to be uniform, but the influence of temperature non-uniformity on ice mechanics parameters has not been explored. In subsequent research, theory should be combined with practice, and the bending strength and fracture toughness obtained from experiments should be used to provide support for the design and construction of ice sports projects and cold-region engineering.
2. In the fracture toughness test, we only analyzed the fracture toughness of granular ice at −15 °C to −6 °C. In the future, fracture toughness tests can be conducted on granular ice at different temperatures, and better analytical methods can be found to analyze its fracture toughness.

Author Contributions: Conceptualization, H.H. and Y.L.; methodology, H.H., W.L. and X.L.; formal analysis, Y.L. and Z.L.; investigation, H.H., W.L., Y.L., Z.L. and X.L.; data curation, H.H. and W.L.; writing—original draft preparation, H.H. and W.L.; writing—review and editing, Y.L. and X.L.; funding acquisition, H.H. and X.L. All authors have read and agreed to the published version of the manuscript.

Funding: This research was supported by the Major Scientific and Technological Projects of the Ministry of Water Resources of China (No. SKS-2022017) and the Project to Support the Development of Young Talent by Northeast Agricultural University.

Data Availability Statement: The data are available upon request.

Conflicts of Interest: The authors declare no conflicts of interest.

References

1. Mironov, E.U.; Egorova, E.S. Seasonal and interannual variations in the Greenland Sea ice age composition in the winter period. *Russ. Meteorol. Hydrol.* **2024**, *49*, 221–229. [CrossRef]
2. Zhang, Y.X.; Qiu, Y.B.; Li, Y.; Leppäranta, M.; Jia, G.; Jiang, Z.X.; Liang, W.S. Spatial-temporal variation of river ice coverage in the Yenisei river from 2002 to 2021. *J. Hydrol.* **2024**, *637*, 131440. [CrossRef]
3. Zhang, L. A Study on the Brazilian Disk Splitting Test of Antarctic Ice and Liaohe Ice. Master's Thesis, Dalian University of Technology, Dalian, China, 2022. (In Chinese)
4. Zhang, Y. Study on the Internal Structure and Surface Crack Characteristics of River Ice. Master's Thesis, Shenyang Agricultural University, Shenyang, China, 2019. (In Chinese)
5. Makkonen, L.; Tikanmäki, M.; Sainio, P. Friction in sliding heavy objects on ice. *J. Glaciol.* **2016**, *62*, 1186. [CrossRef]
6. Cheng, P.; Li, J.; Li, X. Analysis of carrying capacity of ice cover based on Winkler Model. *Yangtze River* **2017**, *48*, 64–68. (In Chinese) [CrossRef]
7. Yan, Y.; Zhang, J.; Wang, Y.; Tao, Y.; Xu, Y.; Gu, W. Spatiotemporal distribution characteristics of sea ice disasters in the Northern China Sea from 2001 to 2020. *Ocean Coast. Manag.* **2023**, *246*, 106889. [CrossRef]
8. Xiu, Y.R.; Li, Z.J.; Wang, Q.K.; Han, H.W. Experimental study on the bending mechanical behavior of granular sea ice in Liaodong Bay. *Mar. Sci. Bull.* **2023**, *42*, 667–676. (In Chinese)
9. Zhang, Y.; Qian, Z.; Lv, S.; Huang, W.; Ren, J.; Fang, Z.; Chen, X. Experimental investigation of uniaxial compressive strength of distilled water ice at different growth temperatures. *Water* **2022**, *14*, 4079. [CrossRef]
10. Long, X.; Liu, S.; Ji, S. Discrete element modelling of relationship between ice breaking length and ice load on conical structure. *Ocean Eng.* **2020**, *201*, 107152. [CrossRef]
11. Yu, J. Experimental Study on Physical and Mechanical Properties of Artificially Frozen Ice Considering Irradiation Effects. Master's Thesis, Northeast Agricultural University, Harbin, China, 2021. (In Chinese)

12. Zhang, H.B. Experimental Study on Tensile Strength and Fracture Toughness of Yellow River Ice by Using Split Test. Master's Thesis, Dalian University of Technology, Dalian, China, 2016. (In Chinese)
13. Zhang, X.P.; Chen, K.; Li, F.; Xing, H.N. Experimental study of the fracture toughness of natural freshwater ice. *J. Glaciol. Geocryol.* **2010**, *32*, 960–963. (In Chinese)
14. Weeks, W.F.; Assur, A. The mechanical properties of sea ice. In *US Army Cold Regions Research and Engineering Laboratory (CRREL) Monograph II-C3*; Cold Regions Research and Engineering Laboratory: Hanover, NH, USA, 1967.
15. Sinha, N.K. Short-term rheology of polycrystalline ice. *J. Glaciol.* **1978**, *21*, 457–474. [CrossRef]
16. Lee, R.W.; Schulson, E.M. The strength and ductility of ice under tension. *J. Offshore Mech. Arct. Eng.* **1988**, *110*, 187–191. [CrossRef]
17. Schulson, E.M.; Renshaw, C.E. Fracture, friction, and permeability of ice. *Annu. Rev. Earth Planet. Sci.* **2022**, *50*, 323–343. [CrossRef]
18. Wang, C.; Han, D.; Wang, Q.; Wang, Y.; Zhang, Y.; Jing, C. Study of elastoplastic deformation and crack evolution mechanism of single-crystal ice during uniaxial compression using 3D digital image correlation. *Eng. Fract. Mech.* **2023**, *293*, 109712. [CrossRef]
19. Chen, X.D.; Wang, A.L.; Ji, S.Y. The study on brittle-ductile transition mechanism and failure mode of sea ice under uniaxial compression. *Sci. Sin.-Phys. Mech. Astron.* **2018**, *48*, 124601. [CrossRef]
20. Zhang, Y.; Li, Z.; Guo, W.; Yu, H.; Liang, W. Method and equipment for three-pint flexural strength test of ice. *Hydro Sci. Cold Zone Eng.* **2018**, *1*, 63–68. (In Chinese)
21. Schulson, E.M. Brittle failure of ice. *Eng. Fract. Mech.* **2001**, *68*, 1839–1887. [CrossRef]
22. Schulson, E.M. An analysis of the brittle to ductile transition in polycrystalline ice under tension. *Cold Reg. Sci. Technol.* **1979**, *1*, 87–91. [CrossRef]
23. Yasui, M.; Schulson, E.M.; Renshaw, C.E. Experimental studies on mechanical properties and ductile-to-brittle transition of ice-silica mixtures: Young's modulus, compressive strength, and fracture toughness. *J. Geophys. Res. Solid Earth* **2017**, *122*, 6014–6030. [CrossRef]
24. Kellner, L.; Stender, M.; Herrnring, H.; Ehlers, S.; Hoffmann, N.; Høyland, K.V. Establishing a common database of ice experiments and using machine learning to understand and predict ice behavior. *Cold Reg. Sci. Technol.* **2019**, *162*, 56–73. [CrossRef]
25. Meng, D.; Chen, X.; Ji, S. Experimental study on the flexural strength and failure process of sea ice. *Mar. Sci. Bull.* **2021**, *40*, 609–620. (In Chinese) [CrossRef]
26. Zhang, J.; Mou, J.; Pan, Z.; Li, Y. Discussion and prospects of the development on measurement while drilling technology in oil and gas wells. *Petrol. Sci. Bull.* **2024**, *9*, 240–259. (In Chinese) [CrossRef]
27. Efimov, V.P. Experimental study on loading rate effects on the tensile strength and fracture toughness of rocks. *Geotech. Geol. Eng.* **2020**, *38*, 1–8. [CrossRef]
28. Yin, T.; Wu, Y.; Wang, C.; Zhuang, D.; Wu, B. Mixed-mode I+II tensile fracture analysis of thermally treated granite using straight-through notch Brazilian disc specimens. *Eng. Fract. Mech.* **2020**, *234*, 107111. [CrossRef]
29. Abdolghanizadeh, K.; Hosseini, M.; Saghafiyazdi, M. Effect of freezing temperature and number of freeze–thaw cycles on mode I and mode II fracture toughness of sandstone. *Theor. Appl. Fract. Mech.* **2020**, *105*, 102428. [CrossRef]
30. Xu, Y.; Yao, W.; Zhao, G.; Xia, K. Evaluation of the short core in compression (SCC) method for measuring mode II fracture toughness of rocks. *Eng. Fract. Mech.* **2020**, *224*, 106747. [CrossRef]
31. Xiao, Z. Reliability Study on the Brazilian Test Method for Determining the Mechanical Properties of Ice. Master's Thesis, Dalian University of Technology, Dalian, China, 2017. (In Chinese).
32. Han, H.W.; Jia, Q.; Huang, W.F.; Li, Z.J. Flexural strength and effective modulus of large columnar-grained freshwater ice. J. *Cold Reg. Eng.* **2016**, *30*, 04015005. [CrossRef]
33. Timco, G.W. Flexural strength and fracture toughness of urea model ice. *J. Energy Resour. Technol.* **1985**, *107*, 498–505. [CrossRef]
34. Petrovic, J.J. Review mechanical properties of ice and snow. *J. Mater. Sci.* **2003**, *38*, 1–6. [CrossRef]
35. Liu, H.W.; Miller, K.J. Fracture toughness of fresh-water ice. *J. Glaciol.* **1979**, *22*, 135–143. [CrossRef]
36. Nixon, W.A.; Schulson, E.M. The fracture toughness of ice over a range of grain sizes. *J. Offshore Mech. Arct. Eng.* **1988**, *110*, 192–196. [CrossRef]
37. Timco, G.W.; Frederking, R.M.W. Flexural strength and fracture toughness of sea ice. *Cold Reg. Sci. Technol.* **1983**, *8*, 35–41. [CrossRef]
38. Wang, L. Experimental Study on the Bending Resistance of Artificial Ice and Numerical Simulation Analysis of Artificial Ice Rinks. Master's Thesis, Harbin Institute of Technology, Harbin, China, 2021. (In Chinese).
39. Gagnon, R.E.; Gammon, P.H. Characterization and flexural strength of iceberg and glacier ice. *J. Glaciol.* **1995**, *41*, 103–111. [CrossRef]
40. Ji, S.; Liu, H.; Xu, N.; Ma, H. Experiments on sea ice fracture toughness in the Bohai Sea. *Adv. Water Sci.* **2013**, *24*, 386–391. (In Chinese) [CrossRef]
41. Zhang, A.D.; Ji, H.L.; Li, Z.J.; Zhang, B.S.; Xu, J. Experimental research on flexural strength of fresh water ice in Hongqipao Reservoir in Da-qing. *Chin. Rural Water Hydropower* **2011**, *3*, 70–72. (In Chinese)
42. Zhang, L.Z. Flexural strength and effective modules of Songhua River ice. *Adv. Mater. Res.* **2013**, *2569*, 783–793. [CrossRef]
43. Xu, X.; Jeronimidis, G.; Atkins, A.G.; Trusty, P.A. Rate-dependent fracture toughness of pure polycrystalline ice. *J. Mater. Sci.* **2004**, *39*, 225–233. [CrossRef]

44. Litwin, K.L.; Zygielbaum, B.R.; Polito, P.J.; Sklar, L.S.; Collins, G.C. Influence of temperature, composition, and grain size on the tensile failure of water ice: Implications for erosion on Titan. *J. Geophys. Res.* **2012**, *117*, E08013. [CrossRef]
45. Mulmule, S.V.; Dempsey, J.P. Scale effects on sea ice fracture. *Mech. Cohesive-Frict. Mater.* **1999**, *4*, 505–524. [CrossRef]
46. Dempsey, J.P.; Adamson, R.M.; Mulmule, S.V. Scale effects on the in-situ tensile strength and fracture of ice. Part II: First-year sea ice at Resolute, NWT. *Int. J. Fract.* **1999**, *95*, 347–366. [CrossRef]
47. Yuan, B.J. Research and Application of Artificial Ice Failure Criteria in Ice Rink. Master's Thesis, Harbin Institute of Technology, Harbin, China, 2021. (In Chinese)
48. Huang, W.F.; Li, Z.J.; Han, H.W.; Jia, Q. Seasonal evolution of static freshwater lake ice microstructures and the effects of growth processes. *J. Glaciol. Geocryol.* **2016**, *38*, 699–707. (In Chinese)
49. Schwarz, J.; Frederking, R.; Gavrillo, V.; Petrov, I.G.; Hirayama, K.I.; Mellor, M.; Tryde, P.; Vaudrey, K.D. Standardized testing methods for measuring mechanical properties of ice. *Cold Reg. Sci. Technol.* **1981**, *4*, 245–253. [CrossRef]
50. Han, H.W. Study on the Spatial and Temporal Distribution of Sea Ice and the Physical, Mechanical Properties of Sea Ice in Polar Routes. Doctoral Thesis, Dalian University of Technology, Dalian, China, 2016. (In Chinese)
51. Gold, L.W. Some observations on the dependence of strain on stress for ice. *Can. J. Phys.* **1958**, *36*, 1265–1275. [CrossRef]
52. Wei, J.; Zhu, W.C.; Li, R.F.; Niu, L.L.; Wang, Q.Y. Experiment of the tensile strength and fracture toughness of rock using notched three point bending test. *J. Water Resour. Archit. Eng.* **2016**, *14*, 128–142. (In Chinese)
53. Huang, Y. Experimental Study on the Fracture Performance of the Yellow River Ice During Freezing Period. Master's Thesis, Zhengzhou University, Zhengzhou, China, 2021. (In Chinese)
54. Han, H.W.; Li, Z.J.; Zhang, L.M.; Qing, X.X.; Zhou, Q. Experimental Study on Three Point Bending Mechanical Properties of Sea Ice in Taiping Bay, Bohai Sea. In Proceedings of the National Conference on Hydraulic and Hydraulic Informatics, Tianjin, China, 13 October 2011. (In Chinese)
55. Blanchet, D.; Abdelnour, R.; Comfort, G. Mechanical properties of first-year sea ice at Tarsiut Island. *J. Cold Reg. Eng.* **1997**, *11*, 59–83. [CrossRef]
56. Cole, D.M. The microstructure of ice and its influence on mechanical properties. *Eng. Fract. Mech.* **2001**, *68*, 1797–1822. [CrossRef]

water

Article

Air–Ice–Water Temperature and Radiation Transfer via Different Surface Coverings in Ice-Covered Qinghai Lake of the Tibetan Plateau

Ruijia Niu [1,2,3], Lijuan Wen [1,2,*], Chan Wang [1], Hong Tang [1,2,3] and Matti Leppäranta [4]

1 Key Laboratory of Cryospheric Science and Frozen Soil Engineering, Northwest Institute of Eco-Environment and Resources, Chinese Academy of Sciences, Lanzhou 730000, China
2 Qinghai Lake Comprehensive Observation and Research Station, Chinese Academy of Sciences, Gangcha 812300, China
3 University of Chinese Academy of Sciences, Beijing 100049, China
4 Institute of Atmospheric and Earth Sciences, University of Helsinki, 00014 Helsinki, Finland
* Correspondence: wlj@lzb.ac.cn

Abstract: There are numerous lakes in the Tibetan Plateau (TP) that significantly impact regional climate and aquatic ecosystems, which often freeze seasonally owing to the high altitude. However, the special warming mechanisms of lake water under ice during the frozen period are poorly understood, particularly in terms of solar radiation penetration through lake ice. The limited understanding of these processes has posed challenges to advancing lake models and improving the understanding of air–lake energy exchange during the ice-covered period. To address this, a field experiment was conducted at Qinghai Lake, the largest lake in China, in February 2022 to systematically examine thermal conditions and radiation transfer across air–ice–water interfaces. High-resolution remote sensing technologies (ultrasonic instrument and acoustic Doppler devices) were used to observe the lake surface changes, and MODIS imagery was also used to validate differences in lake surface conditions. Results showed that the water temperature under the ice warmed steadily before the ice melted. The observation period was divided into three stages based on surface condition: snow stage, sand stage, and bare ice stage. In the snow and sand stages, the lake water temperature was lower due to reduced solar radiation penetration caused by high surface reflectance (61% for 2 cm of snow) and strong absorption by 8 cm of sand (absorption-to-transmission ratio of 0.96). In contrast, during the bare ice stage, a low reflectance rate (17%) and medium absorption-to-transmission ratio (0.86) allowed 11% of solar radiation to penetrate the ice, reaching 11.70 $W \cdot m^{-2}$, which increased the water temperature across the under-ice layer, with an extinction coefficient for lake water of 0.39 ($\pm$0.03) m^{-1}. Surface coverings also significantly influenced ice temperature. During the bare ice stage, the ice exhibited the lowest average temperature and the greatest diurnal variations. This was attributed to the highest daytime radiation absorption, as indicated by a light extinction coefficient of 5.36 ($\pm$0.17) m^{-1}, combined with the absence of insulation properties at night. This study enhances understanding of the characteristics of water/ice temperature and air–ice–water solar radiation transfer through effects of different ice coverings (snow, sand, and ice) in Qinghai Lake and provides key optical radiation parameters and in situ observations for the refinement of TP lake models, especially in the ice-covered period.

Keywords: Qinghai lake; lake ice; snow cover; sand cover; attenuation coefficient

Academic Editor: Hung Tao Shen

Received: 25 November 2024
Revised: 28 December 2024
Accepted: 6 January 2025
Published: 8 January 2025

Citation: Niu, R.; Wen, L.; Wang, C.; Tang, H.; Leppäranta, M. Air–Ice–Water Temperature and Radiation Transfer via Different Surface Coverings in Ice-Covered Qinghai Lake of the Tibetan Plateau. *Water* **2025**, *17*, 142. https://doi.org/10.3390/w17020142

1. Introduction

The Tibetan Plateau (TP), often referred to as the "Asian Water Tower", is home to a vast expanse of high-altitude lakes, collectively covering over 50,000 km^2, constituting more than half of China's total lacustrine area [1–3]. These lakes, perched at elevated altitudes, are prone to seasonal freezing, with ice periods lasting from several months to as long as half a year and maximum ice thicknesses ranging from 0.58 to 0.83 m [4–11]. The formation of lake ice is not merely a natural phenomenon; it significantly influences the local climate by altering the radiation transfer from the air to the water, which in turn affects the thermal dynamics of the lakes [12,13].

The ice cover acts as a barrier to heat exchange, curtailing heat loss and evaporation, and acts as a pivotal component of the climate system on the Tibetan Plateau [14,15]. However, our current understanding of the physical processes and parameters governing lake ice formation is limited, leading to considerable uncertainty in the simulation of the plateau's lake ice period by current climate models. The complexity of these processes, which include heat transfer, the evaporation of moisture, and changes in ice thickness, is compounded by the scarcity of direct observational data due to the challenging observational conditions in the region [15,16]. Uncertainties in parameters such as the optical properties and thermal conductivity of lake ice further increase the complexity of model simulations.

Existing models face challenges in estimating key parameters such as ice thickness, sub-ice water temperatures, and the ice surface albedo, which directly affect the simulation accuracy of the lake's energy balance and the hydrological cycle process [4,15,16]. For instance, the increase in water temperature during the ice period and the rapid rise in water temperature near the ice layer towards the end of the ice melting are phenomena that current models struggle to simulate accurately [15]. Weather phenomena, including strong winds and precipitation, can significantly alter the internal radiation and thermal conditions within frozen lakes by influencing the ice cover's composition and surface properties, such as albedo and light transmittance [17,18]. The reflection of solar radiation by snowfall, in particular, can lead to a decrease in lake temperature [19].

The thermal dynamics of fully ice-covered lakes are predominantly influenced by solar radiation, which can induce melting at various levels of the ice and create heat convection cells beneath the ice [4,20]. Observations indicate that, during the ice-covered period, lake water beneath the ice undergoes a gradual warming process attributed to the penetration of solar radiation through the ice layer [21,22]. Despite the presence of ice cover, a measurable fraction of solar radiation is transmitted, directly contributing to the incremental increase in water temperature throughout the freezing season [22–24]. In contrast, widely applied lake models for Tibetan Plateau (TP) lakes, such as the Freshwater Lake Model (FLake) and Community Land Model (CLM) coupled with lake schemes in the Weather Research and Forecasting Model (WRF), generally incorporate assumptions of negligible or non-existent solar radiation penetration through lake ice after freezing [15,21]. As a result, these models are unable to accurately represent the observed thermal dynamics, particularly the sustained warming of sub-ice water during the ice-covered period. This limitation underscores the need for incorporating more realistic representations of radiation transfer processes into lake models to enhance their capability in simulating energy balance and thermal dynamics under ice-covered conditions.

Field observations on the Tibetan Plateau are challenging due to the harsh climatic conditions, which has led to a focus on lake studies during the ice-free period. The scarcity of observational data during the freezing season, especially in the remote and sparsely populated Qinghai–Tibet Plateau, limits our understanding of the complex interactions between lake water, ice, and the atmosphere. The optical characteristics of lakes, including absorptivity and attenuation coefficients during the freezing period, remain elusive, hindering the advancement of lake modeling in the region [3,7,25,26]. Current models, such as the WRF-Flake [27], Flake [28], and LAKE2.0 [29], tend to overestimate the albedo of lake ice on the Tibetan Plateau, resulting in unsatisfactory simulations of energy balance and ice phenology during the ice-covered period [30]. To further refine global circulation models, regional climate models, and numerical weather forecasting models, it is essential to improve the representation of lake features within these models [31,32]. This highlights the urgent need for a better understanding of lake ice physics and the physical principles governing the seasonal evolution of ice.

This study sought to contribute to the understanding of atmosphere–ice–water interactions in Qinghai Lake through systematic multi-layer observations. Utilizing field observations and meteorological data, the impact of different weather processes on the lake surface and the effects of different ice coverings on radiation and temperature within the system were analyzed. This study represents one of the first attempts to conduct stratified observations of lake ice energy, analyzing how different ice cover materials influence temperatures across various ice layers. The findings contribute to a deeper understanding of how ice cover affects the lake's energy balance and thermodynamic properties, offering valuable insights for refining TP lake model parameters and improving knowledge of the Qinghai Lake basin and the regional aquatic environment.

2. Materials and Methods

2.1. Study Area

The Qinghai Lake (36.53~37.25° N, 99.60~100.78° E), situated at the northeastern edge of the TP with an elevation of 3195 m, is the largest lake in China, which stretches approximately 106 km from east to west and 63 km from north to south (Figure 1a). The lake surface area is 4486.1 km^2, and the average depth is 21 m [33]. The average winter air temperature in the Qinghai Lake Basin ranges from -13.8 to -10.8 °C, with an extreme minimum temperature of approximately -33.4 °C. The lake water, with a salinity of 12.50~12.96 g·L^{-1}, is a weak alkaline solution with a pH ranging from 8.95 to 9.03. The lake's drainage basin, with topography at higher elevations and an area of 2.97 $\times$ 10^4 km^2, forms a closed inland basin [34].

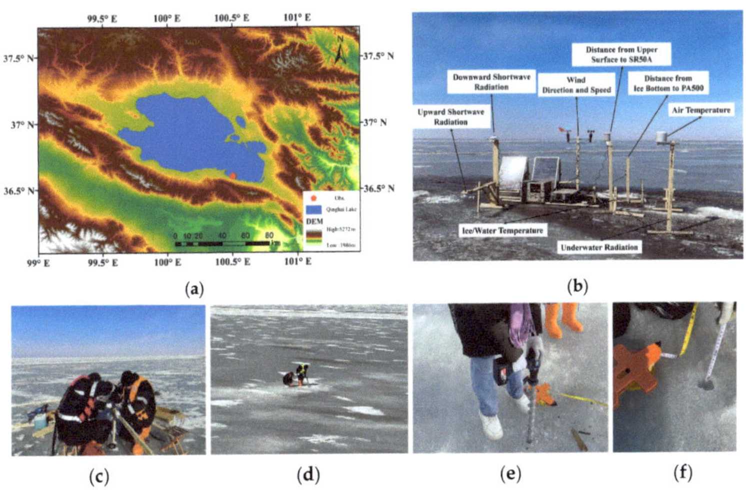

Figure 1. (**a**) Overview of Qinghai Lake, with the observation location marked by a red pentagram. (**b**) Layout of the observational instrumentation. (**c**–**f**) Instrument setup, manual snow thickness measurements, and lake ice thickness measurements via drilling.

Qinghai Lake undergoes a significant seasonal ice cover, with freezing typically commencing in December and culminating in substantial ice layers by January. These ice layers, averaging several tens of centimeters in thickness, gradually melt by March with the lake ice completely thawing by April [7]. Accompanying the ice cover, Qinghai Lake receives winter snowfall predominantly between November and February [35]. The annual precipitation in this period is variable yet generally ranges from low to moderate, influenced by the lake's geographic setting and interannual climatic variability. Surrounding the lake, the dry and semi-arid zones are occasionally subject to dust and sand events. While infrequent, these dust storms, driven by strong winds carrying sand particles over the lake, are indicative of the region's distinctive climatic characteristics, particularly during the winter months [36].

To comprehensively understand the various factors affecting the thermal balance of Qinghai Lake, we have conducted a study examining the optical properties of the lake water and ice, as well as the propagation of solar radiation through the ice layer.

2.2. In Situ Observation

To comprehensively understand the various factors affecting the thermal balance of Qinghai Lake, we conducted a field study examining the optical properties of the lake water and ice, as well as the propagation of solar radiation through the ice layer. An atmosphere–ice–water trinity observation program was conducted in Qinghai Lake in 6–24 February 2022. The observation site (Figure 1b) was located close to the shore in the Erlangjian Scenic Area of Qinghai Lake (36.59° N, 100.50° E), where the water depth is 18.5 meters. Observation data were collected with temporal resolutions of 1 minute for air temperature, wind, and radiation; 10 minutes for the ultrasonic distance meter system; and 30 minutes for underwater irradiance, as detailed in Table 1. The underwater irradiance was measured in lux but converted to $W{\cdot}m^{-2}$ here. The local noon at the site is within ±15 min of 13:30 h CST (China Standard Time; CST = UTC + 8 h). The solar radiation data with a solar elevation angle less than 15° were eliminated due to the weak solar radiation at sunrise and sunset. A practical problem in our setup was that snow accumulated around the instrument platform because of the winds. In the analysis, the lake water temperature data observed from February to April 2023 were also utilized to analyze the long-term characteristics of the water temperature during the ice cover period of Qinghai Lake.

Table 1. Introduction of observation instrument.

Observation Item	Sensor (Manufacturer)	Accuracy	Range	Height (Depth)
Temperature	PTWD (JST, Jinzhou, China)	0.2 °C	−40~80 °C	1.5 m
Wind speed	MaxiMet GMX 501 (Gill Instruments Ltd., Lymington, Hampshire, UK)	$0.1\ m{\cdot}s^{-1}$	$0.1{\sim}60\ m{\cdot}s^{-1}$	1.5 m
Global radiation	TBQ-2 (JST, Jinzhou, China)	<5%	300−3000 nm	1.5 m
Snow/sand depth (ice surface)	SR50A (Campbell Scientific, Logan, UT, USA)	0.01 cm	0.5~10 m	−0.6 m
Ice thickness (ice bottom)	Tritech PA500/6 (Tritech International Ltd., Westhill, Aberdeenshire, UK)	0.1 cm	0.1~10 m	−0.4 m
Ice temperature	PTWD (JST, Jinzhou, Liaoning, China)	<5%	−40~150 °C	−0.05, −0.10, −0.15, −0.20 m
Water temperature	PTWD (JST, Jinzhou, China)	<5%	−40~150 °C	−0.4, −0.5, −2.1, −6.7, −8.7, −12.7 m

Table 1. *Cont.*

Observation Item	Sensor (Manufacturer)	Accuracy	Range	Height (Depth)
Underwater irradiance	HOBO Pendant Temperature/Light 64K Data Logger-UA-002-64 (Onset Computer Corporation, Bourne, MA, USA)		175–1200 nm	−0.7, −2.1 m

The precipitation data were obtained from the National Meteorological Science Data Center (http://data.cma.cn/) (accessed on 10 May 2024). The dataset used was the China Surface Climate Data Daily (V3.0), which includes precipitation data from the Qinghai Lake 151 station with a temporal resolution of 1 hour. This station, the nearest ground meteorological station to the lake, is located in the southern side of the lake (36.58° N, 100.48° E) at an elevation of 3200.8 m. According to field observation, there was a thin layer of sand on ice, and the video surveillance shows a clear sand blowing.

2.3. Remote Sensing Instrumentation for Lake Surface and Ice Bottom Monitoring

In this study, two remote sensing instruments were utilized to monitor changes at the upper lake interface (covering or ice) and the ice bottom during the freezing period at Qinghai Lake. The SR50A, manufactured by Campbell Scientific (Logan, UT, USA), is an ultrasonic sensor that measures changes at the upper lake surface, which may include transitions between snow, sand, and ice. It operates by emitting high-frequency sound pulses and measuring the time it takes for the echo to return, achieving an accuracy of ±0.01 cm over a range of 0.5 to 10 m, with a deployment height of −0.6 m. This instrument's quick response time and high resolution make it particularly effective for continuous monitoring in challenging environments where traditional measurement methods may be impractical or hazardous [37,38].

In parallel, the Tritech PA500/6 (Tritech, UK) measures changes at the underwater ice bottom using acoustic Doppler technology. This instrument provides distance measurements with an accuracy of ± 0.1 cm and operates over a range of 10 to 1000 cm, with a deployment depth of −0.4 m. By analyzing the frequency shifts in the reflected sound waves from particles within the water, the PA500/6 offers detailed insights into the growth and decay of the ice bottom [39,40].

Together, these instruments measure the dynamic changes at the lake surface and the ice bottom to assess the total thickness of the lake cover and ice, contributing to a better understanding of the effects of different surface covers on ice melt.

2.4. Terra/MODIS Remote Sensing Imagery

The Moderate Resolution Imaging Spectroradiometer (MODIS), developed by NASA, was used alongside automatic weather station monitoring images to comprehensively assess the weather conditions and lake surface processes of Qinghai Lake in 6–24 February 2022. The MODIS instrument on the Terra satellite, using corrected reflectance and the Band 3-6-7 combination, provided false-color images that are particularly effective for snow and ice mapping due to the distinct reflective and absorptive properties of these features in different parts of the electromagnetic spectrum. The images, available from NASA's Earth Data site (https://wvs.earthdata.nasa.gov/) (accessed on 10 May 2024), have a spatial resolution of 250 meters and a temporal resolution of 1 day.

2.5. Methodology

2.5.1. Albedo α

The surface albedo is the ratio of the upward solar irradiance E_u (unit: W·m^{-2}) to the downward solar irradiance E_d just above the surface:

$$\alpha = \frac{E_u}{E_d} \quad (1)$$

It is an important parameter for the surface energy balance.

2.5.2. Lake Water Body Attenuation Coefficient K_{dw} and Lake Ice and Covering Attenuation Coefficient K_{di}

Infrared radiation is absorbed in a thin near-surface layer, and only photosynthetically active radiation (PAR) wavelengths (400–700 nm) are present in radiation that travels through ice [41]. Two radiation sensors were used to measure the downward radiation in the under-ice water body: sensor 1 at a depth of 0.7 m and sensor 2 at a depth of 2.1 m. By utilizing these PAR measurements, the attenuation coefficient K_{dw} (unit: m^{-1}) can be calculated using Equation (2). For a vertically optically homogeneous water body, the attenuation of radiation follows the exponential decay law [42]:

$$K_{dw} = -\frac{1}{\Delta z} ln \frac{E_d(z_{0.7})}{E_d(z_{2.1})} \quad (2)$$

Here, $\Delta z = z_{2.1} - z_{0.7}$, $E_d(z_{0.7})$ and $E_d(z_{2.1})$ (unit: W m^{-2}) represent the downward radiation at depths $z_{0.7}$ and $z_{2.1}$, respectively.

The coefficient K_{di} is determined by utilizing the radiation at the ice bottom $E_d(z_{ice-water})$ and the incident PAR on the lake surface. Based on previous research conducted at Xiaopo Lake at the eastern shore of Qinghai Lake, the average value of the PAR coefficient (ηPAR) in February in the Qinghai Lake basin is 0.42 [43]. Combining this with the thickness of the lake ice, K_{di} can be calculated.

2.5.3. Ice–Water Interface PAR $z_i(PAR)$, Euphotic Zone Depth Z_{eu}, and Lake Ice Transmittance

By utilizing the PAR from Sensor 1, the distance between the ice bottom and sensor 1, and K_{dw}, one can estimate the PAR at the ice bottom during the observation period using Equation (3), where $z_{air-ice}$ represents the depth at the air–ice interface, $z_{air-ice}$ represents the depth at the ice–water interface, $E_d(z_{air-ice})$ represents the irradiance at the air–ice interface, and $E_d(z_{ice-water})$ represents the irradiance at the ice–water interface.

$$E_d(z_{ice-water}) = E_d(z_{air-ice}) e^{K_{dw}(z_{air-ice} - z_{ice-water})} \quad (3)$$

The euphotic zone depth is defined as the depth at which the net primary production becomes zero, coincident with the depth of the layer of photosynthetic activity. The euphotic zone depth Z_{eu} is usually defined as the depth at which the irradiance is 1% of the PAR irradiance at the surface.

It can be calculated by incorporating h_i, the ice thickness, and h_w, the depth in water where irradiance reaches 1% of the surface PAR irradiance.

$$Z_{eu} = h_i + h_w \quad (4)$$

$$0.01 = \eta PAR * (1 - \alpha) e^{-(h_w K_{dw} + h_i K_{di})} \quad (5)$$

The transmittance refers to the ratio of downward radiation at the ice layer depth to the PAR on the lake surface. By using the radiation reaching the ice bottom, one can calculate the ice layer transmittance:

$$\tau = \frac{E_d(z_{ice-water})}{\eta PAR * E_d} \quad (6)$$

3. Results

3.1. Background Field

Taking into account the prevailing meteorological conditions and the variability in surface coverage, the study period was delineated into three distinct stages represented by different shades in Figure 2: the snow stage, encompassing 6–11 February; the sand stage, occurring in 13–14 February; and the bare ice stage, observed in 19–24 February. Table 2 provides a summary of major weather phenomena and lake surface features during the study period. For reference, the MODIS remote sensing images and snapshots from the automatic weather station during the observation period are shown in Figure 3.

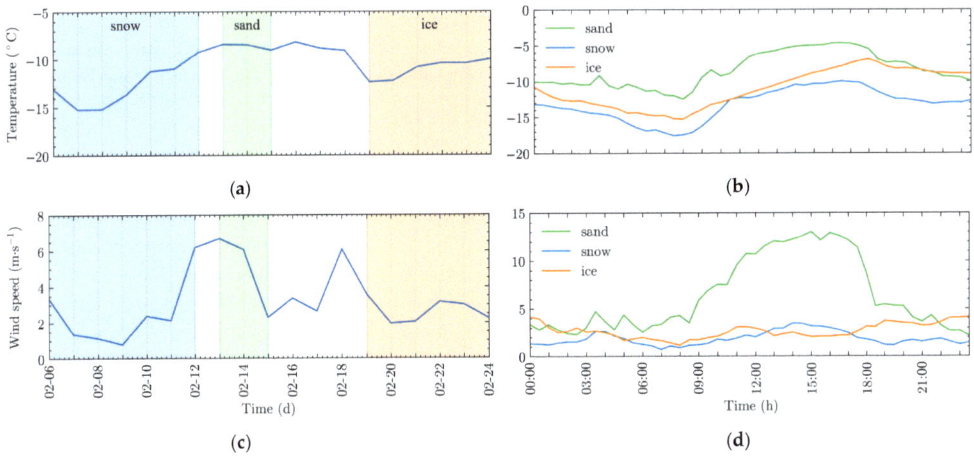

Figure 2. (**a**,**c**) Daily and (**b**,**d**) diurnal variations in (**a**,**b**) temperature and (**c**,**d**) wind speed at Qinghai Lake in 6–24 February 2022. The shaded areas in (**a**,**c**) correspond to the standard stages of lake cover: blue for snow, green for sand, and yellow for bare ice. Panels (**b**,**d**) display stage-averaged data for each variable. Note: Consistent with this article's approach, the color coding in panels (**a**,**c**) is applied across all figures to represent the three distinct stages of the lake's cover.

Table 2. Major weather phenomena and lake surface features.

Date	Weather Phenomena	Lake Surface Features
February 5–6, 10	Snow	Snow cover
February 12–14	Sand blowing	Sand cover
February 18	Strong wind	Bare ice

During the snow stage, the daily mean air temperatures dropped to about −15.23 °C on 7 February and then increased to −8.42 °C in the next stage. The meteorological observatory recorded the weather as snowfall on 5 February, with a total of 3.5 mm of snow falling during two periods: 02:00–03:00 and 23:00–06:00 in 5–6 February. A thin layer of snow forming on the ice surface was observed, with a thickness of about 2.03 cm on 5 February.

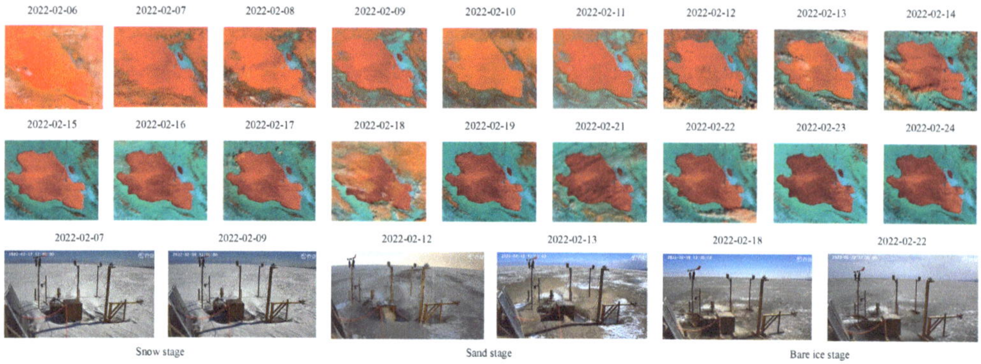

Figure 3. Terra/MODIS images during the stable freezing period of Qinghai Lake in 6–24 February 2022, along with snapshots from the automatic weather station during the snow, sand, and bare ice stages. Two images from the automatic weather station are provided for each stage. The MODIS images are shown daily, except for 20 February, which has been removed due to distortion. Red corresponds to Band 3 (459–479 nm), green corresponds to Band 6 (1628–1652 nm), and blue corresponds to Band 7 (2105–2155 nm). Red areas represent ice and snow, cyan represents exposed soil, and white indicates small liquid water droplets in clouds. The lake surface is covered by a stable frozen ice layer.

The average wind speed was 3.18 m·s^{-1} during the observation period. There were four days experiencing speeds exceeding 6.00 m·s^{-1} accompanying two significant wind events. The first wind event spanned in 12-14 February, with daily average wind speeds ranging from 6.07 to 6.69 m·s^{-1}. The intensity of these winds resulted in the deposition of fine sand particles onto the ice surface and started the sand stage. Subsequently, the second significant wind event was recorded on 18 February, characterized by a daily average wind speed of 6.09 m·s^{-1} and instantaneous winds of up to 17.7 m·s^{-1}. This event led to the dispersion of sand particles, thereby exposing the underlying ice surface. The bare ice period began, which is a common characteristic of TP lakes due to less snow than the low-altitude lakes with high latitudes.

3.2. Lake Surface-Covering Transformation and Ice Thickness

As seen in Figure 4, the distance between the underwater ultrasonic device and the ice base did not change much during the entire observation period, decreasing by only 1.95 cm in 6–24 February. This suggests that the sinking of the lake ice bottom was slow, at a rate of about 0.11 cm·d^{-1}.

The lake boundary comprises the coverings of snow or sand on the top and ice on the bottom. During the snow stage, the ultrasonic device was positioned 60.37~61.15 cm from the surface. The average thickness of ice and snow layers remained constant at 34.30~35.85 cm. The snow cover was relatively thin, around 2 cm.

In the sand stage, with sand covering the surface, the distance of the ultrasonic device from the ice to the sand surface rapidly decreased from 60.59 cm to the minimum value of 50.32 cm. There was a significant increase in the thickness of the ice and sand layer from 35.85 cm to the maximum of 47.53 cm. Wind activity resulted in a measured sand and snow layer thickness of approximately 8 cm in the observation area, where accumulation was notable due to observation structure effects causing substantial deposition. The spatial distribution of sand in the Qinghai Lake is heterogeneous due to its considerable size, leading to thin sand layers in other regions.

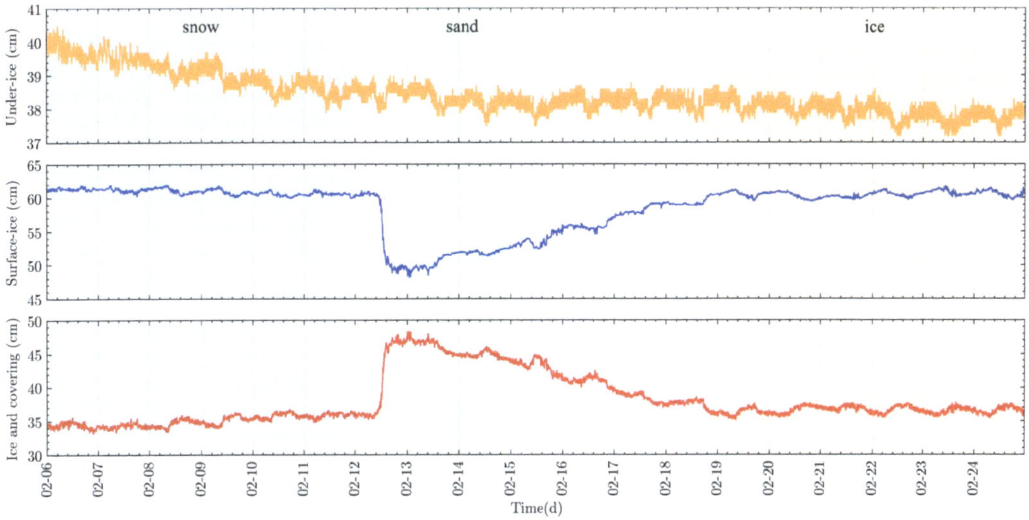

Figure 4. High-precision ultrasonic measurements of lake ice surface distances and thicknesses. The (**top**) graph depicts the distance from the sub-ice ultrasonic sensor to the underside of the ice, referred to as 'Under-ice'. The (**middle**) graph illustrates the distance from the ice surface ultrasonic sensor to the ice surface (or covering surface, if present), referred to as 'Surface-ice'. The (**bottom**) graph presents the combined thickness of the ice and any covering, measured from the top to the bottom surface, referred to as 'Ice and covering'.

In the bare ice stage, only with the bare ice, the daily average lake ice thickness decreased and stabilized at 36.36~36.89 cm. The daily fluctuation in ice thickness ranged from 2.20 cm to 2.70 cm, exhibiting the most significant diurnal variation observed during the study period. This could possibly be attributed to increased ice sublimation and deposition during the daytime and nighttime.

Based on the analysis above, the ice bottom had a minimal variation, and the ice thickness maintained about 36.6 cm during the whole observation period that happened to be the stable ice period. The thickness between the lake surface and ice bottom varied mostly because of the changed coverings.

3.3. Lake Water and Ice Temperature

3.3.1. Lake Water Temperature Under Ice

The lake water hovered around 0 °C with fluctuations not exceeding 0.54 °C (Figure 5a) during the whole observation period. The temperature of the lake water near the bottom of the ice (0.4 m or 0.5 m below the surface of lake ice) was lower than 0 °C, which was because Qinghai Lake is a saline lake (12.50 $g \cdot L^{-1}$), which has about a −0.69 °C negative freezing point [44]. The deep lake water (12.7 m) temperature remained below 0 °C until 18 February, after which it exceeded 0 °C.

The water in all layers with snow and sand coverages was colder than that with only bare ice. The mean temperature at the water depth of 0.4 m (12.7 m) below the ice surface was −0.24, −0.29, and −0.10 °C (−0.18, −0.17, and 0.15 °C) during snow, sand, and ice stages, respectively (Table 3).

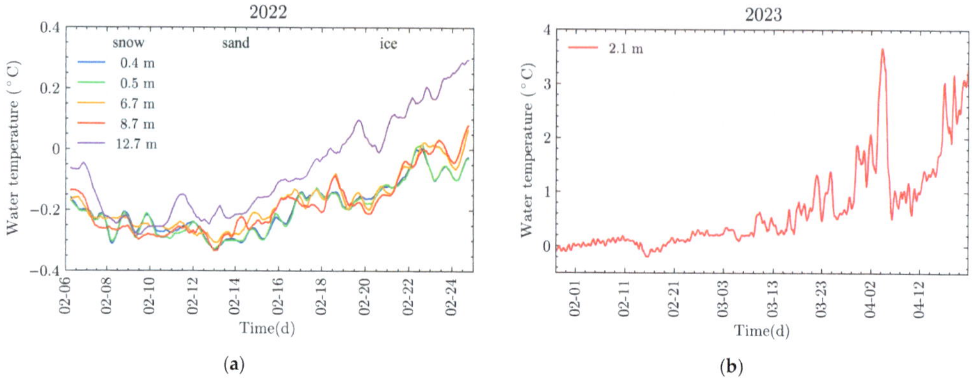

Figure 5. Temporal profiles of water temperature at various depths: (**a**) 12-hourly smoothed temperatures at 0.4 m, 0.5 m, 6.7 m, 8.7 m, and 12.7 m depths in February 2022; (**b**) 12-hourly smoothed temperatures at a depth of 2.1 m from February to April 2023, with the shaded area indicating the ice-covered period.

Table 3. Maximum, minimum, and mean of lake water temperature at different depths below the ice surface in three stages.

Depth	Max (°C)			Min (°C)			Mean (°C)		
	Snow	Sand	Ice	Snow	Sand	Ice	Snow	Sand	Ice
0.4 m	−0.11	−0.22	0.05	−0.33	−0.36	−0.22	−0.24	−0.29	−0.10
0.5 m	−0.11	−0.22	0.06	−0.32	−0.35	−0.20	−0.24	−0.29	−0.10
6.7 m	−0.12	−0.17	0.11	−0.31	−0.34	−0.23	−0.23	−0.27	−0.07
8.7 m	−0.12	−0.17	0.11	−0.31	−0.33	−0.23	−0.24	−0.27	−0.08
12.7 m	0.01	−0.14	0.36	−0.31	−0.28	−0.06	−0.18	−0.17	0.15

The water temperature was almost uniformly mixed throughout the water column in the snow and sand stages, with the vertical gradient within 0.1 °C between the shallow and deep layer. During the bare ice stage, the vertical temperature difference gradually increased to 0.37 °C.

The short duration limited the visibility of this subtle warming trend in Qinghai Lake in the February 2022 observation period (Figure 5a), but the extended observations in 2023 confirmed a gentle warming tendency throughout the ice period and a notable temperature rise of 3.87 °C before ice melting from late March to early April (Figure 5b). Qinghai Lake, characterized as a brackish body and the biggest lake in the TP and China, also exhibited similar typical characteristics with increasing seasonal under-ice water temperatures of the Tibetan Plateau's lakes. The phenomenon had been observed in TP lakes including Ngoring Lake Bangong Co, Zhari Namco, and Dagze Co, etc. [15], but the related conditions of lake ice and air–ice–water radiation transfer have not been studied or supported by the in situ observations.

3.3.2. Ice Temperature

Significant disparities in ice temperature in the three distinct stages were observed (Figure 6). The ice located 5 cm beneath the ice surface had the smallest daily minimum temperature and biggest daily maximum temperature (−10.50 and −0.40 °C) during the bare ice stage compared to that covered with snow (−5.46 and −3.00 °C) and sand (−4.41 and −2.99 °C). Thus, its mean temperature (−5.29 °C) was lower in the bare ice stage than in the other stages (−4.02, and −3.68 °C), while the diurnal ice temperature fluctuations in the bare ice stage (8.52 °C) is 5~8 times higher than the daily temperature fluctuation observed in the snow (1.58 °C) and sand (1.04 °C) stages. The deeper ice layers (10 cm, 15 cm, and 20 cm below the ice surface) shared similar stage temperature characteristics, and they were generally warmer and exhibited smaller temperature ranges (Table 4).

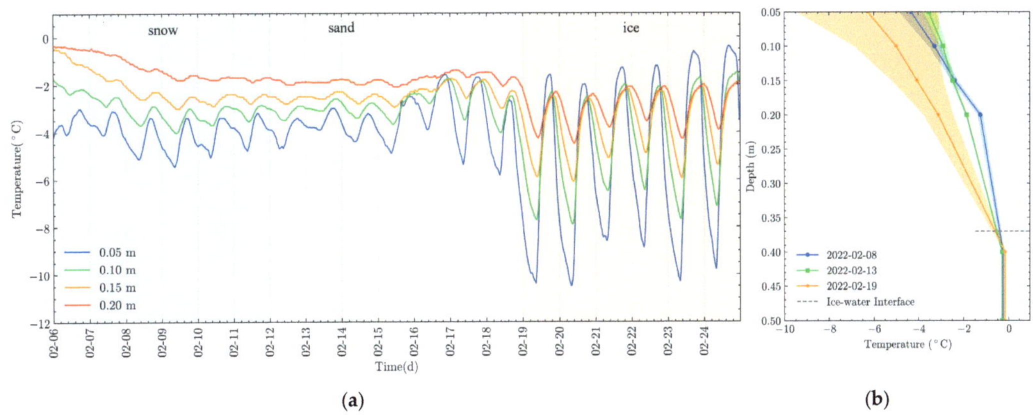

Figure 6. (**a**) Thirty-minute average lake ice temperature and (**b**) vertical temperature profile.

Table 4. Maximum, minimum, and mean of lake ice temperature at different depths below the ice surface in three stages.

Depth	Max (°C)			Min (°C)			Mean (°C)		
	Snow	Sand	Ice	Snow	Sand	Ice	Snow	Sand	Ice
5 cm	−3.00	−2.99	−0.40	−5.46	−4.41	−10.50	−4.02	−3.68	−5.29
10 cm	−1.70	−2.60	−1.51	−4.03	−3.41	−7.90	−3.00	−2.98	−4.39
15 cm	−0.42	−2.30	−1.90	−3.00	−2.80	−6.10	−2.14	−2.52	−3.69
20 cm	−0.32	−1.77	−1.90	−1.90	−2.00	−4.50	−1.25	−1.88	−2.93

During the snow, sand, and bare ice stages, the minimum values of the daily average temperatures recorded beneath the ice surface at 5 cm (−4.76, −4.16 and −9.71 °C) and 20 cm (−1.34, −1.95, and −4.05 °C) were all documented between 07:30 and 08:30. The maximum values of the daily average temperatures at these depths (5 cm: −3.26, −3.11, and −1.19 °C; 20 cm: −1.09, −1.79, and −2.16 °C) were observed between 17:00 and 18:00 during the three stages. The vertical temperature difference between the surface and deeper layers of ice was greater in the morning (−3.41, −2.20, and −5.66 °C) than in the afternoon (−2.17, −1.33, and 0.97 °C). There was a negative vertical temperature difference throughout the observation period, except during the bare ice stage in the afternoon, when a positive vertical temperature difference was observed. The vertical temperature gradient was steeper during the morning of the bare ice stage than during the other two stages.

Observational findings have unveiled the impact of lake ice and its coverings on the temperature variations across different layers of lake water and ice. However, the intrinsic physical mechanism of increasing in water temperature during the freezing period in plateau lakes remains inadequately explained. Subsequently, we will elucidate the causes of these temperature discrepancies by analyzing variations in radiation and the optical properties of both the lake water and ice.

3.3.3. Correlation Between Air Temperature and Ice/Water Temperature at Different Depths

The correlation between air temperature and ice/water temperature at various depths shows significant inconsistency (Table 5). The surface ice temperature almost immediately reflects changes in air temperature. Correlation analysis reveals a high correlation between air temperature and surface ice temperature at depths of 0.05 m and 0.10 m, with values of 0.76 and 0.74, respectively. This suggests that variations in air temperature have a significant impact on surface ice temperature, indicating the sensitivity of surface ice to environmental climate changes.

Table 5. Correlation between air temperature and ice and water temperature at different depths.

	Ice					Water			
Depth (m)	**0.05 m**	**0.10 m**	**0.15 m**	**0.20 m**	**0.4 m**	**0.5 m**	**6.7 m**	**8.7 m**	**12.7 m**
Correlation	0.76 *	0.74 *	0.66 *	0.5 *	0.13	0.14 *	0.21 *	0.16 *	0.11 *

Note: * denotes correlation is significant at the 0.001 level.

As depth increases, the correlation between ice temperature and air temperature decreases significantly. At depths of 0.15 m and 0.20 m, the correlations are 0.66 and 0.50, respectively, indicating that the effect of air temperature on deeper ice temperatures gradually diminishes. At depths of 6.7 m, 8.7 m, and 12.7 m, the correlation with water temperature further decreases to 0.21, 0.16, and 0.11, respectively. This suggests that deeper water temperatures are less influenced by air temperature due to the insulating effect, and the variations in air temperature, unlike solar radiation, are difficult to penetrate and affect a deeper water body.

3.4. Radiation and Optical Parameters

3.4.1. Downward Shortwave Radiation

Throughout the entire observation period (6–24 February 2022), as shown in Figure 7, the downward shortwave radiation reaching the lake surface exhibits a weak upward trend with no significant changes, averaging 235.82 $W{\cdot}m^{-2}$. The exceptions occurred with a decrease during the day in 10, 18 and 20 February, attributed to cloudy weather (with daily means of 191.97, 209.75, and 210.49 $W{\cdot}m^{-2}$, respectively), and sand blowing on the 12 February (daily mean of 193.72 $W{\cdot}m^{-2}$). However, the radiation level remained high during the remaining, mostly sunny observation period. It is worth noting that the snowfall on the 5-6 February occurred at night and did not impact the radiation during the daytime.

3.4.2. Upward Shortwave Radiation and Albedo

In the snow stage, the albedo of the freshly snow-covered surface was 0.68 at maximum, and the upward shortwave radiation was 159.56 $W{\cdot}m^{-2}$ on 6 February. In the case of a thin snow cover here, the underlying medium influences the albedo, resulting in the present value lower than reported for optically thick, new snow (around 0.9) [45,46]. Subsequently, due to snow metamorphosis, the albedo decreased to 0.57. Snowfall occurred on the morning of the 10 February, blanketing the lake surface with fresh snow, which possesses a higher albedo than old snow. This maintained the albedo approximately at 0.55, with the upward shortwave radiation of 124.21 $W{\cdot}m^{-2}$ on 11 February. Throughout the snow stage, the lake surface maintained high reflectivity (albedo ranging from 0.55 to 0.68) and exhibited strong upward shortwave radiation (from 110.58 to 159.56 $W{\cdot}m^{-2}$).

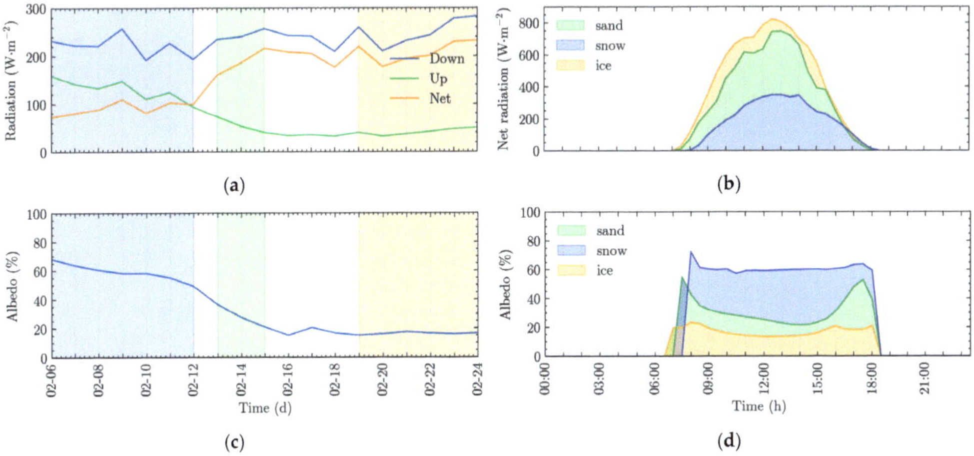

Figure 7. (**a**,**c**) Long-term trends and (**b**,**d**) daily variations in (**a**,**b**) solar shortwave radiation and (**c**,**d**) albedo. The lines in (**a**) denote downward (blue), upward (green), and net (yellow) shortwave radiation The shaded areas in (**b**,**d**) correspond to the snow (blue), sand (green), and bare ice (yellow) periods.

In the sand stage, the albedo of the sand-covered ice was low, leading to a sudden decrease from 0.49 to 0.37 in 12–13 February, while the corresponding upward shortwave radiation decreased from 93.94 to 53.54 W·m^{-2}. By the 16 February, the sand cover remained on the ice surface, and the albedo continued to decrease, stabilizing at 0.15 due to the melting and deterioration of the snow present in the sand. The daily mean upwelling shortwave radiation further declined, reaching a minimum of 32.90 W·m^{-2} on 18 February in the sand stage. The average daily upward shortwave radiation in the sand stage varied greatly (32.90~93.94 W·m^{-2}), and the albedo remained low (0.15~0.49).

During the bare ice stage, much of the particles (snow and sand) on the lake surface were dispersed by strong winds, leaving the ice surface bare. The albedo during 19–24 February at noon remains stable, with a consistent range of 0.15 to 0.18 without any notable fluctuations. The upward shortwave radiation increased slightly during the bare ice stage but still maintained a low level (32.72 to 50.78 W·m^{-2}).

3.4.3. Net Solar Shortwave Radiation

During the snow stage, the net shortwave radiation incident on the lake surface was relatively small (89.46 W·m^{-2}), due to the high reflectivity of the snow. In the sand stage, the sand absorbed a large quantity of radiation (174.00 W·m^{-2}). As the wind blew the sand and snow away, the net shortwave radiation rose quickly and eventually stabilized. Higher net radiation (209.39 W·m^{-2}) was absorbed by the ice and the under-ice water column during the bare ice stage due to the increased solar radiation intensity and decreased albedo. The net shortwave radiation incident on the lake surface during the bare ice stage was larger than that during the snow and sand stages due to the higher reflectance of snow and the snow–sand mixture compared to bare ice. The snow stage and the sand stage accounted for only 43% and 62%, respectively, of the net radiation in the bare ice stage.

3.4.4. Underwater Radiation of 0.7 m

Underwater radiation refers to solar shortwave radiation that penetrates to liquid water through the ice cover. During the snow stage, there was an increasing trend of underwater radiation at 0.7 m depth, rising from 4.94 W·m^{-2} to 7.89 W·m^{-2} during

6–9 February (Figure 8). This corresponds to the decreased surface albedo due to the aging of snow and to the increased incident radiation. The high albedo of new snow in the early morning of the 10 February resulted in a reduction in underwater radiation to 6.41 W·m^{-2} on the 10 February. Subsequently, with the aging of snow, the radiation increased to 8.10 W·m^{-2} on the 11 February.

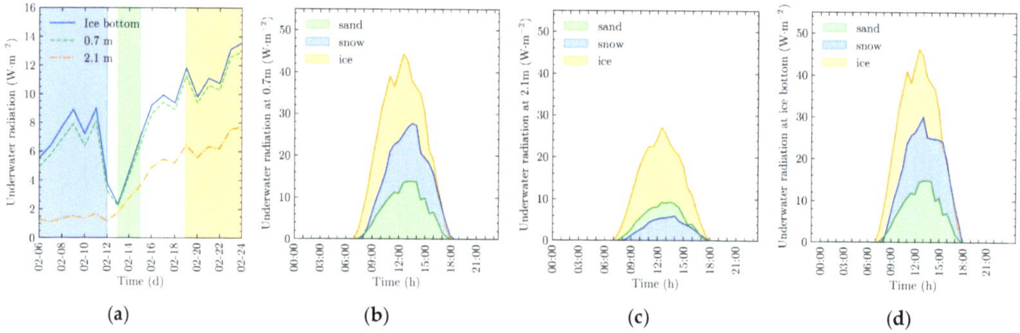

Figure 8. (**a**) Long-term trend and (**b–d**) daily variations in underwater radiation at depths of 0.7 m, 2.1 m, and the ice bottom.

The radiation level at the depth of 0.7 m substantially reduced during the sand blowing on the 12 February, reaching a minimum of 2.22 W·m^{-2} on the 13 February. The reduction in radiation was attributed to the surface coverage around the site by an approximately 8 cm thick layer of sand, decreasing the penetration of radiation through the sand layer. Thus, the sand largely changed the solar forcing of ice into a surface boundary condition rather than a distributed source term. As the wind dispersed the sand particles over the lake, the thickness of the sand layer gradually decreased. This gave rise to a continuous increase in the underwater radiation, which reached 8.62 W·m^{-2} at 0.7 m depth on the 16 February, exceeding the maximum of the snow stage. During the bare ice stage, the radiation at the depth of 0.7 m showed a gradual increase from 9.39 W·m^{-2} to 12.97 W·m^{-2} during the 20–24 February. Without the absorption of a deposited covering, the shortwave solar radiation penetrates the ice surface in large quantities.

Comparatively, the mean radiation at 0.7 m depth was significantly higher in the bare ice stage (11.70 W·m^{-2}) than during the snow cover stage (7.44 W·m^{-2}), with the sand stage recording the lowest values (3.42 W·m^{-2}). The diurnal variation in underwater radiation peaked just after local noon (12:30–13:30), with maxima recorded at 27.71, 13.91, and 44.33 W·m^{-2} for each stage, respectively. The radiation reaching the water is influenced by the thickness and optical properties of snow, sand, and ice. The underwater radiation is significantly affected by the thickness and optical attributes of the overlying snow, sand, and ice. Snow, even when thin, provides exceptional reflection, while sand absorbs considerable incoming solar radiation. The variation in underwater radiation is primarily driven by snow's high albedo and sand's absorptivity of solar shortwave radiation.

3.4.5. Underwater Radiation of 2.1 m

The mean underwater radiation at 2.1 m depth during the bare ice stage (6.63 W·m^{-2}; range: 5.58~7.71 W·m^{-2}) was higher compared to that covered with snow (average: 1.39 W·m^{-2}; range: 1.10~1.66 W·m^{-2}) and sand (average: 2.25 W·m^{-2}; range: 1.75~2.74 W·m^{-2}). The diurnal variation in underwater radiation at 2.1 m depth peaked just after local noon (12:30–14:00), with maxima recorded at 5.97, 9.33, and 27.14 W·m^{-2} for each stage, respectively (Figure 8c). During the snow stage, the radiation exhibited stable fluctuations (0.56 W·m^{-2}), while, in the sand and bare ice stages, it showed a clear increasing trend, with rates of 2.03 W·m^{-2}·d^{-1} and 0.34 W·m^{-2}·d^{-1}, respectively. The increase during the bare ice stage was due to the gradual rise in solar radiation and thinning ice, leading to more incoming radiation, whereas the rapid rise during the sand stage was primarily caused by wind thinning the surface sand, allowing quicker radiation penetration. Additionally, radiation at 2.1 m was consistently lower than that at 0.7 m in all three stages, which can be attributable to the substantial absorption of radiation by the water layer, with the respective differences of 5.27, 0.99, and 4.56 W·m^{-2}. The disparity in the magnitude underscores impacts of surface conditions on the penetration of solar radiation in the deeper water.

3.4.6. K_{dw} and Ice Bottom Radiation

The diurnal variation of K_d for both the lake water body and ice was computed using Equation (2) (Figure 9). Throughout the snow stage, the average K_{dw} was 1.17 (±0.06) m^{-1}, decreasing sharply to a minimum of 0.27 (±0.05) m^{-1} during the sand stage. In the bare ice stage, K_{dw} was a consistent average value of 0.39 (±0.03) m^{-1} for an extended period, which was lower than during the snow stage but higher than during the sand stage. There are no data of the water quality in the three stages; however, the differences in K_{dw} may be due to variations in the spectral distribution of underwater radiation. The attenuation coefficient of snow, ice, and water, along with the albedo, shows diurnal cycles from dawn to sunset, particularly under clear skies. Generally, both K_{dw} and albedo are higher in the morning and afternoon compared to the solar noon largely due to the solar elevation angle. This fluctuation is most prominent during the bare ice stage, while the snow and sand stages tend to exhibit milder variations, as the covering materials (snow and sand) block solar radiation, allowing it to penetrate into the ice layer only when the radiation is sufficiently strong. Another potential factor behind the daily variation is the melting and the consequent presence of liquid water, even in small amounts.

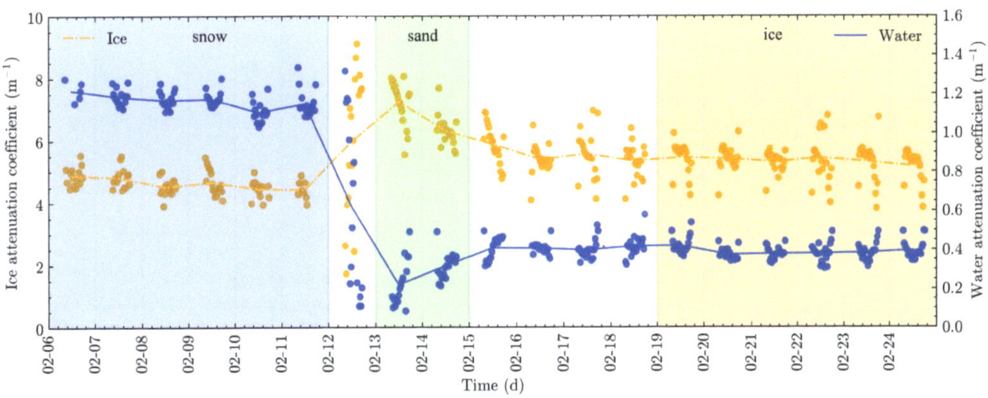

Figure 9. Temporal variation in the attenuation coefficients of the lake water (blue) and lake ice (yellow). Dots represent values at 10-minute intervals, and lines represent the daily average.

Equation (2) was utilized to calculate the radiation at the ice bottom considering the underwater radiation at 0.7 m depth, K_{dw}, and the distance from the ice bottom. The results show changes consistent with the underwater radiation level. The daily peak values of the radiation at the ice bottom are 30.01, 14.77, and 46.24 W·m^{-2} in the three stages (Figure 8d). The mean radiation is higher in the bare ice stage (11.70 W·m^{-2}) than in the snow cover (7.44 W·m^{-2}) and sand (3.42 W·m^{-2}) stages. The determined attenuation coefficient of the ice is greater than that often predicted for bare ice due to the small quantity of sand present on the ice; the analysis is unable to differentiate between the attenuation in ice and sand due to the lack of data.

3.4.7. K_{di} and Lake Ice Transmittance

Equation (3) was employed to calculate the K_d of lake ice with its coverings, considering the radiation at the ice bottom, the radiation entering the lake surface, and the thickness of the ice. In the snow stage, the K_{di} of the snow-covered ice layer is 4.63 ($\pm$0.25) m^{-1}, while the K_{di} of bare ice is 5.36 ($\pm$0.17) m^{-1}, which is higher than that of the ice–snow layer but lower than the sand-covered ice layer, which is 6.78 ($\pm$0.47) m^{-1}. Even a small amount of sand on ice will increase the K_{di}, reflecting the strong absorption characteristics of sand. This is also one reason why the K_{di} of the bare ice stage is larger than that of the snow stage.

Figure 10 illustrates the evolution of light transmittance in February together with the daily variation described by Equation (5). The total transmittance in the snow, sand, and bare ice stages are 8%, 3%, and 11%, respectively. Combined with the reflectance values of each stage, the absorption rates of the ice with the different coverings are 31%, 65%, and 72%, respectively. In terms of the daily variation, light transmission first appears at 07:00 (08:00) CST during the sand and bare ice stages (but was delayed by one hour in the snow stage). Thus, not only do the coverings influence the intensity of light transmission but also the period with underwater light is shorter, thereby reducing the heating impact of radiation. All physical property parameters of lake ice and coverings mentioned in this chapter are shown in Table 6.

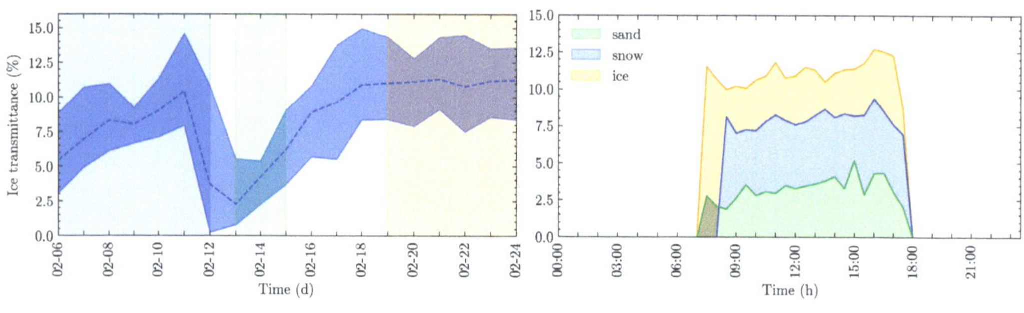

Figure 10. (**a**) Long-term trend and (**b**) daily variation in lake ice transmittance.

Table 6. Daily average of main physical parameters in each stage.

	Snow	Sand	Bare Ice
Albedo of coverings or ice	61%	32%	17%
Absorptivity of coverings and ice	31%	65%	72%
Transmittance of coverings and ice	8%	3%	11%
Absorption-to-transmission ratio (β)	0.79	0.96	0.86
K_{dw} of lake water (m^{-1})	1.17	0.27	0.39
Ice bottom radiation (W·m^{-2})	7.44	3.42	11.70
K_{di} of coverings and ice (m^{-1})	4.63	6.78	5.36
$Z_{eu}(PAR)$ (m)	-	-	11.50

4. Discussion

4.1. Mechanism of Lake Ice Temperature Variations Across Three Stages

Lake water temperature remained near the freezing point during the period of this experiment and hardly changed, indicating that heat transfer from the water body to ice was small and the lake ice heat content is not much controlled by the water layer below the ice. Meanwhile, the growth of ice at the bottom was small (1.95 cm) over the 19-day study period; thus, the latent heat released by the stage transitions was small. The transfer of strong solar radiation on the TP can be taken as the foremost factor to explain the lake ice temperature differences in the three stages (−4.02, −3.68, and −5.29 °C at a depth of 0.05 m in snow, sand, and bare ice stages, respectively).

The net solar radiation over the lake surface during the snow stage was much lower than in sand and bare ice stages (89.46, 174.00, and 209.39 W·m^{-2}, respectively) due to the higher reflectance of snow (61%) to that of sand and bare ice (32%, 17%). Furthermore, the net solar radiation was lessened more by the higher absorptivity of sand (the absorption-to-transmission ratio: 0.96) during the sand stage than in snow and bare ice stages (0.79, 0.86). Thus, as depicted in Figure 11, the net radiation penetrating from surface into water during the snow and sand stages was much lower than in the bare ice stage (7.44, 3.42, and 11.70 W·m^{-2}, respectively) owing to the strong absorptivity of sand and the strong reflectivity of snow, which was consistent with the lake ice temperature.

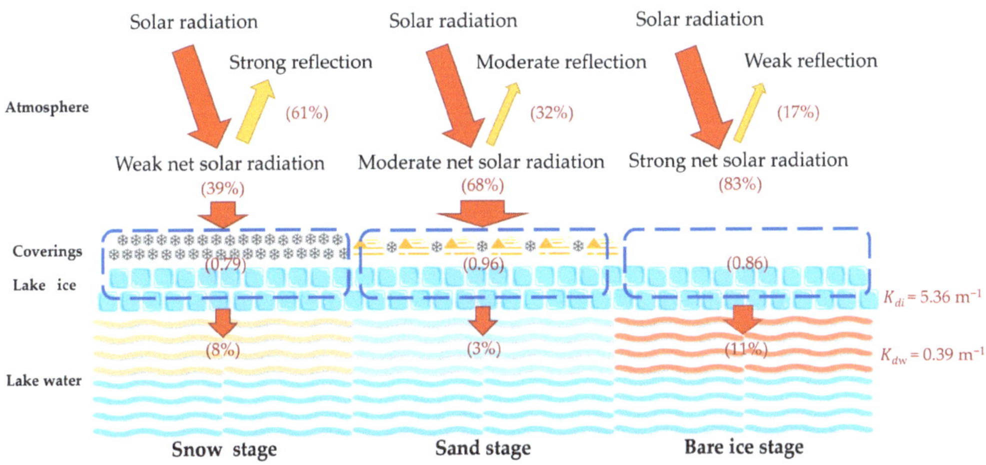

Figure 11. Schematic diagram depicting radiation transfer within the air–ice–water system of Qinghai Lake. The blue dashed box shows the absorption-to-transmission ratio.

Another noteworthy point is the diurnal variation in the three stages. The snow layer that contained air served as a blanket due to the very low thermal conductivity. Therefore, the diurnal range of ice temperature was only −5.46~−3.00 °C in the snow stage, much less compared with the bare ice stage (−10.50~−0.40 °C). The thick and high-density sand layer almost totally blocked the ice from radiation and made the ice temperature diurnal range (−4.41~−2.99 °C) even less than in the snow stage.

In the bare ice stage, the incoming solar radiation could directly reach the ice without any loss, and, at night, due to the lack of covering, the long-wave radiation loss was enhanced. This caused dramatic fluctuations of lake ice temperature and the lowest average temperature, unlike in the surface-covering stages.

4.2. Mechanism of Lake Water Temperature Variations Across Three Stages

The ice cover acts as a barrier to wind-driven momentum and prevents the wind-induced vertical mixing within the lake [17], which would be contributed by the density-driven mixing owing to the gradient of salinity and temperature of lake water. Observations showed that the salinity gradient was only 0.33 g·L^{-1}·m^{-1} and had a small influence on vertical density difference. All the observed water temperatures were below the maximum density temperature (274.43 K) [47,48]. Thus, the density would always increase with the increasing temperature warming due to solar heating and daytime convection events in interaction with the evolution of the vertical temperature–salinity distribution. Throughout the observation period, the lake's water body persistently conveyed heat to the ice at the ice/water interface, driven by the ice's lower temperature. The water maintained a relatively stable temperature, which can be partially credited to the solar radiation that compensated for the heat absorbed by the ice, sustaining the water's temperature around 0 °C, with only subtle variations observed across three distinct stages. Specifically, at a depth of 0.4 meters, the water temperature registered at −0.24, −0.29, and −0.10 °C during the snow-covered, sand-covered, and bare ice stages, respectively (Table 3). In these stages, underwater radiation predominantly dictated the thermal exchange within the lake. To elucidate these temperature discrepancies, an analysis of the ice bottom radiation was conducted to ascertain the phenomenon of energy penetration into the aqueous medium.

During the sand stage, the sand and superficial snow layer significantly reflected the incident solar radiation (32%), with the residual radiation absorbed by the intervening snow, sand, and ice layers (65%), resulting in minimal energy (3%, equivalent to 3.42 $W \cdot m^{-2}$) permeating the water body, aligning with the lowest observed lake water temperatures. In the snow-covered stage, the high reflectance of the snow (61%) led to a substantial reflection of downward radiation, coupled with absorption by the snow and ice layers (31%), thereby reducing the radiation reaching the water. Consequently, the underwater radiation during this stage was relatively diminutive, amounting to 7.44 $W \cdot m^{-2}$ (8%), which sustained the lake's lower water temperature. In contrast, during the bare ice stage, the absence of coverings allowed solar radiation to be minimally reflected by the ice surface (17%) before being substantially absorbed by the ice layer (72%), thereby heating the shallow water adjacent to the underside of the ice. Ultimately, the incident radiation penetrated the water layer with the greatest intensity (11%, equivalent to 11.70 $W \cdot m^{-2}$), with this direct heating effect and enhanced radiation absorption contributing to a comparatively elevated lake water temperature, marking the highest temperatures during the bare ice stage.

The transmittance of solar radiation through the ice layer and coverings plays a decisive role in determining the lake's water temperature. Given that the ice during the bare ice stage is not entirely transparent, it exhibits a strong absorptive capacity for solar radiation, particularly in the long-wave spectrum, with this capacity intensifying as the ice thickness increases [42]. In Qinghai Lake, with an ice thickness of approximately 36.6 cm, the transmittance rates for solar radiation are as follows: 11% for the bare ice stage, 8% for the snow-covered ice stage, and 3% for the sand-covered ice stage. These data underscore the significant impact of the ice layer and coverings' characteristics on the lake's water temperature, with the ice layer's absorptive and transmittance capabilities of solar radiation as pivotal factors influencing the lake's thermal regime.

4.3. K_{di}, K_{dw}, and Euphotic Zone Depth in the Qinghai Lake

The mean (±standard deviation) K_{di} in Qinghai Lake is 5.36 (±0.17) m^{-1} calculated with the observation data, which is higher than that in nine Estonian and Finnish boreal lakes and the brackish Santala Bay of the Baltic Sea (0.51~3.54 m^{-1}) and in the central Asian arid climate zone of the Wuliangsuhai (0.21 m^{-1}) [49–51]. Higher K_{di} in Qinghai Lake could be explained from three aspects. Lake ice contains impurities that consist of gas bubbles, liquid inclusions, and particles, which originate from the water body, bottom sediments, or atmospheric deposition [52]. The gas bubbles in the ice have a great impact on the scattering of light and in the liquid inclusions of brackish ice eventual CDOM (chromogenic dissolved organic matter), and algae can absorb light [53]. Additionally, sand particles brought by strong winds would thinly cover the wrinkled lake ice surface in the TP [27], and sand particles would be stuck in the ice by the freeze–thaw process, forming sand layers that absorb much light. Although the K_{di} in Qinghai Lake is quite big, the incident radiation is large due to its high altitude and low latitude, and sunlight does have an important role in heating the water and in providing photons for primary production beneath the ice. During the ice stable stage, with only bare ice that is the normal status for the TP lakes, there are 11.70 $W \cdot m^{-2}$ of photosynthetically active radiation penetrating into the water in Qinghai Lake, much higher than that in northern lakes with high latitude. With the consideration of penetrated solar radiation from the lake ice bottom in lake models, the originally simulated flat under-ice water temperature in TP lakes will be improved.

Compared to the rather clustered K_{di}, the K_{dw} ranges widely from 3% to 15 m^{-1}. Also, water bodies could lead to disparate levels of attenuation of solar radiation, and the euphotic zone depth varies (0.3 to 60 m), with significant discrepancies in deep water temperatures [38,49,54]. It is very necessary to have the accurate K_{dw} in lake models. According to the observed PAR during the freezing period in the Qinghai Lake, the average K_{dw} is 0.39 (±0.03) m^{-1}, and the euphotic zone depth is estimated to be 11.50 m, which is smaller than the average water depth (21 m). In fact, many lakes on the TP have a low K_{dw} of 0.10~1.17 m^{-1} (the mean is 0.26 m^{-1}) [55]. For example, the K_{dw} of Namco Lake in the northwestern region of the TP is only 0.14 m^{-1} [51]. Meanwhile, many lakes in Eastern China exhibit a relatively high K_{dw}, such as Chaohu Lake (1.56~18.01 m^{-1} in winter), West Lake (0.49~2.25 m^{-1}), Taihu Lake (2.45~10.42 m^{-1}), and Longgan Lake (0.71~3.72 m^{-1}). That is because these lakes are more eutrophic and shallower than the TP lakes. The dynamic effect of wind-induced turbulent eddies gives rise to the resuspension of inorganic particles from the sediment in shallow lakes of Eastern China, while the ice cover and large depth damp the resuspension in the Qinghai Lake, even under strong wind conditions.

5. Conclusions and Future Works

Based on the systematic field experiment on air–ice–water temperature and radiation transfer conducted during the ice-covered period of Qinghai Lake in February 2022, combined with high-resolution remote sensing technology (ultrasonic instruments, acoustic Doppler devices) and MODIS imagery to analyze changes in ice thickness and surface conditions, this study examined characteristics of water and ice temperatures, air–ice–water radiation transfer, and corresponding optical properties, as well as the effects of different coverings (snow, sand, and ice) on temperature and radiation transfer.

Common weather processes (e.g., snowfall, sand blowing, and strong winds) on the Tibetan Plateau can significantly alter the surface conditions of the ice cover. These coverings play an important role in the variations in ice and water temperatures. The mean ice temperature at 0.05 m beneath the ice surface for the three stages—snow cover, sand cover, and bare ice—was found to be −4.02, −3.68, and −5.29 °C, respectively. The daily ice temperature variation ranges were smaller in the snow and sand stages (1.58 and 1.04 °C) compared to the bare ice stage (8.52 °C). In contrast, the water temperature slightly increased around 0 °C (the water depth of 12.7 m: −0.18, −0.17, and 0.15 °C) with fluctuations not exceeding 0.54 °C during the entire observed ice-stable period. The lake water temperature at the depth of 2.1 m in ice-covered Qinghai Lake continued to increase to 3.87 °C before ice melted and was similar to most Tibetan Plateau lakes.

The different coverings (snow layer and sand layer) on ice exhibit distinct properties, dividing the incident radiation into reflected, absorbed, and transmitted components. For bare ice, the contributions of these three components were 17%, 72%, and 11%, respectively. In contrast, a thin 2 cm snow cover resulted in corresponding values of 61%, 31%, and 8%, while an 8 cm sand cover yielded values of 32%, 65%, and 3%. These differences explain why the ice (water) temperatures in the snow and sand stages were similar but influenced by different mechanisms: the high reflectivity of snow (61%) and strong absorption-to-transmission ratio of sand (0.96). During these stages, less solar radiation penetrated into the water (8% for snow and 3% for sand), resulting in lower water temperatures (−0.18, −0.17 °C). Additionally, the reduced absorbed solar radiation by ice (31% for snow and 65% for sand) limited diurnal temperature variations (1.58, 1.04 °C) due to the insulation of snow and sand. In contrast, during the bare ice stage, lake ice had the lowest temperature (−5.29 °C) and the greatest diurnal variations (8.52 °C). This was attributed to the absorption of 72% of solar radiation (171 $W \cdot m^{-2}$) by 37 cm of ice, which had a light attenuation coefficient of 5.36 (±0.17) m^{-1}.

Percentages of 8%, 3% and 11% of solar radiation penetrated into the lake water in snow, sand, and bare ice stages, respectively, which resulted in colder lake water in the first two stages. The averaged radiation of approximately 11.70 $W \cdot m^{-2}$ penetrating through the ice layer during the bare ice stage primarily contributed to the warming of lake water during this specific period. The averaged light attenuation coefficient of water K_{dw} was 0.39 (±0.03) m^{-1}, which corresponded to a euphotic zone depth of 11.50 m and influenced the special thermal conditions of water temperature.

Author Contributions: Conceptualization, R.N. and L.W.; validation, H.T. and M.L.; writing—original draft preparation, R.N.; writing—review and editing, L.W., M.L. and C.W.; visualization, R.N.; project administration, L.W.; funding acquisition, L.W. All authors have read and agreed to the published version of the manuscript.

Funding: This research was funded by the National Natural Science Foundation of China (Grant No. 42275044), the program of the Key Laboratory of Cryospheric Science and Frozen Soil Engineering, Chinese Academy of Sciences (Grant No. CSFSE-ZZ-2410; CSFSE-TZ-2405), the Chinese Academy of Sciences President's International Fellowship Initiative (Grant No. 2021VTA0007), and the Gansu Provincial Postdoctoral Funding Program (Grant No. E439880118).

Data Availability Statement: Data utilized in this research were obtained from http://www.ncdc.ac.cn/portal/organization/8078220d-03f2-49ae-b81f-ae7cb30adede (accessed on 10 May 2024).

Acknowledgments: The authors would like to thank Xiang Fu and Shuchao Chen from Dalian University of Technology for his support with the in situ measurements.

Conflicts of Interest: The authors declare no conflicts of interest. The funders had no role in the design of this study; in the collection, analyses, or interpretation of data; in the writing of the manuscript; or in the decision to publish the results.

References

1. Zhang, G.; Luo, W.; Chen, W.; Zheng, G. A Robust but Variable Lake Expansion on the Tibetan Plateau. *Sci. Bull.* **2019**, *64*, 1306–1309. [CrossRef] [PubMed]
2. Wan, W.; Long, D.; Hong, Y.; Ma, Y.; Yuan, Y.; Xiao, P.; Duan, H.; Han, Z.; Gu, X. A Lake Data Set for the Tibetan Plateau from the 1960s, 2005, and 2014. *Sci. Data* **2016**, *3*, 160039. [CrossRef] [PubMed]
3. Ma, Y.; Hu, Z.; Wang, B.; Ma, W.; Chen, X.; Han, C.; Li, M.; Zhong, L.; Sun, F.; Lai, Y.; et al. The review of the observation experiments on land-atmosphere interaction progress on the Qinghai-Xizang (Tibetan) Plateau. *Plateau Meteorol.* **2021**, *40*, 1241–1262.
4. Kirillin, G.; Leppäranta, M.; Terzhevik, A.; Granin, N.; Bernhardt, J.; Engelhardt, C.; Efremova, T.; Golosov, S.; Palshin, N.; Sherstyankin, P.; et al. Physics of Seasonally Ice-Covered Lakes: A Review. *Aquat. Sci.* **2012**, *74*, 659–682. [CrossRef]
5. George, G. Modelling the Formation and Decay of Lake Ice. In *The Impact of Climate Change on European Lakes*; Springer: Dordrecht, The Netherlands, 2010.
6. Huang, W.; Han, H.; Niu, F.; Li, R. Field observations on water temperature and stratification in a seasonally ice-covered shallow thermokarst lake. *Adv. Water Sci.* **2016**, *27*, 280–289. [CrossRef]
7. Wang, G. Lake Ice Characteristics and Changes in Qinghai Lake. Master's Thesis, Lanzhou University, Lanzhou, China, 2021. [CrossRef]

8. Chen, X.; Wang, G.; Li, W.; Zeng, Q.; Jin, D.; Wang, L. Lake ice and its remote sensing monitoring in the Tibetan Plateau. *J. Glaciol. Geocryol.* **1995**, *17*, 241–246.
9. Qu, B.; Kang, S.; Chen, F.; Zhang, Y.; Zhang, G. Lake ice and its effect factors in the Nam Co Basin, Tibetan Plateau. *Progress. Inquisitiones De Mutat. Clim.* **2012**, *8*, 18–24.
10. Kirillin, G.; Wen, L.; Shatwell, T. Seasonal Thermal Regime and Climatic Trends in Lakes of the Tibetan Highlands. *Hydrol. Earth Syst. Sci.* **2017**, *21*, 1895–1909. [CrossRef]
11. Sharma, S.; Magnuson, J.J.; Batt, R.D.; Winslow, L.A.; Korhonen, J.; Aono, Y. Direct Observations of Ice Seasonality Reveal Changes in Climate over the Past 320–570 Years. *Sci. Rep.* **2016**, *6*, 25061. [CrossRef]
12. Zhang, J. Solar Radiation Transfer in Lake Ice/Snow Cover and Its Impacts on Lake Heat Balance During Ice-Covered Period. Master's Thesis, Chang'an University, Xi'an, China, 2021. [CrossRef]
13. Song, S. Optics and Thermal Characteristics of Wuliangsuhai During the Frozen Period and the Primary Productivity in the Water Under Ice. Ph.D. Thesis, Inner Mongolia Agricultural University, Nei Monggol, China, 2020.
14. Wu, Y.; Guo, L.; Zhang, B.; Zheng, H.; Fan, L.; Chi, H.; Li, J.; Wang, S. Ice phenology dataset reconstructed from remote sensing and modelling for lakes over the Tibetan Plateau. *Sci. Data* **2022**, *9*, 743. [CrossRef]
15. Wen, L.; Lv, S.; Li, Z.; Zhao, L.; Nagabhatla, N. Impacts of the Two Biggest Lakes on Local Temperature and Precipitation in the Yellow River Source Region of the Tibetan Plateau. *Adv. Meteorol.* **2015**, *2015*, 248031. [CrossRef]
16. Wang, G.; Zhang, T.; Yang, R.; Zhong, X.; Li, X. Lake ice changes in the Third Pole and the Arctic. *J. Glaciol. Geocryol.* **2020**, *42*, 124–139. [CrossRef]
17. Duguay, C.R.; Flato, G.M.; Jeffries, M.O.; Ménard, P.; Morris, K.; Rouse, W.R. Ice-cover Variability on Shallow Lakes at High Latitudes: Model Simulations and Observations. *Hydrol. Process.* **2003**, *17*, 3465–3483. [CrossRef]
18. Wu, Q.; Li, C.; Sun, B.; Shi, X.; Zhao, S.; Han, Z. Change of ice phenology in the Hulun Lake from 1986 to 2017. *Prog. Geogr.* **2019**, *38*, 1933–1943. [CrossRef]
19. Cheng, Y.; Cheng, B.; Zheng, F.; Vihma, T.; Kontu, A.; Yang, Q.; Liao, Z. Air/Snow, Snow/Ice and Ice/Water Interfaces Detection from High-Resolution Vertical Temperature Profiles Measured by Ice Mass-Balance Buoys on an Arctic Lake. *Ann. Glaciol.* **2020**, *61*, 309–319. [CrossRef]
20. Leppäranta, M. *Freezing of Lakes and the Evolution of Their Ice Cover*, 2nd ed.; Springer: Berlin/Heidelberg, Germany, 2023.
21. Lazhu; Yang, K.; Wang, J.; Lei, Y.; Chen, Y.; Zhu, L.; Ding, B.; Qin, J. Quantifying Evaporation and Its Decadal Change for Lake Nam Co, Central Tibetan Plateau: Lake Evaporation and Its Change. *J. Geophys. Res. Atmos.* **2016**, *121*, 7578–7591. [CrossRef]
22. Lazhu; Yang, K.; Hou, J.; Wang, J.; Lei, Y.; Zhu, L.; Chen, Y.; Wang, M.; He, X. A new finding on the Prevalence of Rapid Water warming during Lake Ice Melting on the Tibetan Plateau. *Sci. Bull.* **2021**, *66*, 2358–2361. [CrossRef]
23. Kirillin, G.B.; Shatwell, T.; Wen, L. Ice-Covered Lakes of Tibetan Plateau as Solar Heat Collectors. *Geophys. Res. Lett.* **2021**, *48*, e2021GL093429. [CrossRef]
24. Wang, M.; Wen, L.; Li, Z.; Leppäranta, M.; Stepanenko, V.; Zhao, Y.; Niu, R.; Yang, L.; Kirillin, G. Mechanisms and Effects of Under-Ice Warming Water in Ngoring Lake of Qinghai–Tibet Plateau. *Cryosphere* **2022**, *16*, 3635–3648. [CrossRef]
25. Liu, Q. Research on Characteristics and Influencing Factors of Qinghai Lake Water Body Variation Based on Remote Sensing Data. Master's Thesis, China University of Geosciences, Beijing, China, 2021. [CrossRef]
26. Huang, W.; Li, R.; Han, H.; Niu, F.; Wu, Q.; Wang, W. Ice Processes and Surface Ablation in a Shallow Thermokarst Lake in the Central Qinghai–Tibetan Plateau. *Ann. Glaciol.* **2016**, *57*, 20–28. [CrossRef]
27. Li, Z.; Ao, Y.; Lyu, S.; Lang, J.; Wen, L.; Stepanenko, V.; Meng, X.; Zhao, L.I.N. Investigation of the Ice Surface Albedo in the Tibetan Plateau Lakes Based on the Field Observation and MODIS Products. *J. Glaciol.* **2018**, *64*, 506–516. [CrossRef]
28. Lang, J.; Lyu, S.; Li, Z.; Ma, Y.; Su, D. An Investigation of Ice Surface Albedo and Its Influence on the High-Altitude Lakes of the Tibetan Plateau. *Remote Sens.* **2018**, *10*, 218. [CrossRef]
29. Li, Z.; Lyu, S.; Wen, L.; Zhao, L.; Ao, Y.; Meng, X. Study of Freeze-Thaw Cycle and Key Radiation Transfer Parameters in a Tibetan Plateau Lake Using LAKE2.0 Model and Field Observations. *J. Glaciol.* **2021**, *67*, 91–106. [CrossRef]
30. Fang, N.; Yang, K.; Zhu, L.; Chen, Y.; Wang, J.; Zhu, L. Research on the Application of WRF-Lake Modeling at Nam Co Lake on the Qinghai-Tibetan Plateau. *Plateau Meteorol.* **2017**, *36*, 610–618.
31. Gou, P.; Ye, Q.; Che, T.; Feng, Q.; Ding, B.; Lin, C.; Zong, J. Lake Ice Phenology of Nam Co, Central Tibetan Plateau, China, Derived from Multiple MODIS Data Products. *J. Great Lakes Res.* **2017**, *43*, 989–998. [CrossRef]
32. Brown, L.C.; Duguay, C.R. The Response and Role of Ice Cover in Lake-Climate Interactions. *Prog. Phys. Geogr. Earth Environ.* **2010**, *34*, 671–704. [CrossRef]
33. Sun, Y.; Li, Q.; Liu, T.; Li, Y.; Yang, Y. Analysis of Qinghai Lake's water level changes and water balance analysis from 1956 to 2019. *J. China Hydrol.* **2021**, *41*, 91–96. [CrossRef]
34. Wang, Z.; Zhang, F.; Li, X.; Cao, Y.; Hu, J.; Wang, H.; Lu, H.; Li, T.; Liu, W. Changes in the Depth of Lake Qinghai since the Last Deglaciation and Asynchrony between Lake Depth and Precipitation over the Northeastern Tibetan Plateau. *Glob. Planet. Change* **2020**, *188*, 103156. [CrossRef]

35. Shen, J.; Shen, Y.; Ouyang, Z.; Guo, H.; Wang, X. Study on the Phenological Changes of Snow and Lake Ice in Qinghai Lake Basin based on MODIS Data. *Plateau Meteorol.* **2024**, *43*, 1177–1189.
36. Zhu, C.; Sun, S.; Xiao, J.; Qiao, B. Spatial-temporal pattern analysis of vegetation restoration in sandy areas of northeast Qinghai Lake. *J. Arid. Land Resour. Environ.* **2023**, *37*, 169–177. [CrossRef]
37. Jiang, L.; Xu, W.; Zhang, J.; Wang, G.; Liu, X.; Zhao, S. A Dataset for Automatically Measured Snow Depth on the Tibetan Plateau (2015–2016). *China Sci. Data* **2017**, *2*, 51–58. [CrossRef]
38. Xie, F.; Lu, P.; Leppäranta, M.; Cheng, B.; Li, Z.; Zhang, Y.; Zhang, H.; Zhou, J. Heat Budget of Lake Ice during a Complete Seasonal Cycle in Lake Hanzhang, Northeast China. *J. Hydrol.* **2023**, *620*, 129461. [CrossRef]
39. Zhao, G.; Yu, X.; Li, X.; Sun, J.; Zhang, X.; Zhang, S.; Wang, X.; Li, Z.; Cao, R.; Xu, C.; et al. Benvir: A in situ Deep-sea observation system for Benthic enviromental monitoring. *Chin. High Technol. Lett.* **2015**, *25*, 54–60.
40. Xie, F.; Lu, P.; Li, Z.; Wang, Q.; Zhang, H.; Zhang, Y. A Floating Remote Observation System (FROS) for Full Seasonal Lake Ice Evolution Studies. *Cold Reg. Sci. Technol.* **2022**, *199*, 103557. [CrossRef]
41. Cao, X. Observation and Simulation Research on Growth and Decay Processes of Ice Cover on Lake Wuliangsuhai. Ph.D. Thesis, Dalian University of Technology, Dalian, China, 2022. [CrossRef]
42. Kirk, J.T. *Light and Photosynthesis in Aquatic Ecosystems*; Cambridge University Press: Cambridge, UK, 2011.
43. Gao, L.; Zhang, L.; Chen, K.; Mao, Y. Photosynthetically Active Radiation in Alpine Wetland in the Qinghai Lake Watershed. *Arid Zone Res.* **2018**, *35*, 50–56. [CrossRef]
44. Deng, S.C.; Dong, H.L.; Lv, G.; Jiang, H.C.; Yu, B.S.; Bishop, M.E. Microbial dolomite precipitation using sulfate reducing and halophilic bacteria: Results from Qinghai Lake, Tibetan Plateau, NW China. *Chem. Geol.* **2010**, *278*, 151–159. [CrossRef]
45. Leppäranta, M.; Terzhevik, A.; Shirasawa, K. Solar Radiation and Ice Melting in Lake Vendyurskoe, Russian Karelia. *Hydrol. Res.* **2010**, *41*, 50–62. [CrossRef]
46. Xie, F.; Zhang, Y.; Lu, P.; Cao, X.; Zu, Y.; Li, Z. Characteristics and influencing factors of lake ice growth and decay in a shallow lake from a cold region. *J. Lake Sci.* **2021**, *33*, 1552–1563.
47. Caldwell, D.R. The Maximum Density Points of Pure and Saline Water. *Deep Sea Res.* **1978**, *25*, 175–181. [CrossRef]
48. Su, D.; Wen, L.; Gao, X.; Leppäranta, M.; Song, X.; Shi, Q.; Kirillin, G. Effects of the Largest Lake of the Tibetan Plateau on the Regional Climate. *J. Geophys. Res. Atmos.* **2020**, *125*, e2020JD033396. [CrossRef]
49. Arst, H.; Erm, A.; Leppäranta, M.; Reinart, A. Radiative Characteristics of Ice-Covered Fresh- and Brackish-Water Bodies. *Proc. Est. Acad. Sciences. Geol.* **2006**, *55*, 3–23. [CrossRef]
50. Cao, X.; Lu, P.; Leppäranta, M.; Arvola, L.; Huotari, J.; Shi, X.; Li, G.; Li, Z. Solar Radiation Transfer for an Ice-Covered Lake in the Central Asian Arid Climate Zone. *Inland Waters* **2021**, *11*, 89–103. [CrossRef]
51. Nima, C.; Zhuo, G. Preliminary Research of Attenuation of Radiation in the Lake Namtso. *J. Tibet. Univ.* **2012**, *27*, 11–14. [CrossRef]
52. Notaro, M.; Bennington, V.; Vavrus, S. Dynamically Downscaled Projections of Lake-Effect Snow in the Great Lakes Basin. *J. Clim.* **2015**, *28*, 1661–1684. [CrossRef]
53. Bi, R.; Zhang, H.; Li, H.; Chang, F.; Duan, L.; He, Y.; Zhang, H.; Wen, X.; Zhou, Y. Characteristics and Changes of Water Quality Parameters of Qinghai Lake in 2015. *J. Water Resour. Res.* **2018**, *7*, 74–83. [CrossRef]
54. Goldman, C.R.; Mason, D.T.; Hobbie, J.E. Two Antarctic Desert Lakes1. *Limnol. Oceanogr.* **1967**, *12*, 295–310. [CrossRef]
55. Shang, Y.; Song, K.; Jiang, P.; Ma, J.; Wen, Z.; Zhao, Y. Optical absorption properties and diffuse attenuation of photosynthetic active radiation for inland waters across the Tibetan Plateau. *J. Lake Sci.* **2018**, *30*, 802–811. [CrossRef]

water

Article

Investigation on the Excitation Force and Cavitation Evolution of an Ice-Class Propeller in Ice Blockage †

Qiaogao Huang, Sijie Zheng, Han Li *, Xing He and Xinming Li

School of Marine Science and Technology, Northwestern Polytechnical University, Xi'an 710072, China; huangqiaogao@nwpu.edu.cn (Q.H.); zhengsijie@mail.nwpu.edu.cn (S.Z.); hexing@mail.nwpu.edu.cn (X.H.); lixinming@mail.nwpu.edu.cn (X.L.)

* Correspondence: lihan_hhxy@nwpu.edu.cn

† This paper is an extended version of our paper published in Zheng, S.; Huang, Q.; Li, H.; He, X.; Li, X. Research on the Excitation Force and Cavitation Evolution of an Ice-Class Propeller in Ice Blockage Environments. In Proceedings of the 12th National Ice Engineering Conference, Zhuhai, China, 25–28 October 2024.

Abstract: When an ice-class propeller is operating in an ice-covered environment, as some ice blocks slide along the ship hull in front of the propeller blades, the inflow ahead of the propeller will become non-uniform. Consequently, the excitation force applied to the blades will increase and massive cavitation bubbles will be generated. In this paper, a hybrid Reynolds-Averaged Navier–Stokes/Large Eddy Simulation method and Schnerr–Sauer cavitation model are used to investigate the hydrodynamics, excitation force, cavitation evolution and flow field characteristics of the propeller in ice blockage conditions. The results show that the numerical method adopted has a relatively high accuracy and the hydrodynamic error is controlled within 3.0%. At low cavitation numbers, although the blockage distance decreases, the cavitation phenomenon is still severe and the hydrodynamic coefficients hardly increase accordingly. Ice blockage causes a sharp increase in cavitation. When the distance is 0.15 times the diameter, the cavitation area amounts to 20% of the propeller blades. As the advance coefficient grows, the total cavitation area diminishes, while the cavitation area of the blade behind ice does not decrease, resulting in an increment in excitation force. Ice blockage also causes backflow in the wake. At this time, the largest backflow appears at the tip of the blade behind the ice. The higher the advance coefficient, the more significant the high-pressure area of the pressure side and the greater the pressure difference, causing the excitation force to rise sharply. This work offers a positive theoretical basis for the anti-cavitation design and excitation force suppression of propellers operating in icy regions.

Keywords: propeller; ice blockage; excitation force; cavitation evolution

Academic Editor: Bommanna Krishnappan

Received: 21 November 2024
Revised: 10 January 2025
Accepted: 17 January 2025
Published: 22 January 2025

Citation: Huang, Q.; Zheng, S.; Li, H.; He, X.; Li, X. Investigation on the Excitation Force and Cavitation Evolution of an Ice-Class Propeller in Ice Blockage. *Water* **2025**, *17*, 295. https://doi.org/10.3390/w17030295

1. Introduction

The opening and utilization of Arctic Sea routes have shortened the voyage of merchant ships and promoted the development and utilization of Arctic resources. Whether a vessel is an icebreaker for opening a route or an ice-strengthened ship in the route, in an ice-covered environment, it will be affected by ice resistance [1]. However, the proportion of ice resistance is often more than half of the total resistance [2]. It is necessary for an ice-class propeller to raise its rotational speed, aiming to obtain greater thrust and maintain its ship's navigating efficiency [3]. However, since the rotational speed rises and the navigating speed maintains, the ice-class propeller will undergo heavy loads. Meanwhile, the pressure on the propeller blade drops to the saturated vapor pressure, leading to the appearance of cavitation. In particular, when an ice-blockage phenomenon occurs, the ice-blockage will intensify the cavitation phenomenon [4,5].

Due to the complexity and importance of the hydrodynamic and cavitation evolution of propellers in ice blockage and cavitation environments, some scholars carry out research in terms of experiments, theories and numerical simulations. Sampson et al. carried out model tests in ice blockage and milling environments in the Emerson cavitation tunnel, and it was confirmed that as ice and propeller collided, accompanied by an increment in cavitation bubbles, it not only hindered propulsion but also caused structure damage [6,7]. In an environment without cavitation bubbles, the ice block was close to the propeller, causing its thrust to grow by 40%. However, in an environment with serious cavitation, the change in propulsion performance was not significant when the ice–propeller distance was reduced [8]. Since the ice block moved closer to the suction surface of the blade, an induced conjoined vortex was observed, along with vortex cavitation [9,10]. Simultaneously, this led to a significant increase in the excitation force, especially as the blade behind the blockage was subjected to a heavy load and the low cavitation number brought a drastic growth in the amplitude of the pressure fluctuation in some high-order frequencies [10]. We measured the propulsion performance of an ice-class propeller, considering the effect of different ice–propeller distances, advance coefficients and cavitation numbers in the cavitation tunnel, which indicated that ice blockage intensified the cavitation phenomenon and severe cavitation will reduce the growth in thrust caused by the ice block [5].

Benefiting from the development of theoretical methods and computer technology, the assessment of the hydrodynamics and cavitation evolution of propellers according to viscous flow theories has achieved a swift development. Under an ice blockage condition with no cavitation bubbles, Wang adopted the panel method, Reynolds-Averaged Navier–Stokes equations (RANS) and model tests to research the load on the propeller surface and its hydrodynamic performance in icy conditions [11]. Moreover, the RANS numerical method and overset grid were also applied to simulate the hydrodynamics as the ice block neared the propeller and the hydrodynamics of a single blade with the circumferential position and the pressure distribution were obtained [12]. Ice blockage induced great oscillations in the thrust and torque by obstructing the incoming flow of the propeller, thereby increasing the loads on the propeller [13]. Particularly when cavitation occurred, its excitation force increased rapidly. For this reason, Sun et al. used the RANS method in order to analysis the excitation force caused by ice–propeller interaction and mainly analyzed the cavitation effect on the propeller's excitation force and its evolution process when the ice block approached the propeller [14]. The ice–propeller distance affected the propeller's propulsion and cavitation performance; this numerical method was able to effectively predict the hydrodynamics as the blockage and cavitation appeared simultaneously [15]. We also adopted the RANS method to carry out a numerical simulation study. The hydrodynamic coefficients and the cavitation phenomena were in accordance with the experimental results in the cavitation tunnel. The flow characteristics showed that the ice blockage resulted in the turning of the direction of the propeller wake, cavitation appeared for the low pressure area on the blade's suction surface, and cavitation decreased the vorticity on the suction surface of the propeller blade. Further, a hybrid method combining Reynolds-Averaged Navier–Stokes equations and Large Eddy Simulation (LES) methods was adopted to obtain detailed flow field information. The findings revealed that the ice blockage gave rise to a significant excitation force and delayed the excitation force occurrence at closer distances [16].

The model tests and theoretical and numerical studies carried out by previous scholars are all of great guiding significance for the improvement of the propulsion performance of ice-class propellers and the design of anti-cavitation performance. However, there are still deficiencies in the research on the evolution process of cavitation in an ice-blockage environment. In this work, the hybrid RANS/LES method combined with the Schnerr–Sauer cavitation model is used to explore the excitation force characteristics and the evolution process of cavitation in an ice blockage. First, the precision of the numerical method is validated through comparing numerical and experimental hydrodynamic coefficients; then, the hydrodynamic, excitation force, cavitation performance and flow field are analyzed; finally, the influencing mechanisms of the ice–propeller distance and advance coefficient on the excitation force and cavitation evolution are summarized, as well as their internal relations.

2. Numerical Theory

2.1. Governing Equations

In fluid dynamics, even in the case of multiphase flow with cavitation, the continuum hypothesis is still employed to simplify the problem, enabling the equations of continuum mechanics to be utilized for describing the fluid behavior. Under the continuum hypothesis, the physical properties of the fluid are considered as quantities that vary continuously in space, which allows for the use of partial differential equations to depict the flow motion. The Navier–Stokes equations are the fundamental sets of equations for describing fluid motion and the governing equations are presented as follows:

$$\frac{\partial \rho_m}{\partial t} + \frac{\partial(\rho_m u_j)}{\partial x_j} = 0 \quad (1)$$

$$\frac{\partial(\rho_m u_i)}{\partial t} + \frac{\partial(\rho_m u_i u_j)}{\partial x_j} = -\frac{\partial p}{\partial x_i} + \frac{\partial}{\partial x_j}\left[(\mu_m + \mu_t)\left(\frac{\partial u_i}{\partial x_j} + \frac{\partial u_j}{\partial x_i} - \frac{2}{3}\frac{\partial u_k}{\partial x_k}\delta_{ij}\right)\right] \quad (2)$$

where u_i and u_j represent velocity vectors, p stands for static pressure, μ_t denotes turbulent viscosity and δ_{ij} means the Kronecker function. ρ_m and μ_m represent the density and viscosity coefficients and $\rho_m = \rho_l\alpha_l + \rho_v\alpha_v$, $\mu_m = \mu_l\alpha_l + \mu_v\alpha_v$ and $\alpha_l + \alpha_v = 1$. m, l and v stand for the mixture flow, water and cavitation, respectively.

2.2. Turbulence Model

In this paper, the hybrid RANS/LES method is employed, aiming to solve the Navier–Stokes equations. This method combines the high precision of the LES method and the high efficiency of the RANS method and realizes the transformation between the LES and RANS methods by controlling the physical quantities. The Improved Delayed DES method (IDDES) on the basis of the hybrid RANS/LES concept is widely used in the calculations of traditional propellers [17], podded propellers [18], pump-jet propulsors [19,20] and shaftless rim-driven propulsors [21]. The IDDES expression is as follows:

$$\frac{\partial \rho_m k}{\partial t} + \nabla \cdot (\rho_m \vec{U} k) = \nabla \cdot [(\mu + \sigma_k \mu_t)\nabla k] + P_k - \rho_m \sqrt{k^3}/l_{IDDES} \quad (3)$$

$$\frac{\partial \rho_m \omega}{\partial t} + \nabla \cdot (\rho_m \vec{U} \omega) = \nabla \cdot [(\mu + \sigma_\omega \mu_t)\nabla \omega] + 2(1 - F_1)\rho_m \sigma_{\omega 2}\frac{\nabla k \cdot \nabla \omega}{\omega} + \alpha\frac{\rho_m}{\mu_t}P_k - \beta\rho_m\omega^2 \quad (4)$$

$$\mu_t = \rho_m \frac{a_1 \cdot k}{\max(a_1 \cdot \omega, F_2 \cdot S)} \quad (5)$$

where k represents turbulent kinetic energy, ω stands for turbulent dissipation rate, σ_k and σ_ω donate the turbulent Prandtl numbers, P_k represents the productions of k and ω, F_1

and F_2 stand for the SST blending functions and l_{IDDES} donates the length scale for the transition from RANS to LES, which is as follows:

$$\begin{aligned} l_{IDDES} &= \widetilde{f}_d \cdot (1 + f_e) \cdot l_{RANS} + (1 - \widetilde{f}_d) l_{LES} \\ l_{RANS} &= \sqrt{k}/(C_\mu \omega), l_{LES} = C_{DES}\Delta \\ C_{DES} &= C_{DES1} \cdot F_1 + C_{DES2} \cdot (1 - F_1) \end{aligned} \tag{6}$$

where $\widetilde{f}_d$ represents the empiric blending function and f_e stands for the elevating function. C_{DES1}, C_{DES2} and C_μ are the constants, which are 0.78, 0.61 and 0.09, respectively.

2.3. Cavitation Model

In this paper, the cavitation phenomenon is described by the Schnerr–Sauer cavitation model. The relational formula between the cavitation mass conversion rate and volume fraction is

$$\dot{m} = \frac{\rho_v \rho_l}{\rho_m} \frac{d\alpha}{dt}, \ \alpha = \frac{\frac{4}{3}\pi R_B^3 n_0}{1 + \frac{4}{3}\pi R_B^3} n_0 \tag{7}$$

where R_B represents the radius of the cavitation bubble and n_0 stands for its number. The cavitation mass change rate can be expressed as

$$\dot{m} = sign \frac{3\alpha_v(1-\alpha_v)}{R_B} \frac{\rho_l \rho_v}{\rho_m} \sqrt{\frac{2}{3} \frac{|P_v - P|}{\rho_l}} \tag{8}$$

where sign represents the sign function and P_v stands for the saturated vapor pressure.

2.4. Hydrodynamic Coefficients

For hydrodynamic performance tests of model-scale propellers, it is required to meet the similarity criteria of dimensionless coefficients stipulated by the International Towing Tank Conference (ITTC), which include

$$J = \frac{V}{nD}, \ K_T = \frac{T}{\rho n^2 D^4}, \ K_Q = \frac{Q}{\rho n^2 D^5}, \ \eta_0 = \frac{J K_T}{2\pi K_Q} \tag{9}$$

where V represents inflow velocity, n donates rotational speed, D stands for the diameter, T represents thrust, Q donates torque, J stands for the advance coefficient, K_T represents the thrust coefficient, K_Q stands for the torque coefficient and η_0 donates the open-water efficiency.

In addition, the similarity of the cavitation numbers also needs to be satisfied. Since the method of changing the inflow velocity at a fixed rotational speed is used to adjust the advance coefficient, a rotational-speed cavitation number σ_n is defined and the expression is presented as follows:

$$\sigma_n = \frac{P - P_v}{\frac{1}{2}\rho(nD)^2} \tag{10}$$

3. Numerical Strategy

3.1. Models

The research object is an ice-class propeller with 4 blades. In order to study the performance of the propeller with an ice blockage and cavitation, a model test is conducted in the cavitation tunnel of China Ship Scientific Research Center. The scale ratio of the model is 1:28 and the scaled-down model diameter D is 0.25 m. The geometric model of the propeller is depicted in Figure 1 and its main parameters are presented in Table 1. To simulate the ice-blockage environment, a block is installed in front of the propeller. The ice block has a length of 1.72D, a width of D and a height of 0.5D and the distance between it and the propeller is L. The propeller model is fixed inside the cavitation tunnel through the central propeller shaft, while the ice block is suspended and fixed through the upper fixing bracket. The change in the distance between the ice block and the propeller can be achieved by adjusting the fixing bracket. The experiment of the hydrodynamic performance is carried out by the method of fixed rotating speed and variable inflow velocity. The rotating speed of the propeller n = 35 rps. The inflow velocity and the pressure are adjusted to the advance coefficient and cavitation number. This ice-class propeller is a right-hand propeller and the definitions of the propeller rotation direction and the coordinate system are presented in Figure 1.

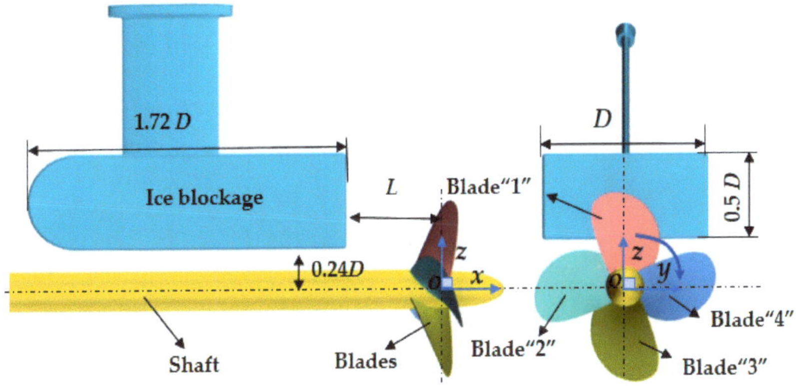

Figure 1. The geometric models of the propeller and ice block.

Table 1. The main parameters of the propeller.

Blade Number	Scale Ratio	Diameter	Pitch Ratio	Disc Ratio	Hub Diameter Ratio
Z	Λ	D/m	$(P/D)_{0.7R}$	A_E/A_0	d_h/D
4	1:28	0.25	0.84	0.75	0.21

3.2. Computational Domain

Figure 2 illustrates the computational domain and its boundary condition settings. The dimensions of the computational domain are identical to that of the cavitation tunnel, having a length of 12.8D and a diameter of 3.2D. On its left side is set the velocity inlet, while on the right side is set the pressure outlet. The distance from both sides to the center of the propeller is 6.4D and the cylindrical side is a slip wall. The boundary conditions of both the propeller and ice are non-slip walls. Given that the propeller needs to rotate within the computational domain, the computational domain is therefore divided into a stationary domain and a rotating domain and the interface between them is an internal interface. The rotating domain encloses the propeller blades and has a diameter of 1.2D.

The grid generation of the computational domain model is displayed in Figure 3. For the intricate geometry of this propeller, a cut-body grid with high stability is utilized to achieve the spatial discretization of the computational domain. In order to ensure the calculation accuracy, a greater grid concentration is provided for the propeller, the ice blockage and the wake of the propeller. The basic grid size of the propeller blade is 0.001 m and further refinement is carried out at the blade edges. The basic grid size of the ice blockage is 0.002 m and its edges are also refined. Because of the high-speed rotation of the propeller, a high-gradient boundary layer is formed around it. To ensure the accuracy of the flow calculation within the boundary layer, a prismatic layer grid is used on the propeller and the ice blockage. The height of first prismatic layer of the ice-class propeller is

2.8×10^{-5} m, the extension ratio of the prismatic layer is 1.2 and there are 16 layers in total. The height of the first prismatic layer of the ice blockage is 4.48×10^{-5} m, its extension ratio is 1.2 and there are 16 layers in total. There are a total of 1.26×10^{7} element grids in the computational domain.

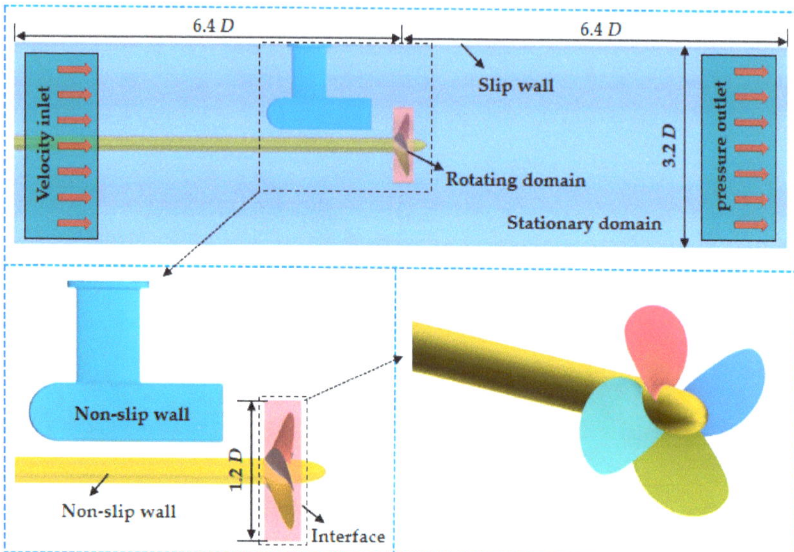

Figure 2. The computational domain and its boundary settings.

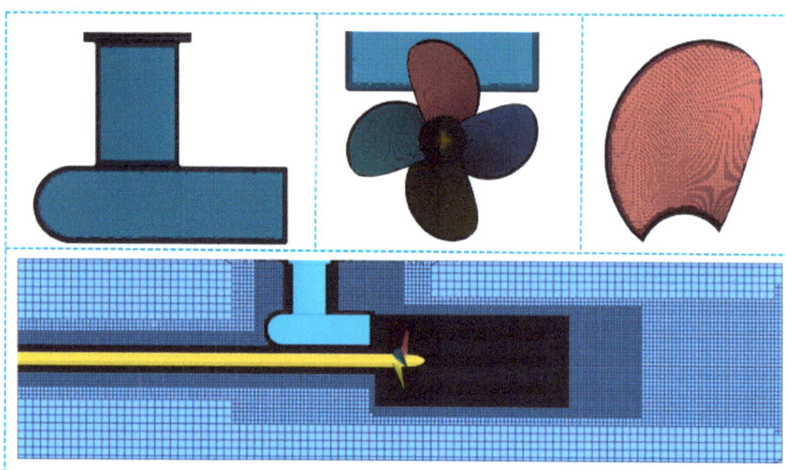

Figure 3. The grids of the computational domain.

3.3. Settings

Based on the previous work on the hydrodynamics and cavitation of the propeller within the ice-blockage environment, we further investigate the excitation force, cavitation evolution and flow field characteristics of the propeller when $\sigma_n = 1.5$, $J = 0.35$, 0.45 and 0.55 and $L/D = 0.15$ and 0.5. The Reynolds number is 2.46×10^{6}. The calculation is carried out on a 48 core workstation using STAR-CCM+ software version 2302 and a calculation duration of 120 h. The calculation is divided into two steps. Firstly, the steady-state flow field is solved based on the RANS method. The SST k-ω turbulence model is employed to close the N-S equations and the SIMPLEC numerical algorithm is utilized to solve the discrete difference equations of the velocity and pressure terms. At this moment, the moving reference frame (MRF) method is adopted to simulate the rotation of the propeller. After 1000 iteration steps, a stable flow field in a steady state is obtained. Then, the hybrid RANS/LES method is used to precisely solve the unsteady turbulent flow field, with the IDDES as the turbulence model. The Schnerr–Sauer cavitation model is introduced to simulate the cavitation. In this working condition, the propeller rotates at a constant speed of 35 rps, the rotation period T_n is 1/35 s and the inflow velocity is determined by the advance coefficient. The rotational movement of the rotor is achieved by the sliding grid. The time step is 6.25×10^{-5} s, which is equivalent to 0.7875° per time step, and the total calculation is 35 T_n.

4. Validation

Grid uncertainty validation is the evaluation and quantification of errors caused by grid division and numerical methods. This paper uses three sets of grids with different sizes; the total grid quantities are 6.8 million, 12.6 million and 22.5 million, respectively. The grid convergence index (GCI) is introduced to analyze grid errors, as shown in Table 2. ϕ represents the hydrodynamic parameter and φ_1, φ_2 and φ_3 represent the hydrodynamic values under the grid of 22.5 million, 12.6 million and 6.8 million, respectively. e_a^{21}, e_{ext}^{21} and GCI_{fine}^{21} stand for the relative error, extrapolation error and grid convergence indicator, respectively. Table 2 reveals that the values of e_a^{21}, e_{ext}^{21} and GCI_{fine}^{21} are less than 0.8%, indicating that the grid is converged and can be used for numerical calculations.

Table 2. Grid convergence validation (σ_n = 1.5, J = 0.35 and L/D = 0.15).

ϕ	φ_1	φ_2	φ_3	e_a^{21}	e_{ext}^{21}	GCI_{fine}^{21}
K_T	0.2755	0.2751	0.2746	0.145%	0.009%	0.011%
$10K_Q$	0.3732	0.3729	0.3725	0.080%	0.601%	0.756%
η_0	0.4112	0.4109	0.4106	0.073%	0.028%	0.036%

Aiming to validate the accuracy of the simulation results, a validation is made between the computational fluid dynamics (CFD) and experimental fluid dynamics (EFD) when σ_n = 1.5 and J = 0.35, as illustrated in Figure 4. It can be observed that the trends of the hydrodynamic coefficients K_T, K_Q and η_0 in CFD and EFD are in agreement and the errors are within 3.0%. As L/D rises from 0.15 to 0.5, K_T drops from 0.2744 to 0.2726 and $10K_Q$ declines from 0.3723 to 0.3654, while η_0 ascends from 0.4108 to 0.4173. The variation of the hydrodynamic coefficients does not exceed 2.0% and is relatively steady. When σ_n = 4.0, K_T and K_Q decrease rapidly as L/D increases from 0.15 to 0.25; when L/D is within the range of 0.25~0.5, K_T and K_Q gradually decrease but are always larger than the hydrodynamic coefficients as σ_n = 1.5 [5]. This indicates that a decrease in the ice–propeller distance under a higher cavitation number will lead to an increase in K_T and K_Q, while a low cavitation number will raise the cavitation on the suction surface, thereby leading to a reduction in the hydrodynamic coefficients. Particularly when L/D = 0.15, the hydrodynamic coefficients are more markedly affected by the ice blockage and cavitation performance.

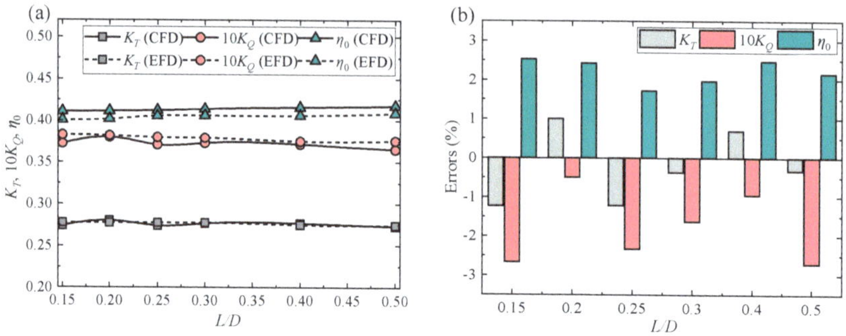

Figure 4. Validation of hydrodynamics between CFD and EFD (σ_n = 1.5 and J = 0.35): (**a**) K_T, $10K_Q$, η_0; (**b**) Errors.

5. Results and Analyses

5.1. Hydrodynamics

Figure 5 presents the comparison of the hydrodynamic coefficients K_T, K_Q and η_0 of the ice-class propeller under the ice-blockage condition at σ_n = 1.5. It can be observed that the CFD results are in line with the EFD results, having the same trends and keeping the error within 3.0%. As J rises from 0.35 to 0.55, both K_T and K_Q for L/D = 0.15 and 0.50 decrease accordingly, while η_0 increases. The rotational speed of the propeller is fixed. The increase in the advance coefficient causes the inflow velocity to increase. The increase in the inflow velocity adds the angle of attack between the propeller blade and the inflow, which in turn results in the decrease of K_T and K_Q. For L/D = 0.15, K_T drops from 0.2751 to 0.2076 and $10K_Q$ drops from 0.3729 to 0.3004; for L/D = 0.50, K_T drops from 0.2736 to 0.2078 and $10K_Q$ drops from 0.3654 to 0.2948. Because of the severe cavitation when σ_n = 1.5, the differences in K_T, K_Q and η_0 for L/D = 0.15 and 0.50 are very small and the hydrodynamic coefficients scarcely increase as the blockage distance decreases.

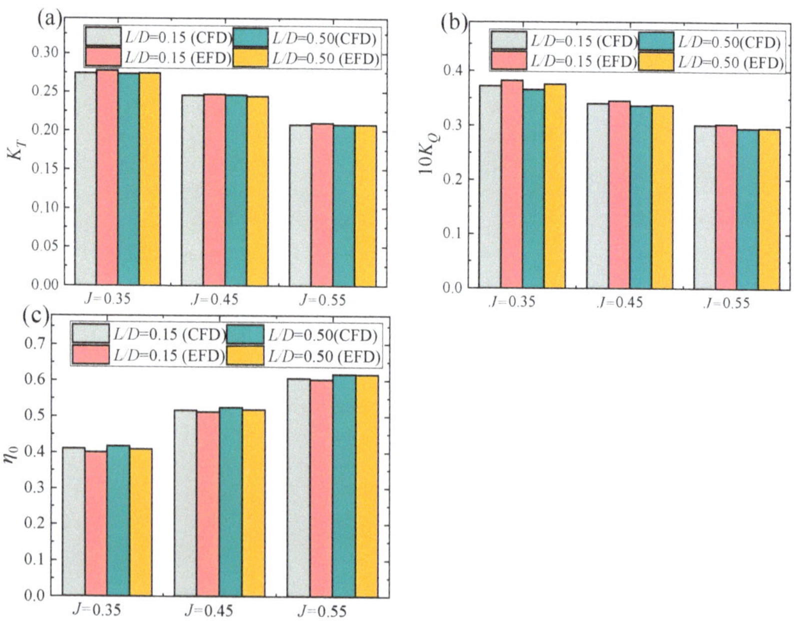

Figure 5. Comparison of hydrodynamic coefficients: (**a**) K_T; (**b**) $10K_Q$; (**c**) η_0.

5.2. Excitation Force

Figure 6 displays the time history curves of four blades' excitation force superposition within the last 7 T_n. The area below the line represents the excitation force of each blade. It can be found that the K_T exhibits periodic excitation. In each rotation period, a single propeller blade undergoes one excitation behind the ice blockage. Before entering the ice blockage's wake, the blade's excitation force rises abruptly, reaches its maximum behind the ice blockage and then decreases gradually. Owing to the effect of the distance between the ice blockage and propeller, the excitation force with L/D = 0.15 occurs later than that with L/D = 0.50 and its excitation force is more prominent. During this, the advance coefficient climbs from 0.35 to 0.55 and the average value of the K_T for a single blade decreases, while the peak value of the excitation force increases. When J = 0.35 and L/D = 0.15 and 0.5, the maximum excitation force is around 0.3. As J increases, the mean K_T decreases but its excitation force increases. When J = 0.55 and L/D = 0.15, the excitation force even reaches 0.43, as shown in Figure 6e. The excitation force at J = 0.55 and L/D = 0.50 does not exceed 0.25, as presented in Figure 6f. The propeller blade generates excitation behind the ice block and there is a phase difference in the occurrence of the excitation forces among different blades. The superposition of the excitation forces of the four blades leads to four excitations of the total excitation force within one rotation period. The greater the excitation force exerted by a single propeller blade, the larger the total excitation force.

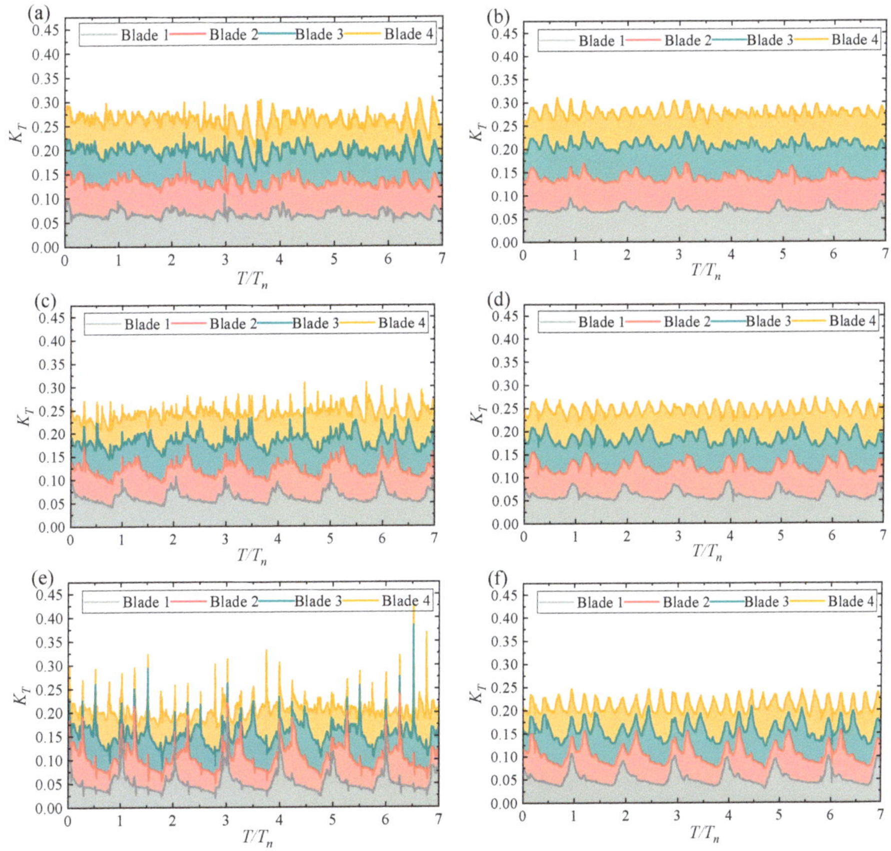

Figure 6. Time history curves of excitation force superposition: (**a**,**c**,**e**) L/D = 0.15 and J = 0.35, 0.45 and 0.55, respectively; (**b**,**d**,**f**) L/D = 0.50 and J = 0.35, 0.45 and 0.55, respectively.

5.3. Cavitation Evolution

Figure 7 presents the evolution of cavitation shapes, indicating that the cavitation on blade "1" is the most prominent. Moreover, the cavitation on blade "4", blade "3" and blade "2" decreases successively and gradually collapses. The ice blockage causes a growth in the cavitation on the blade. The smaller the L/D, the larger the cavitation coverage area will be. The evolution of the cavitation pattern in one rotation period accounts for the change in excitation force. Differing from the cavitation patterns at other propeller blades, multiple protrusions appear at the lower edge of the cavitation on blade "1", which indicates that cavitation occurs in the vortex, as showed in black circle. Owing to the blockage effect, an induced vortex is formed in its wake. When the induced vortex approaches the blade surface, the pressure in the vortex tube decreases and cavitation occurs when it is lower than the saturated vapor pressure. As the J increases, the cavitation patterns on blade "1" vary from each other, but its volume change is minimal. Meanwhile, the cavitation on blade "4", blade "3" and blade "2" decreases significantly, which reveals the mechanism of the more significant excitation forces caused by the growth of J. As the L/D increases, the blockage effect decreases, resulting in a reduction in cavitation. The decrease in cavitation at blade "1" is more evident and the reduction in cavitation volume brings about a decline in the excitation forces.

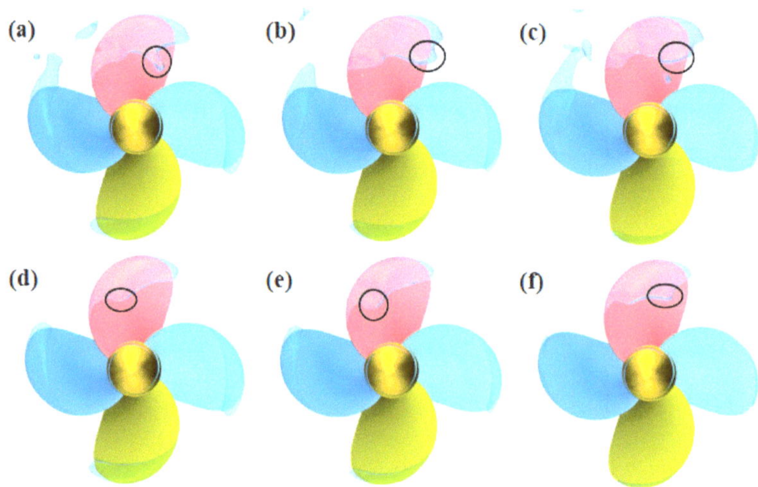

Figure 7. Evolution of cavitation shapes: (**a–c**) $L/D = 0.15$ and $J = 0.35$, 0.45 and 0.55, respectively; (**d–f**) $L/D = 0.50$ and $J = 0.35$, 0.45 and 0.55, respectively.

To explore the trend of the cavitation coverage area on the propeller blades, the fraction C_s is defined as $C_s = S_c/S$, where S_c stands for the cavitation coverage area and S is the total area of the blades. The time-history curves of the cavitation coverage areas superposition is presented in Figure 8. The area below the line represents the C_s of each blade. It can be discovered that the trends of the cavitation coverage area on the propeller blades are nearly identical to that of the excitation force. Both reach their maximum values behind the ice blockage, but the peak value of the cavitation coverage area is relatively stable. When $L/D = 0.15$ and $J = 0.35$, the C_s peak value of a single blade reaches 8.0% and the total C_s is approximately 20%. While the advance coefficient rises, the total C_s decreases. However, the C_s peak value of the blade behind the ice block remains almost the same and the change amount of the cavitation coverage area increases, which leads to a growth in the excitation force. When the distance between the ice block and propeller rises, the change in the amount of the cavitation coverage area decreases and the excitation force decreases to a certain extent. When $L/D = 0.50$ and $J = 0.35$, the total cavitation coverage area fraction fluctuates around 20.0%. When the advance coefficient grows, the larger the distance between the ice block and propeller is, the smaller the total cavitation coverage area will be, yet the peak of the cavitation coverage area of a single propeller blade is always around 7.5%.

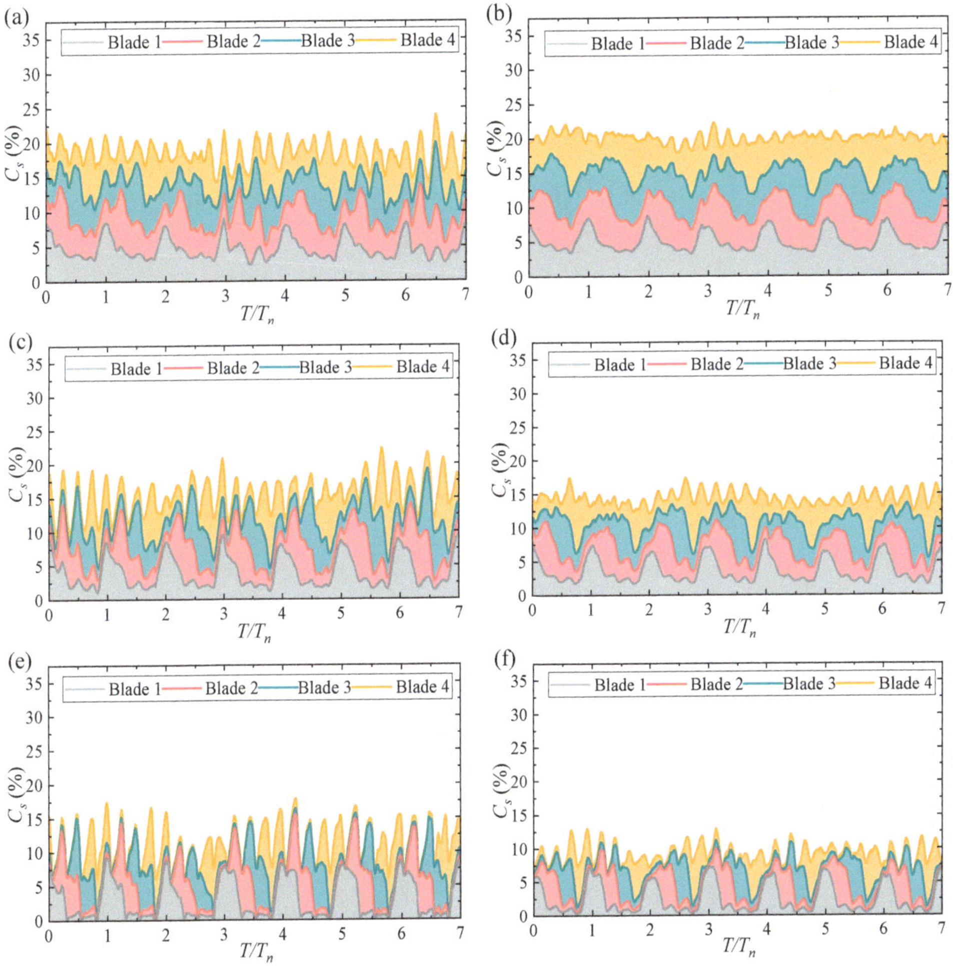

Figure 8. Time history curves of cavitation areas' superposition: (**a**,**c**,**e**) $L/D = 0.15$ and $J = 0.35$, 0.45 and 0.55, respectively; (**b**,**d**,**f**) $L/D = 0.50$ and $J = 0.35$, 0.45 and 0.55, respectively.

5.4. Flow Field Characteristics

Figure 9 illustrates the axial velocity V_x/nD in the flow field surrounding the ice-class propeller. It is obvious that the suction effect of the propeller makes the velocity of the surrounding flow increase. However, the ice blockage hinders the incoming flow, reduces the wake velocity of the ice block and gives rise to backflow, which enhances the chaos of the incoming flow to the propeller blades. When $L/D = 0.15$, the maximum backflow takes place at the blade tip behind the ice blockage and has an impact on the velocity field behind the propeller. The greater the advance coefficient is, the stronger the backflow will be and the greater the influence on the velocity field is, thereby leading to more significant excitation forces. As the distance of the ice blockage increases, the obstruction effect of the ice block on the propeller diminishes, the backflow behind the blockage weakens and the effect of the backflow on the velocity field of the ice-class propeller reduces.

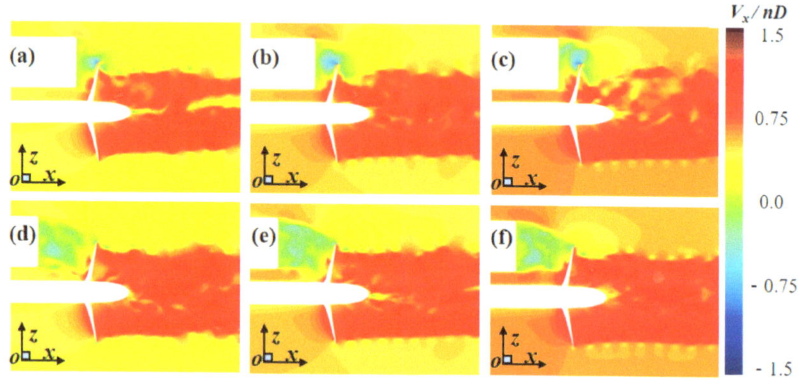

Figure 9. Axial velocity V_x/nD in flow field: (**a–c**) $L/D = 0.15$ and $J = 0.35$, 0.45 and 0.55, respectively; (**d–f**) $L/D = 0.50$ and $J = 0.35$, 0.45 and 0.55, respectively.

Figure 10 presents the pressure C_p distribution in the ice-class propeller flow field, with $C_p = (P - P_0)/0.5\rho n^2 D^2$. It can be found that the suction effect of the ice-class propeller on the flow creates a pressure difference on the propeller blades and then generates thrust. In comparison with the pressure of the unblocked propeller blade, the obstruction effect of the ice block on the incoming flow leads to an increase in both the low-pressure area and the high-pressure area on the suction surface behind the ice blockage. When $L/D = 0.15$, the low-pressure area of suction surface lies within the low-pressure area behind the blockage, which makes the low-pressure even lower and the coverage area larger; the high-pressure area on the pressure surface is influenced by the dynamic pressure of the backflow, which causes the pressure to increase further, and the larger the advance coefficient is, the greater the increase in pressure will be. As shown in Figure 10c, the pressure difference on the propeller blades is the largest, which causes the excitation force to rise sharply.

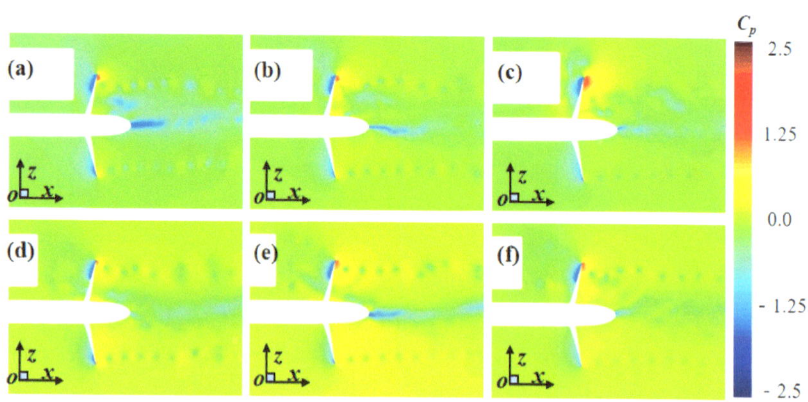

Figure 10. Pressure C_p in flow field: (**a–c**) $L/D = 0.15$ and $J = 0.35$, 0.45 and 0.55, respectively; (**d–f**) $L/D = 0.50$ and $J = 0.35$, 0.45 and 0.55, respectively.

Figure 11 presents the C_p on the suction and pressure sides of the blade, which discloses the mechanism of cavitation evolution on the blade. The suction effect of the blade results in a low-pressure area on its suction side and a high-pressure area on its pressure side, and thrust is generated under the pressure difference. When the pressure on the suction side is lower than the saturation vapor pressure, water vaporizes to form cavitation. Under the influence of ice blockage, the low-pressure area on the suction side of blade "1" becomes even lower and the distribution range of this low-pressure area is broader. At the same time, the leading edge pressure on the pressure side of blade "1" rises and a low-pressure area emerges at the top of the trailing edge due to the impact of the suction side. When J increases, the low-pressure ranges on the suction sides of blades "2", "3" and "4" decrease, thereby reducing the cavitation area. However, the low-pressure range on the suction side of blade 1 scarcely changes, leading to a rapid development of cavitation here. The increase in J makes the blade subject to a spatially non-uniformly distributed pressure load, which gives rise to severe cavitation evolution and thus generates an excitation force. Moreover, the increase in J also causes a sharp rise in the leading-edge pressure on the pressure side of blade "1", as shown in Figure 11e. Additionally, when $L/D = 0.15$, the low-pressure area on the suction side also extends to the pressure side, making the blade load more complicated. As L/D increases, the influence of the ice blockage decreases accordingly.

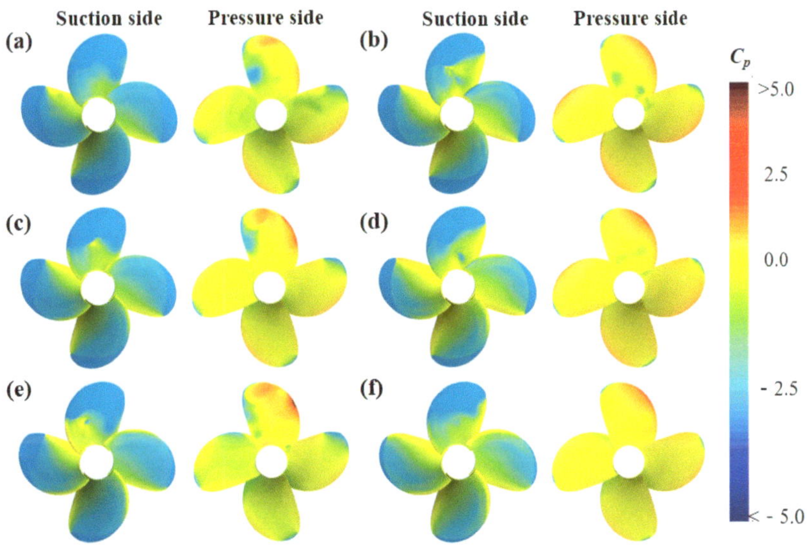

Figure 11. Pressure C_p on blades: (**a**,**c**,**e**) $L/D = 0.15$ and $J = 0.35$, 0.45 and 0.55, respectively; (**b**,**d**,**f**) $L/D = 0.50$ and $J = 0.35$, 0.45 and 0.55, respectively.

6. Conclusions

In this work, the hybrid RANS/LES method combined with the Schnerr–Sauer cavitation model is adopted to study the hydrodynamics, excitation force, cavitation evolution and the flow field characteristics in an ice blockage environment with a low cavitation number. The conclusions are listed as follows:

(1) The hybrid RANS/LES method and Schnerr–Sauer cavitation model possess good numerical accuracy, with the error between the numerical value and the experimental result being within 3.0%. When the advance coefficient rises, the angle of attack between the propeller blade and the incoming flow also increases, which leads to a reduction in thrust and torque. In the case of a low cavitation number, severe cavitation

makes the hydrodynamic coefficient scarcely increase as the distance between the ice and propeller decreases.

(2) The obstruction effect of the ice block on the incoming flow leads to a great increase in cavitation on the blade behind it. Especially when $L/D = 0.15$, the total cavitation coverage area reaches 20% and the cavitation-covered area of a single blade reaches 8.0%. As the advance coefficient increases, the total cavitation coverage area decreases, but as the blade locates behind the ice blockage its cavitation coverage area hardly reduces, causing rapid cavitation evolution and an increase in the excitation force. Especially when $J = 0.55$, the excitation force is twice its average value.

(3) The ice block gives rise to a backflow behind it. When $L/D = 0.15$, the maximum backflow takes place at the blade tip behind the ice blockage, which results in an increase in the low-pressure zone on the suction surface and the high-pressure zone on the pressure surface. The greater the advance coefficient is, the more the high pressure rises and the larger the pressure difference is, thereby causing the excitation force to increase sharply. The increase in J makes the blade subject to a spatially non-uniformly distributed pressure load, which gives rise to severe cavitation evolution and thus generates an excitation force.

Author Contributions: Methodology, Q.H.; simulation, S.Z. and H.L.; validation, X.H. and X.L.; analysis, Q.H. and S.Z.; investigation, S.Z., X.H. and X.L.; resources, Q.H. and H.L.; data curation, X.L.; writing—original draft preparation, Q.H. and S.Z.; writing—review and editing, H.L.; visualization, X.H. and X.L.; supervision, H.L.; project administration, X.H. and X.L.; funding acquisition, Q.H. All authors have read and agreed to the published version of the manuscript.

Funding: This study is supported by the Fundamental Research Funds for the Central Universities (HYGJXM202319, D5000240318), and National Natural Science Foundation of China (U2341285).

Data Availability Statement: All analyzed data in this study have been included in the manuscript.

Acknowledgments: The authors would like to thank the 12th National Ice Engineering Conference.

Conflicts of Interest: The authors declare no conflicts of interest.

Nomenclature

D	Diameter of the propeller	L	Ice–propeller distance
V	Inflow velocity	n	Rotating speed
T	Thrust	Q	Torque
J	Advance coefficient	K_T	Thrust coefficient
η_0	Open-water efficiency	K_Q	Torque coefficient
σ_n	Cavitation number	C_p	Pressure coefficient

References

1. Zheng, S.; Huang, Q.; Li, H.; He, X.; Li, X. Research on the Excitation Force and Cavitation Evolution of an Ice-Class Propeller in Ice Blockage Environments. In Proceedings of the 12th National Ice Engineering Conference, Zhuhai, China, 25–28 October 2024.
2. Zhang, J.; Zhang, Y.; Shang, Y.; Jin, Q.; Zhang, L. CFD-DEM based full-scale ship-ice interaction research under FSICR ice condition in restricted brash ice channel. *Cold Reg. Sci. Technol.* **2022**, *194*, 103454. [CrossRef]
3. Pustoshny, A.V.; Darchiev, G.K.; Frolova, I.G. The problem of propeller design for high ice class transportation ships. In Proceeding of Fifth International Symposium on Marine Propulsors, Espoo, Finland, 12–15 June 2017; pp. 390–397.
4. Sun, H.S.; Ni, B.Y.; Zhang, Y.X.; Fan, X. A case study on propeller-hull vortex cavitation elimination based on CFD simulation and model test. *Ships Offshore Struct.* **2024**, *19*, 1762–1775. [CrossRef]
5. Zhou, L.; Zheng, S.J.; Diao, F.; Ding, S.F.; Gao, J.L. Experimental and numerical study on ice blockage performance of propeller in cavitation flow. *Water* **2022**, *14*, 1060. [CrossRef]
6. Sampson, R.; Atlar, M.; Sasaki, N. Ice blockage tests with a DAT tanker podded propulsor. In Proceedings of the 2nd International Conference on Technological Advances in Podded Propulsion, Brest, France, 3–5 October 2006.

7. Sampson, R.; Atlar, M.; Sasaki, N. Propulsor ice interaction-does cavitation matter? In Proceedings of the 6th International Symposium on Cavitation (CAV), Wageningen, The Netherlands, 11–15 September 2006.
8. Wu, S.; Zeng, Z.B.; Zhang, G.P. Experimental research on the influence of ice block parameters on propeller hydrodynamic performance. *Ship Mech.* **2018**, *22*, 156–164.
9. Xue, Q.Y.; WU, S.; Gu, X.N.; Zhai, S.C. Analysis on the mechanisms of propeller-hull vortex cavitation induced in ice block en-vironments. *Chin. J. Ship Res.* **2020**, *15*, 90–95.
10. Xu, P.; Wang, C.; Ye, L.Y.; Guo, C.Y.; Xiong, W.P.; Wu, S. Cavitation and induced excitation force of ice-class propeller blocked by ice. *J. Mar. Sci. Eng.* **2021**, *9*, 674. [CrossRef]
11. Wang, G.L. Study of Propeller Hydrodynamic Performance Under Ice-Propeller-Flow Interaction. Ph.D. Thesis, Harbin Engineering University, Harbin, China, 2016.
12. Wang, C.; Sun, S.; Chang, X.; Ye, L.Y. Numerical simulation of hydrodynamic performance of ice class propeller in blocked flow-using overlapping grids method. *Ocean Eng.* **2017**, *141*, 418–426.
13. Hu, S.L.; Diao, F.; Wu, S. The numerical study of the propeller performance with ice blockage. In Proceedings of the ISOPE Pacific/Asia Offshore Mechanics Symposium, Dalian, China, 22–25 November 2020. ISOPE-P-20-192.
14. Sun, S.; Li, L.; Wang, C.; Zhang, H. Numerical prediction analysis of propeller exciting force for hull–propeller–rudder system in oblique flow. *Int. J. Nav. Archit. Ocean Eng.* **2018**, *10*, 69–84. [CrossRef]
15. Wu, S.; Liu, Y.; Zeng, Z.B. Influence of ice block on hydrodynamic performance and cavitation of propeller. *Shipbuild. China* **2018**, *59*, 110–121.
16. Zheng, S.J.; Huang, Q.G.; Zhou, L.; Li, H. Numerical investigation on high-fidelity flow field of an ice-class propeller in ice blockage and cavitation flow. *Ocean Eng.* **2024**, *301*, 117626. [CrossRef]
17. Wang, L.; Wu, T.; Gong, J.; Yang, Y. Numerical simulation of the wake instabilities of a propeller. *Phys. Fluids* **2021**, *33*, 125125. [CrossRef]
18. Nie, Y.; Ouyang, W.; Li, G.; Zhang, C.; Zhou, X. Simulation study on hydrodynamic performance of podded propulsor for curise ship. *Chin. J. Ship Res* **2022**, *17*, 170–177.
19. Zheng, S.; Huang, Q.G.; Zhou, L.; Li, H.; He, X.; Li, X.M. Effects of propeller boss cap fins on hydrodynamics and flow noise of a pump-jet propulsor. *Phys. Fluids* **2024**, *36*, 095104. [CrossRef]
20. Lidtke, A.K.; Lloyd, T.; Lafeber, F.H.; Bosscher, J. Predicting cavitating propeller noise in off-design conditions using scale-resolving CFD simulations. *Ocean Eng.* **2022**, *254*, 111176. [CrossRef]
21. Gaggero, S. A Study on the Wake Evolution of a Set of RIM-Driven Thrusters. *J. Mar. Sci. Eng.* **2023**, *11*, 1659. [CrossRef]

water

MDPI

Article

An Investigation of the Thickness of Huhenuoer Lake Ice and Its Potential as a Temporary Ice Runway

Ying Wang [1], Qiuming Zhao [2], Bo Zhang [1], Qingjiang Wang [1], Peng Lu [2,*], Qingkai Wang [2], Xinghua Bao [1] and Jiahuan He [1]

1 Hulun Buir Meteorological Bureau, Hulun Buir 021008, China; wygohappy@163.com (Y.W.); bzhang6789@163.com (B.Z.); qjwang70857@126.com (Q.W.); baoxinghua2024@sina.com (X.B.); hejiahuan50425@163.com (J.H.)

2 State Key Laboratory of Coastal and Offshore Engineering, Dalian University of Technology, Dalian 116024, China; qiumingzhao@mail.dlut.edu.cn (Q.Z.); wangqingkai@dlut.edu.cn (Q.W.)

* Correspondence: lupeng@dlut.edu.cn

Abstract: The study of ice runways has significant practical importance. Regarding inland lake ice, while little of the practicality of ice runways during the ice formation period was explored in the published articles, the analysis of the time period and suitable locations may be used. This study focused on Huhenuoer Lake, located in Chen Barag Banner in northeastern China. The time-dependent law of ice growth in this lake has been investigated over a study period from 2023 to 2024. Utilizing the drilling approach, the ice thickness, recorded at each site on 29 February 2024, has surpassed 100 cm. On 14 March 2024, the recorded ice thickness at site #2 reached a record high of 139 cm. Second, to assess the project's ease of use and safety, we used the Stefan equation to model the lake's ice growth processes, resulting in a fitted Stefan coefficient of 2.202. For safety considerations, the Stefan coefficient used for the construction of the ice runway was set at 1.870. We investigated the distribution of lake ice and concluded that the lake ice runway should be established in the north. We established the relationship between ice thickness, cumulative snowfall, and negative accumulated temperature by integrating the fitting technique with the Stefan model. Utilizing the P-III method, the minimum value of the maximum negative accumulated temperature for the 50-year return period is 2092.46 °C·d, while the maximum cumulative snowfall for the 50-year period is 58.4 mm. We can apply these values to the aforementioned relationship to derive the ice thickness patterns across varying return periods. Finally, the study provides recommendations for the construction of the ice runway at Huhenuoer Lake. This study introduces ice field research and an ice growth model into the analysis of lake ice runway operations to provide technical assistance for ice runways.

Keywords: ice field investigation; ice thickness; ice runway; Huhenuoer Lake; wind rose; negative accumulated temperature; cumulative precipitation; feasibility analysis

Academic Editor: Jueyi Sui

Received: 9 December 2024
Revised: 13 January 2025
Accepted: 28 January 2025
Published: 31 January 2025

Citation: Wang, Y.; Zhao, Q.; Zhang, B.; Wang, Q.; Lu, P.; Wang, Q.; Bao, X.; He, J. An Investigation of the Thickness of Huhenuoer Lake Ice and Its Potential as a Temporary Ice Runway. *Water* **2025**, *17*, 400. https://doi.org/10.3390/w17030400

1. Introduction

Recent trends in cold places and polar scientific research development have led to a proliferation of studies that focus on ice runways. In polar regions, scientific research heavily relies on air transportation for both personnel and equipment. The presence of polar conditions limits conventional runway construction [1]. The ice runway is crucial infrastructure for supporting polar research [2].

For airplanes to take off and land on ice, the runway must be considered a kind of ice engineering. The evaluation of the ice's carrying capability is important, primarily determined by its thickness and strength [3,4]. Although there have been successful instances of ice takeoff and landing airplanes overseas, the runway requirements for various aircraft types are different. Sharp proposed the thickness of ice required for aircraft landings on skis, with the assumption Sharp assumed that the plane will land on lake ice, river ice, and sea ice. These calculations are suitable for frozen ice at temperatures below −9 °C. If the temperature of the frozen ice exceeds −9 °C, it is necessary to increase the required thickness by 25%. Wheeled aircraft require 20% more ice thickness than aircraft landing on skis [5]. Blaisdell et al. indicated that the compressive strength of ice must be evaluated, with the strength required to exceed the maximum contact stress produced by the target aircraft by at least 25% [4]. Under external force, the floating ice has great deformation and cross-section flexural stress. Due to ice's weak tensile strength, the critical stress represents the peak tensile stress at the bottom of the ice under the load [3]. This resembles the conventional rigid pavement design methodology. Consequently, when the bending strength of ice improves, the load-bearing capacity of the ice track will be enhanced. Given that a layer of compacted snow is often placed over the ice runway, which has a low coefficient of friction, the landing distance on compacted snow is 1.6 times the required length of the land runway [6]. Swithinbank presented the results of the Antarctic ice runway experiment [7]. The ice runways at two Antarctic sites accommodate various wheeled aircraft, enabling the takeoff and landing of C-130, C-141, and C-5B aircraft. Squire et al. [8] performed studies on the sea ice next to Tent Island in McMurdo Strait. In the experiment, the sea ice thickness measured 1.60 m and remained consistent throughout. The experiment used a pickup vehicle with a mass of 2100 kg and an LC-130 aircraft with a mass of about 50,000 kg, measuring the strain rate that airplane landings and vehicular activity impose on the ice. McCallum [9] utilized the Casey ice runway constructed by the Antarctic Division in Australia as a case study, clarifying the impact of glacier movement, including movement and rotation, on the positioning and deformation of the ice runway and subsequently analyzing the implications of glacial dynamics on the long-term viability of the ice runway.

Ice runways promote the advancement of transportation and tourism in colder places; however, there are currently few actual uses of ice runways in northeast China. This paper selects Huhenuoer Lake for an ice field investigation to assess the feasibility of lake ice runways. This article aims to investigate the feasibility of using Huhenuoer Lake ice as a winter ice runway, particularly focusing on the optimal position and period for its operation. To accomplish the study objectives mentioned above, this article includes the following work: Utilizing historical meteorological data and ice field research methodologies, we opted to perform an analysis of the Huhenuoer Lake ice from 5 December 2023 to 21 March 2024. We used the ice drilling technique to assess the variation in thickness at each location. This research presents our findings on the ice formation in Huhenuoer Lake and examines the orientation, location, and operational periods of several aircraft types used on the temporary runway. The research methods and concepts presented in this paper are applicable to the design of many different lake ice runways, beyond just Huhenuoer Lake.

2. Basic Natural Conditions at Huhenuoer Lake

Lake ice formation is influenced by cold air, making local temperature and precipitation the primary factors. As for temporary ice runways, the orientation ofthe runway is related to wind speed and direction during the winter, while the runway length is related to the lake's boundaries. Huhenuoer Lake (49°15′~49°20′ N, 119°11′~119°17′ E) lies in the western plain of the Greater Hinggan Mountains, approximately 14 km west of Chen Barag Banner district, Hulun Buir City, and close to the confluence of the Hailar River and Morigele River [10]. The length is 7.6 km from north to south, the max width is 3.8 km from east to west, and the area covers approximately 21 km^2. The lake's height is 587 m. Refer to Figure 1 for the schematic geographical position of Huhenuoer Lake.

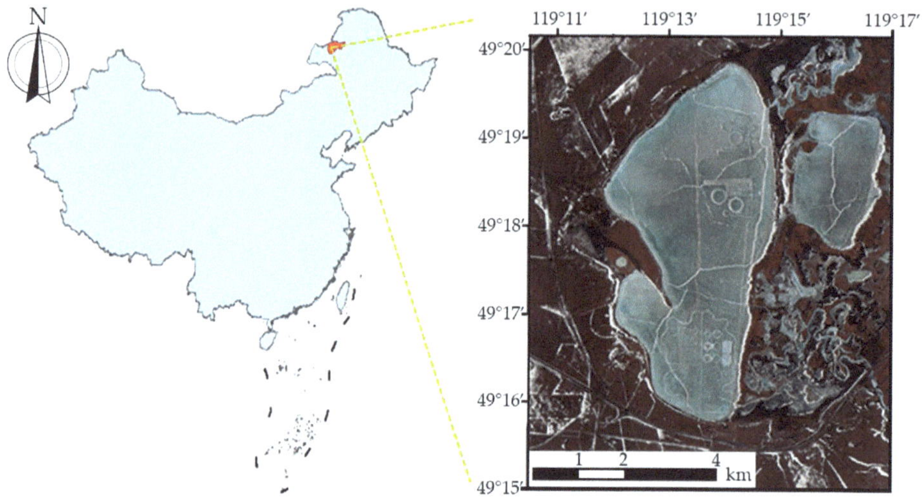

Figure 1. Schematic diagram of the geographical position of Huhenuoer Lake. China is shown in blue on the left, whereas Chen Barag Banner is shown in red. The image on the right displays a high-definition map of Huhenuoer Lake.

3. Wind Rose Analysis and Assessment of the Ice Thickness Research Site

Lake ice formation is influenced by cold air, making local temperature and precipitation the primary factors. As for temporary ice runways, the orientation of the runway is related to wind speed and direction during the winter, while the runway length is related to the lake's boundaries.

From the above, the runway orientation is directly linked to the local wind data. Next, we must ascertain the period of time for gathering wind data. The winter ice season in eastern Inner Mongolia typically spans from late October to late April of the following year [11]. The ice-freezing process starts when the temperature falls below 0 °C [12]. This study defines the research period from one year to the next as the time period during which the daily average temperature remains below 0 °C, based on an examination of the winter icetime from 1993 to 2023. The peak wind speed and direction for all study periods from 1993 to 2023 are summarized in Appendix A, with the wind rose diagram shown in Figure 2.

The prevailing wind enables the aircraft to take off and land safely. The International Civil Aviation Organization (ICAO) mandates that the airport aligns the runway to guarantee a usability factor of 95%, meaning that an excessive crosswind component restricts use of the runway system to no more than 5% of the time. The orientation of the airport runway is determined by visual vector analysis known as the wind rose technique. The typical wind rose comprises a series of concentric circles divided by radial lines on polar coordinate paper. The components of wind data include wind speed, wind direction, and frequency of occurrence. The wind rose displays the percentage and wind speed range in this orientation. The typical wind rose described earlier can offer detailed wind data; however, a specialized template described as follows is required to ascertain the runway orientation.

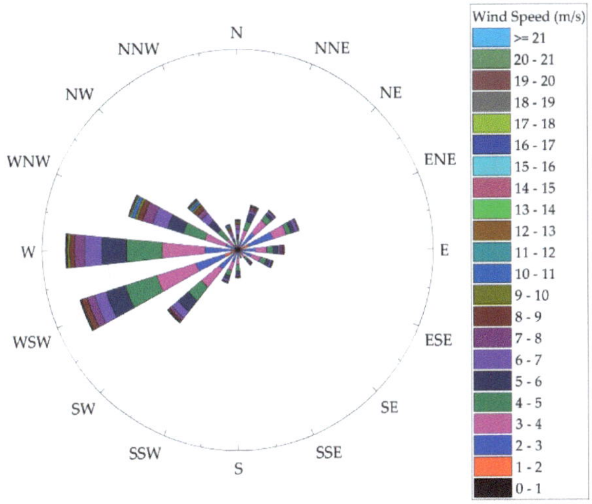

Figure 2. Wind roses for the 2023–2024 research period at Chen Barag Banner.

Three parallel, equally spaced lines were employed with the wind rose. The center line signifies the runway's midpoint, while the span between the center line and each outside line represents the allowable crosswind threshold, often 15 mph, about 7 m/s [13,14].

We have reassessed the data shown in Figure 2 based on the criteria for determining runway direction [15] and generated Table 1.

Table 1. Wind frequency (%) of wind roses used to determine runway orientation in all research periods of Huhenuoer Lake in the past 30 years.

Orientation	0~7 m/s	7~14 m/s	14~21 m/s	Orientation	0~7 m/s	7~14 m/s	14~21 m/s
N	2.09	0.87		S	2.61	0.15	
NNE	4.11	0.46		SSW	3.28	0.13	
NE	4.87	0.11		SW	8.63	0.57	
ENE	6.33	0.13		WSW	15.31	1.24	
E	4.33	0.39		W	15.50	1.72	0.02
ESE	3.59	0.17		WNW	9.26	2.17	0.04
SE	1.72	0.24		NW	4.81	1.61	0.06
SSE	0.76	0.09		NNW	1.61	1.02	

Table 1 suggests creating a wind rose chart to determine the runway orientation, as shown in Figure 3. Afterwards, the maximum permissible crosswind of 7 m/s for manual testing must be employed to maintain the runway's usability factor at or above 95%. The optimal runway orientations for Huhenuoer Lake are 90–270°, with a usabil-ity factor of 95.6%, and 130–310°, with a usability factor of 95.6%.

There are often at least two runways (the primary runway and the secondary runway) at locations where the wind direction varies frequently, making it difficult to choose only one runway if the runway is being utilized to its fullest potential [16]. Given that Huhenuoer Lake is remote from towns and lacks large mountains, these two directions, 90–270° and 130–310°, are appropriate. Simultaneously, these two directions are integer multiples of 10°, facilitating positioning during construction. In general, the ice in the lake's center is not very thick, so we choose the directions of 90–270° and 130–310°. The lakeshore distances of these two plans are 3400 m (solid line) and 4100 m (dashed line), respectively.

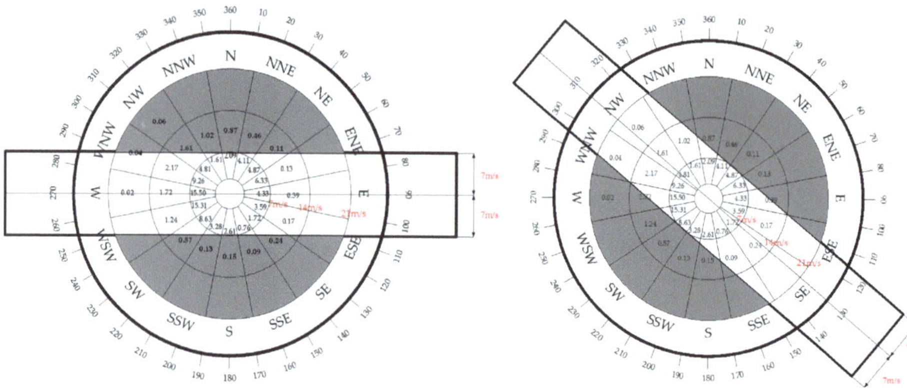

(a)

(b)

Figure 3. The wind rose diagram for two arrangements: (**a**) 90–270°; (**b**) 130–310°.

Table 2 displays the ice thickness investigation sites for Huhenuoer Lake during the study period of 2023–2024, taking into account the thin ice conditions in the lake's center and the goal of maximizing coverage throughout the entire lake. Figure 4 also displays the specified sites.

Table 2. The latitude and longitude of Huhenuoer Lake ice survey sites.

Site Number	Longitude	Latitude
#1	119°12′30″ E	49°18′30″ N
#2	119°13′30″ E	49°18′30″ N
#3	119°14′30″ E	49°18′30″ N
#4	119°13′30″ E	49°19′00″ N
#5	119°13′30″ E	49°17′30″ N
#6	119°13′30″ E	49°16′30″ N

Figure 4. The high-definition map of the proposed route for the optional ice runway and the ice research sites. The solid line represents the 3400 m scheme oriented from 90–270°, and the starting point on the western side is marked with a green pentagon. The dashed line indicates the 4100 m scheme oriented from 130–310°, and the starting point on the western side is marked with a green triangle.

4. Measurement of Ice Thickness Distribution and Its Relationship with Negative Accumulated Temperature

The growth and melting of lake ice mostly rely on temperature and the accumulation of snow on the ice surface. Winter snowfall in Chen Barag Banner affects the processes of lake ice growth and melting. One of the simplest evaluation methods is the mathematical model that relates ice thickness with negative accumulated temperature throughout the ice growth period, known as the degree-day model [17–20]. The statistical coefficient indicates the influence of snow cover, local hydrology, geographical information, and other factors. The Chen Barag Banner meteorological station recorded a daily mean temperature below 0 °C on 30 October 2023, which began to rise above 0 °C on 27 March 2024. Thus, the negative accumulated temperature at the Chen Barag Banner meteorological station over the study period from 2023 to 2024 may be calculated. Huhenuoer Lake ice is mainly constituted of a columnar-grained ice layer. The wetness of the snow is not significant, and the impact of wind erosion on the snow is limited. We performed six measurements of ice thickness between 5 December 2023 and 21 March 2024. We collected ice thickness data from six sites (#1–#6), as depicted in Figure 4. Table 3 displays the measurements of ice thickness and snow depth collected by drilling holes in Huhenuoer Lake during the study period from 2023 to 2024. The data in Table 3 are stored as integers. We use them to analyze the ice-growing process at each site.

Table 3. Ice thickness and snow depth measured by hole-drilling in Huhenuoer Lake during the study period from 2023 to 2024.

Date of Measurement	Snow Depth (cm)	Ice Thickness (cm)						
		#1	#2	#3	#4	#5	#6	Average Value
5 December 2023	13	48	46	42	51	50	48	48
15 December 2023	15					55		55
29 December 2023	19					70	71	71
2 January 2024	18	95	103	107	109	72	117	101
18 January 2024	20	85	98	89	89	84	99	91
26 January 2024	19	89	99	96	97	86	100	95
2 February 2024	20	96	105	102	102	87	105	100
29 February 2024	22	116	112	108	121	107	103	111
7 March 2024	21	119	132	108	119	105	103	114
14 March 2024	19	121	139	108	127	105	103	117
21 March 2024	16	131	139	108	129	125	103	123

This section provides further analysis using a model for ice growth. In the end, we decided to use the Stefan model.

The reason for selecting this approach is to achieve an acceptable compromise between technical ease of use and safety. The Stefan model expression is simple to understand and user-friendly. There is limited literature regarding the application of this strategy to ice runway design, but the subsequent analytical method employs various levels of safety factors to ensure safety.

Stefan developed an analytical formula for calculating ice thickness in 1891 [21]. The surface temperature of the ice corresponds to the air temperature, while the temperature at the ice bottom aligns with the freezing point. Heat transmission within the ice only aligns with its growth direction and spreads linearly throughout. As a result, the ice thickness varies throughout time.

The Stefan model can be shown as follows:

$$H_i = a\sqrt{FDD} \tag{1}$$

where H_i is the ice thickness, and *FDD* is cumulative freezing degree days. The *FDD*, also known as the sum of negative degree days, provides the cumulative total of below-zero temperatures for each day that follows, making the calculation for determining ice thickness simpler [22]. a represents Stefan's coefficient in the degree-day approach. It relates not only to the physical characteristics of ice but also to the average depth of snow above the ice.

Ice analysis extensively uses the previously mentioned Stefan equation. Subsequently, we plan to describe the application of the Stefan equation within this investigation. For the fitting operation, we exclude the measurements of ice thickness (*H*) values from the six sites (i.e., #1–#6) where ice growth has reached the lake bottom. Then, the measured ice thickness and negative accumulated temperature by the freezing degree day (*FDD*) are fitted using the Stefan curve, resulting in the respective statistical coefficients for the six sites. Figure 5 displays the Stefan fitting results for the six sites.

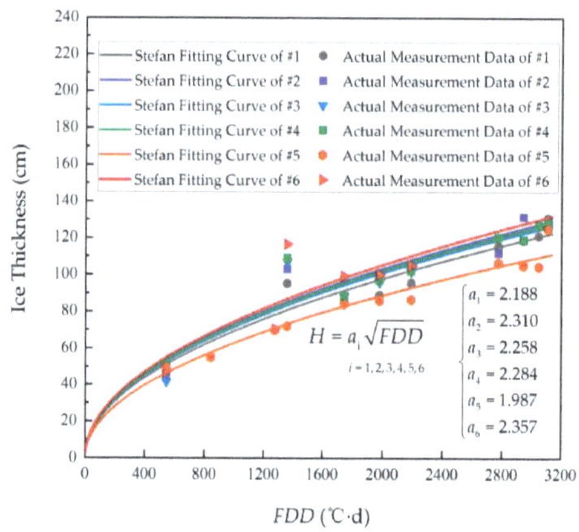

Figure 5. Stefan fitting curves and parameters of negative accumulated temperature and ice thickness of sites #1–#6.

Figure 5 illustrates variations of ice thickness at various places throughout the lake. To describe the findings about the ice thickness at different times more clearly, we created an ice thickness diagram indicating the relative distance from sites #1 (the route oriented towards #1–#2–#3) and #4 (the route oriented towards #4–#2–#5–#6), with the findings shown in Figure 6.

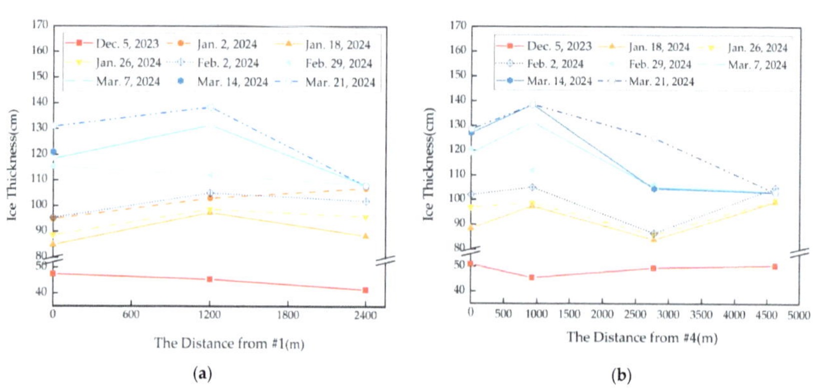

Figure 6. Measured ice thickness at six sites at different times: (**a**) 90–270°; (**b**) 130–310°.

Table 3 indicates that during the initial phase (2 January 2024), ice formations developed more quickly in the southern region than in the northern region, with a small area of thin ice present in the lake's center. During the subsequent period (21 March 2024), the ice in the southern region of the lake grew slowly and even touched the lake bottom due to the shallow water conditions. Conversely, the ice in the northern section of the lake was comparatively thick.

5. Feasibility Analysis of Huhenuoer Lake Ice as a Temporary Runway

5.1. The Runway Position and Ice Thickness Distribution

Analysis of wind speed and direction data at the Chen Barag Banner meteorological station from 1993 to 2023 indicates that the optimal runway orientations for the Huhenuoer Lake temporary airport are 90–270° and 130–310°. Given that the water in the southern region of Huhenuoer Lake is shallow and the ice in the central area is thin, we utilize the 90–270° line and the 130–310° line to identify the broadest location in the northern section of the lake as the proposed runway position. The distances between the two lakesides measured in this way are 3400 m and 4100 m, respectively.

Due to the low friction coefficient of the ice runway, it is necessary to maximize its length. Additionally, the surface of the ice runway may be covered with a layer of compacted snow, as in the case of the Pegasus runway [4], to mitigate melting caused by solar radiation and other factors. Consequently, based on the assumption that the landing distance of a compacted snow runway requires not less than 1.6 times the length of a land runway [6], the 3400 m length of Huhenuoer Lake ice is comparable to approximately 2000 m of a land runway, which is suitable for the takeoff and landing of lightweight aircraft. The C130 requires a land runway length of 1830 m for ground operations. Without a complete examination of the ice thickness, the ice of the northern part of Huhenuoer Lake is capable of supporting the takeoff and landing of aircraft weighing less than the C130, ensuring safety.

Figure 7 identifies two envelope lines, showing the upper- and lower-limit envelopes of ice thickness throughout the study period of 2023–2024 at Huhenuoer Lake. Different research backgrounds, including purposes and locations, require the adoption of different envelopes. Previous research has applied similar concepts across different fields [23]. This study indicates that adopting the lower envelope (i.e., a = 1.870) implies the presence of some safety redundancy. This is safe for the design of ice runways.

Figure 7 illustrates that the Stefan coefficient of Huhenuoer Lake ice is 2.202. The coefficient for the upper envelope is 2.400, while that for the lower envelope is 1.870. For the design of anti-ice structures, such as the anti-ice design of a bridge pier over a reservoir in cold regions studied in reference [24], the upper-bound envelope considered safe for the project should be used. We construct the airport runway on ice to support the aircraft's weight and withstand the dynamic loads during takeoff and landing. The lower-limit envelope, whose fitting coefficient of ice thickness–negative accumulated temperature is 1.870, should be chosen in accordance with the engineering safety criteria.

The survey on ice surface variation has not been carried out, and theoretically, it will not exceed the magnitude of ice thickness variation.

Based on our newly established route, the thickness variation can be assessed as follows:

1. The non-uniformity should be minimized, where non-uniformity = (maximum ice thickness along the route − minimum ice thickness along the route)/maximum ice thickness along the route.
2. The maximum change rate should be minimized, where the maximum change rate refers to the highest value of non-uniformity per unit length over the entire route.

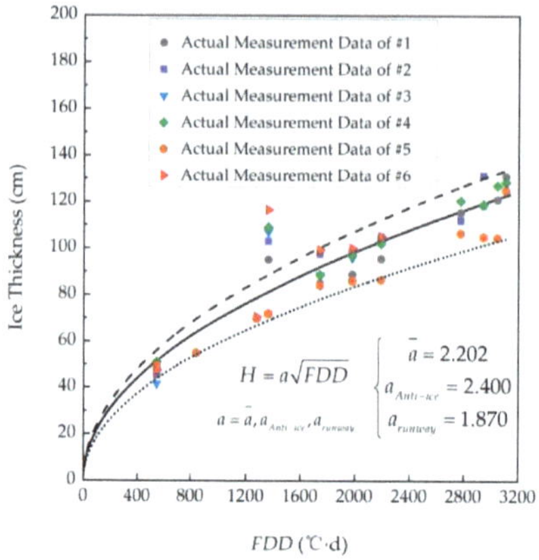

Figure 7. The average Stefan curve (solid line) representing all the ice thickness data from Huhenuoer Lake during the study period of 2023–2024, together with the upper-limit envelope (dashed line) for anti-ice structure design and the lower-limit envelope (dotted line) for ice runway design.

Table 4 presents the ice thickness values for the 90–270° and 130–310° routes, calculated using interpolation, allowing for the assessment of non-uniformity and the maximum change rate of ice thickness for both routes.

Table 4. The maximum value of the ice thickness, non-uniformity, and change rate of ice thickness for runway routes 90–270° and 130–310° at different times.

Date of Measurement	Runway Orientation	Ice Thickness on The Route (cm)		Non-Uniformity (%)	Maximum Change Rate (‰)
		Min Value	Max Value		
5 December 2023	90–270°	39	48	18.2	−0.1
	130–310°	45	50	8.5	−0.0
2 January 2024	90–270°	92	109	15.4	0.1
	130–310°	72	120	39.7	0.7
18 January 2024	90–270°	81	98	16.9	0.1
	130–310°	78	96	19.1	0.2
26 January 2024	90–270°	85	99	13.6	0.1
	130–310°	83	101	18.2	0.2
2 February 2024	90–270°	92	105	12.0	0.1
	130–310°	87	106	18.4	0.3
29 February 2024	90–270°	106	117	9.1	−0.0
	130–310°	101	117	13.4	−0.1
7 March 2024	90–270°	98	132	25.4	−0.2
	130–310°	105	114	8.0	−0.3
14 March 2024	90–270°	94	139	32.4	−0.3
	130–310°	105	119	11.8	−0.6
21 March 2024	90–270°	96	139	31.0	−0.3
	130–310°	90	127	28.9	−0.3

We show the variation in ice thickness for the two routes at each ice measuring time using an interpolation method. The results are shown in Figure 8.

Under adverse conditions, assuming that the ice temperature is higher than −9 °C and that all of the aircraft are wheeled, the necessary lake ice thickness may be determined by calculating for the various aircraft types [5,25–30]. Next, we can determine the beginning of the fundamental time using the negative accumulated temperature associated with the lake ice thickness and designate 26 March 2024 as the end of the fundamental period.

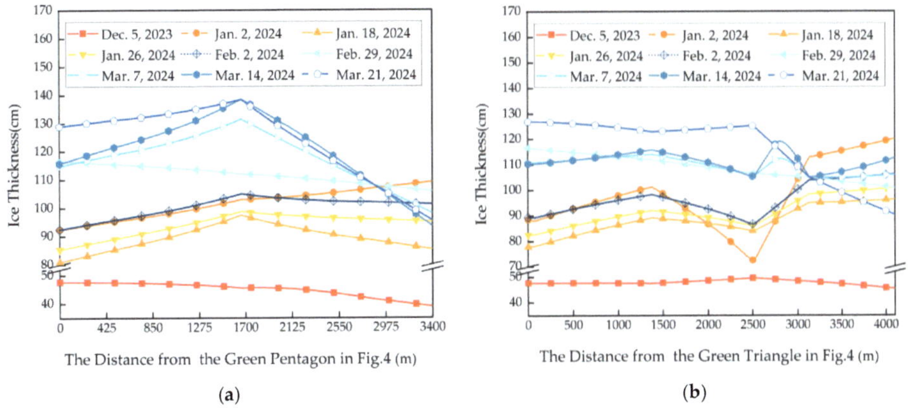

Figure 8. Ice thickness of 90–270° and 130–310° runway routes at different measured times: (**a**) 90–270°; (**b**) 130–310°.

From a safety point of view, the ice thickness may change before and after the operational period. This means that dividing the length of the fundamental period by the safety factor (set at 1.5 in this study) gives us the conservative operational period for the aircraft during the winter, assuming that the middle time point of both the fundamental and conservative periods is the same. Blaisdell et al. indicate that the safety factor for ice runways should be a minimum of 1.25 and a maximum of 1.5 [4]. A safety factor beyond 1.5 diminishes confidence in the runway's capacity to support aircraft. This section establishes the safety factor at 1.5. At the same time, we also observe that in [4], tire pressure has a safety factor of 1.6 for a specific configuration. The Stefan coefficient has been established as 1.870 in this section. The fitted Stefan coefficient is 2.202, resulting in a calculation of $1.5 \times 2.202/1.870 = 1.77$, which exceeds 1.6.

We can mitigate the inaccurate prediction of the ice formation and melting process to the greatest extent feasible regarding operational duration, thereby addressing the deficiency in ice thickness. Using the study period from 2023 to 2024 as an example, Figure 9 can be developed by displaying the data of different aircraft.

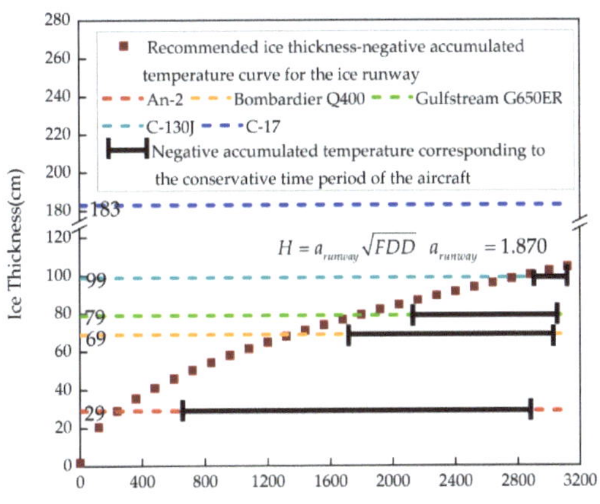

Figure 9. Using the study period from 2023 to 2024 as an example, the Stefan curve depicting the relationship between ice thickness and negative cumulated temperature for Huhenuoer Lake, together with the operating conservative period for each aircraft.

Table 5 presents the above findings in more detail.

Table 5. Using the research period from 2023 to 2024 as an example, the conservative operational period of different aircraft types and the necessary length of the ice runway are assessed. The needed length of the ice runway is 1.6 times the greater of the aircraft's take-off distance and landing distance.

Aircraft Type	Maximum Landing Weight (t)	Start of the Conservative Period	End of the Conservative Period	Conservative Period (d)	Needed Length (m)
An-2	7	10 December 2023	4 March 2024	85	400
Bombardier Q400	29	17 January 2024	12 March 2024	55	2240
Gulfstream G650ER	38	31 January 2024	15 March 2024	44	3070
C-130J	59	5 March 2024	23 March 2024	18	2520
C-17	203	—	—	—	3780

Table 5 indicates that small aircraft like the An-2 will have approximately 85 conservative days during the study period of 2023–2024. However, as aircraft weight increases, the conservative operational period will progressively diminish, making it unsuitable for large transport aircraft such as the C-17. Appendix B [31–34] shows the required ice thickness and runway length associated with the maximum landing weight of other aircraft.

5.2. Feasibility of Using Huhenuoer Lake Ice as the Temporary Runway Every Winter in the Future

In the research period of 2023–2024, a fitting Stefan coefficient of 1.870 may be used to assess the relationship between ice thickness and negative accumulated temperature, supporting the evaluation of ice thickness on various days for the takeoff and landing of different aircraft types. Will we be able to use this coefficient for assessment in future study periods? The assumption is impractical theoretically. The variation of snow depth on the ice surface of Huhenuoer Lake each year affects ice formation differently. This impact is seen in the magnitude of the Stefan coefficient in the relationship between ice thickness and negative accumulated temperature. Generally, a thinner layer of snow corresponds to a higher Stefan coefficient, and vice versa.

Given the above-mentioned impact of snow on the Stefan coefficient, the following will go over how to ascertain the scientific relationship between the two variables. If the snow depth varies, we must use statistics to determine the value of coefficient A. Given that the latitude of Hongqi Pao Reservoir closely resembles that of Huhenuoer Lake (Hongqi Pao Reservoir at 46°36′ N and Huhenuoer Lake at 49°18′ N) [35,36], we refer to Wang's study of Hongqi Pao Reservoir to determine the Stefan coefficient under conditions of maximum snow depth. Wang investigated the empirical formula that identifies the correlation between the thickness of the ice sheet and the negative accumulated temperature beneath the snow cover in Hongqi Pao Reservoir [37]. We note that the average snow depth at Huhenuoer Lake during the 2011–2012 study period reached 24.1 cm, the highest recorded in the past three decades. Through a fitting procedure, we obtain a Stefan coefficient of 1.842 from the Zubov model data [37], which corresponds to a snow depth of 24.1 cm. Using the logistic fitting method, we can find the link between the depth of the snow and the Stefan coefficient at Huhenuoer Lake by combining data from all study sites during the study periods of 2022–2023 and 2023–2024, along with data from other sources [38] that show the Stefan coefficient of windy lake ice with no snow to be 2.700. The primary advantage of this statistical formula is its ability to show that the Stefan coefficient corresponds to the values reported in the literature under bare ice conditions while also indicating that it diminishes as snow depth increases, up to a certain limit. The results are shown in Figure 10.

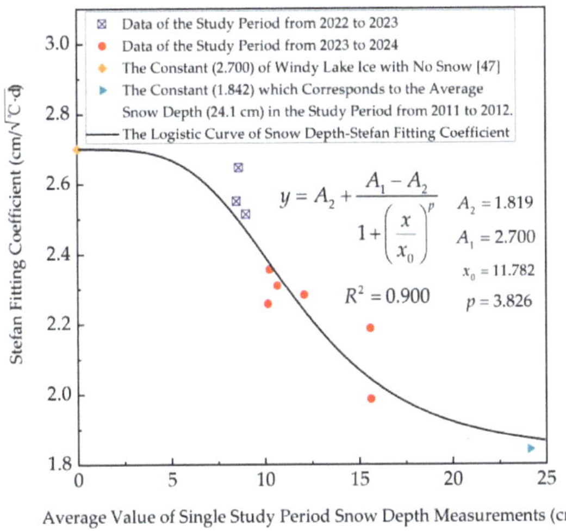

Figure 10. Logistic fitting relationship between snow depth and Stefan coefficient of Huhenuoer Lake ice.

The Chen Barag Banner meteorological station has historically recorded winter snowfall, but it does not record the depth of the snow on the ice surface. How do we measure the depth of snow? This research employs a statistical analysis of the cumulative snowfall and snow depth during the negative accumulated temperature statistical period recorded by the Chen Barag Banner meteorological station. The low temperature and high humidity in Chen Barag Banner, Inner Mongolia, lead to the conclusion that snow undergoes a natural metamorphism process without sublimation. Figure 11 shows a fitting method that can only be used to model the relationship between cumulative snowfall and average snow depth over the study period of the last 30 years in Chen Barag Banner. With an R^2 value of 0.832, this method is good for modeling this relationship.

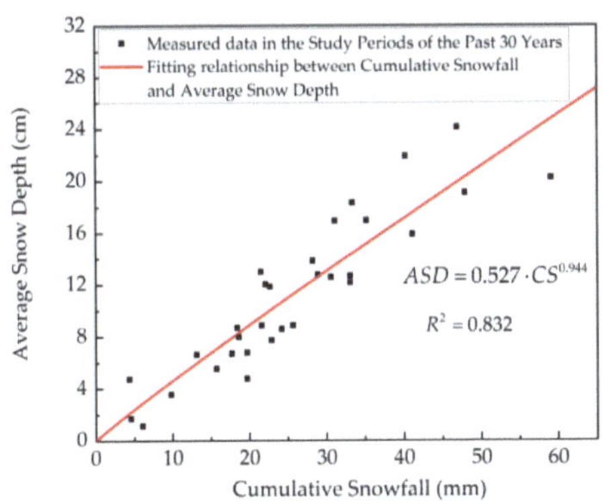

Figure 11. The relationship between cumulative snowfall and average snow depth over the study period of the last 30 years in Chen Barag Banner.

Following these operations, the correlation among ice thickness, negative accumulated temperature, and cumulative snowfall for the temporary runway at Huhenuoer Lake can be expressed as indicated in Equation (2):

$$H_i = \left[1.819 + \frac{2.700 - 1.819}{1 + \left(\frac{0.527 \bullet CS^{0.944}}{11.782}\right)^{3.826}}\right]\sqrt{FDD} = \left[1.819 + \frac{0.881}{1 + \frac{CS^{3.612}}{145491.141}}\right]\sqrt{FDD} \quad (2)$$

where CS is the cumulative snowfall. Can the findings obtained during the study period of 2023–2024 be applicable in the winter of the following 10, 15, 20, 25, or even 50 years? How should we assess the return period? In the return period analysis, although global warming cannot be included, the existing meteorological data already contain information from the time of climatic warming. Moreover, the ice thickness, influenced by harmful climatic circumstances, with a return period of 50 years, could serve as a criterion for other winters that do not exceed 50 years.

The Pearson type three (P-III) distribution curve technique [39–46], often used in hydrometeorological studies, is utilized to derive the P-III distribution curve of cumulative snowfall during the study periods of the last 30 years in Chen Barag Banner, shown as Figure 12. Please refer to Appendix C for details of the P-III method.

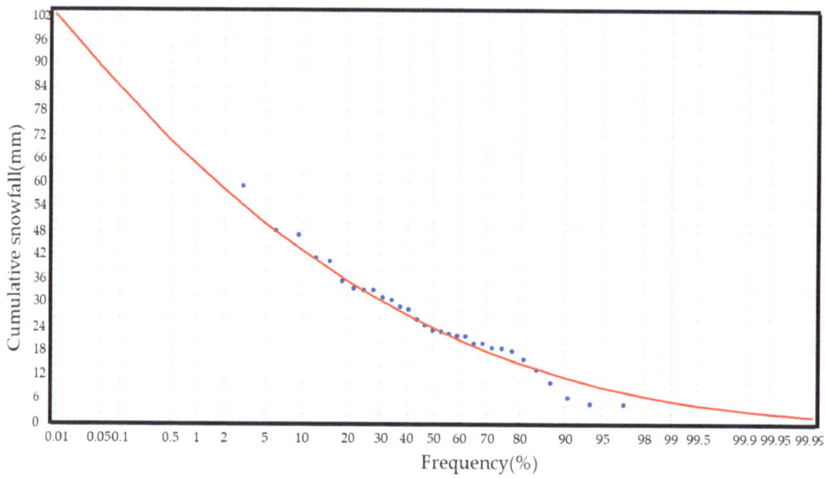

Figure 12. P-III curve representing cumulative snowfall over the study periods of the last 30 years. The blue dots denote the actual data, while the red line illustrates the fitting cumulative snowfall P-III curve for the study period.

A similar P-III distribution analyzes the negative accumulated temperature at the Chen Barag Banner meteorological station during winter for the same time, as shown in Figure 13.

Equation (2) indicates that increased cumulative snowfall results in impacts on the ice becoming thinner, with the 2% frequency in Figure 12 denoting an occurrence once every 50 years. As for FDD, a frequency of 98% in Figure 13 is used to denote the 50-year return period, while Table 6 is designed based on the combination of the thick snow layer on the ice surface and the low negative accumulated temperature throughout various return periods.

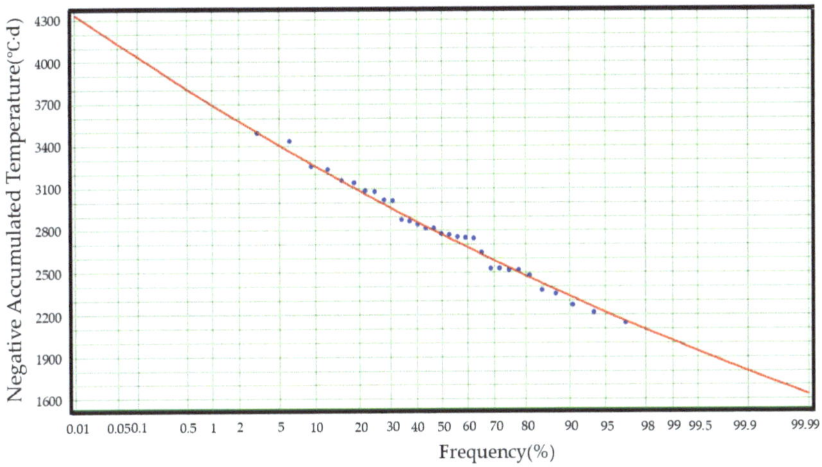

Figure 13. P-III curve representing negative accumulated temperature over the study periods of the last 30 years. The blue dots denote the actual data, while the red line illustrates the fitting negative accumulated temperature P-III curve for the study period.

Table 6. The values of cumulative snowfall and negative accumulated temperature from the P-III curve, together with the corresponding ice thickness over various return periods.

Return Periods (Years)	Value of the Corresponding Return Period by Cumulative Snowfall P-III Curve (mm)	Value of the Corresponding Return Period by Negative Accumulated Temperature P-III Curve (°C·d)
50	58.4	2092.5
45	57.5	2105.5
30	53.7	2158.7
25	52.0	2184.3
20	49.9	2217.0
15	47.0	2262.1
10	43.0	3577.4

We insert the input values into Equation (2), which establishes a relationship between cumulative snowfall, negative accumulated temperature, and ice thickness. This connection allows us to calculate the ice thickness throughout various return periods. We can determine the take-off and landing periods for different aircraft types during different return periods by combining the ice thickness calculated by Equation (2) with the minimum ice thickness required for different aircraft types, taking into account the longest temporary runway length. The most adverse effect of combination might serve as the minimal use period during the planning stage of a temporary airport runway.

6. Conclusions and Suggestions

This paper develops six fixed ice thickness monitoring sites based on limited measurements of ice thickness during the study period of 2022–2023, considering the meteorological data near Huhenuoer Lake and the feasibility of utilizing it as a temporary take-off and landing runway, in accordance with the general characteristics of lake ice thickness distribution. The ice thickness was measured at different times throughout the study period of 2023–2024 at the six sites. The actual requirements of the runway, along with the measured data regarding length, ice thickness, and service period of the temporary runway, as well as the analysis of various return periods, demonstrate the feasibility of the Huhenuoer

Lake ice supporting a temporary runway during different winter periods. The specific conclusions are as follows:

1. Investigations have shown that the ice in the lake's center is thin, particularly in the southwest of Huhenuoer Lake. Furthermore, the southern region of Huhenuoer Lake is shallow, and in this region, the lake ice can reach the bottom of the lake. The northern section of Huhenuoer Lake may serve as the temporary take-off and landing runway for aircraft on the lake ice during winter.
2. The 30-year historical data on wind speed and direction from the Chen Barag Banner meteorological station near Huhenuoer Lake indicate the predominant wind direction and frequency throughout the winter ice season in this region. The criteria requires a 95% usability factor for aircraft takeoffs and landings in wind conditions, unaffected by crosswinds. The aircraft's runway orientations on the Huhenuoer Lake ice are 90–270° and 130–310°.
3. In the northern section of the lake, the straight-line distances of 90–270° and 130–310° indicate that the greatest lengths are 3400 m and 4100 m, respectively. Due to the low friction coefficient of an ice surface, the recommended length of an ice runway with a compacted snow layer is 1.6 times that of land. The minimum operational runway length for Huhenuoer Lake is established at 3100 m. Considering the requirements of maximum ice thickness and irregular ice distribution, it is capable of facilitating the takeoff and landing of aircraft such as the C130.
4. We establish the fitting relationship between ice thickness, negative accumulated temperature, and cumulative snowfall at Huhenuoer Lake, using the safety of aircraft takeoff and landing as an engineering design requirement. We provide the designated service period for takeoff and landing for several aircraft types. An analysis of the shortest application period for the next 50 years is conducted.

This workprovides a comprehensive examination of aircraft takeoffs and landings at Huhenuoer Lake. However, there are parts that require enhancement and supplementation. We recommend that future field investigations and studies incorporate the following components:

1. The findings of this article provide support for the design of ice aircraft takeoff and landing at Huhenuoer Lake; nonetheless, a comprehensive survey is necessary for practical implementation. This research suggests that the northern region of Huhenuoer Lake should be the primary location for a limited number of ice thickness monitoring sites, due to the thick ice, low temperatures, and challenges associated with field operations.
2. For the potential runway ice and compacted snow layer, it is crucial to evaluate not only the variability in ice thickness on the proposed runway but also the undulations of the ice surface. Additionally, it is crucial to conduct research aimed at enhancing the friction coefficient and minimizing surface undulation of the compacted snow.
3. The investigation assessing the bearing capability of the temporary ice runway at Huhenuoer Lake will be integrated into the field of ice engineering. The effective maintenance of the ice runway is in the construction and application stages. It requires measurement of not only ice thickness but also mechanical characteristics, including ice bending strength, elastic modulus, Poisson's ratio, and compressive strength. Therefore, in order to implement the ice runway, it is crucial to conduct experimental research on the physical characteristics of the ice layer during future investigations of Huhenuoer Lake.
4. The Stefan model employed in this research is user-friendly, and various ice thickness models, such as the Ashton model [47,48], are also appropriate for ice runway construction. This represents a significant area for future investigation.

Author Contributions: Conceptualization, P.L.; methodology, Q.Z. and P.L.; investigation, Y.W., Q.W. (Qingjiang Wang) and Q.W. (Qingkai Wang); data curation., B.Z. and X.B.; formal analysis, Q.Z., Q.W. (Qingjiang Wang) and Q.W. (Qingkai Wang); visualization, Y.W., Q.W. (Qingjiang Wang) and Q.W. (Qingkai Wang); writing—original draft preparation, Q.Z. and J.H.; writing—review and editing, Y.W., B.Z., P.L. and X.B.; project administration, Y.W.; funding acquisition, Y.W. All authors have read and agreed to the published version of the manuscript.

Funding: This work was supported by a project of the Inner Mongolia Autonomous Region Natural Science Foundation (grant number 2023MS04006).

Data Availability Statement: The original contributions presented in the study are included in the article; for a more detailed data request, please contact the corresponding author directly.

Conflicts of Interest: The authors declare no conflicts of interest.

Appendix A. Beginning Time, Ending Time, Days, Negative Accumulated Temperature, and Cumulative Snowfall for Each Research Period from 1993 to 2023

Number	Beginning Time	Ending Time	Days	Negative Accumulated Temperature (°C·d)	Cumulative Snowfall (mm)
1993–1994	28 October 1993	29 March 1994	152	2749.2	29.5
1994–1995	6 November 1994	27 March 1995	141	2211.2	13.6
1995–1996	29 October 1995	5 April 1996	159	2511.2	19.7
1996–1997	23 October 1996	25 March 1997	153	3009.9	30.6
1997–1998	20 October 1997	25 March 1998	156	2509.2	17.7
1998–1999	30 October 1998	5 April 1999	157	2746.4	21.6
1999–2000	20 October 1999	6 April 2000	169	3228.8	47.9
2000–2001	2 November 2000	3 April 2001	152	3150.8	15.7
2001–2002	30 October 2001	10 March 2002	131	2367.5	22.6
2002–2003	17 October 2002	29 March 2003	163	3076.2	24.2
2003–2004	1 November 2003	25 March 2004	145	2811.3	28.2
2004–2005	4 November 2004	1 April 2005	148	2869.3	22.1
2005–2006	6 November 2005	31 March 2006	145	2811.2	28.9
2006–2007	4 November 2006	6 April 2007	153	2770.8	33.1
2007–2008	25 October 2007	7 March 2008	134	2524.6	13.1
2008–2009	23 October 2008	1 April 2009	160	2858.5	41.1
2009–2010	6 November 2009	17 April 2010	162	3249.3	35.2
2010–2011	6 November 2010	2 April 2011	147	3016.1	40.2
2011–2012	2 November 2011	27 March 2012	146	3433.6	46.9
2012–2013	27 October 2012	19 April 2013	174	3490.7	59.1
2013–2014	5 November 2013	22 March 2014	137	2519.5	33.4
2014–2015	30 October 2014	24 March 2015	145	2474.3	25.7
2015–2016	14 November 2015	19 March 2016	126	2636.3	18.6
2016–2017	17 October 2016	27 March 2017	161	2837.5	21.5
2017–2018	28 October 2017	23 March 2018	146	3072.4	18.4
2018–2019	3 November 2018	16 March 2019	133	2135.0	4.4
2019–2020	1 November 2019	23 March 2020	143	2739.9	19.7
2020–2021	12 November 2020	12 March 2021	120	2263.0	4.6
2021–2022	4 November 2021	8 March 2022	124	2344.9	9.8
2022–2023	31 October 2022	27 March 2023	147	2766.3	22.9
2023–2024	30 October 2023	26 March 2024	148	3135.0	34.9

Appendix B. The Maximum Landing Weight (MLW) for Some Aircraft Types, the Needed Ice Thickness, and the Needed Length on an Ice Runway (Covered with Compacted Snow)

Aircraft Types	The MLW (t)	Needed Ice Thickness (cm)	Needed Length on Ice Runway Covered with Compacted Snow (m)
CZAW SportCruiser LSA	0.6	10	200
Cessna 172N	1	13	470
Cessna Caravan	4	24	1000
Cessna Citation CJ4 Gen2	7	34	1660
Saab 340	13	46	2060
Bombardier CRJ700	30	71	2500
Embraer 170	33	74	2630
Bombardier Global Express	36	77	2700
ARJ21-700	38	79	2720
ERJ 190	43	84	3290
Airbus A220-100	51	92	2340
Boeing 737-700	58	98	2930
Airbus A319neo	63	102	2960

Appendix C. The Details of P-III Analysis

The probability density function of P-III distribution is:

$$f(x) = \frac{\beta^{\alpha}}{\Gamma(\alpha)}(x - x_0)^{\alpha-1}\exp[-\beta(x - x_0)] \tag{3}$$

where $\Gamma(\alpha)$ is the gamma function, and α, β, and x_0 are the shape, scale, and position coefficients, respectively. Refer to the below for details.

$$\begin{cases} \alpha = 4/C_s{}^2 \\ \beta = 2/\overline{x}C_vC_s \\ x_0 = \overline{x}(1 - 2C_v/C_s) \end{cases} \tag{4}$$

where Cs is skewness coefficient, Cv is variation coefficient, and $\overline{x}$ is the average.

With a 50-year return period—that is, a frequency of 2%—Figure 12 illustrates the upper-limit cumulative snowfall throughout winter: 58.4 mm. This number represents the maximum snowfall that occurs once every 50 years, with the associated snow cover serving as the greatest limit. Snowfall can negatively impact the minimum ice thickness and limit aircraft operations on ice surfaces. Figure 12 delineates the cumulative snowfall in the study period over return periods of 45 years, 30 years, 25 years, 20 years, 15 years, and 10 years.

References

1. Abele, G.; Frankenstein, G.E. *Snow and Ice Properties as Related to Roads and Runways in Antarctica*; US Army Cold Regions Research & Engineering Laboratory: Hanover, NH, USA, 1967; Volume 176.
2. White, G.; McCallum, A. Review of ice and snow runway pavements. *Int. J. Pavement Res. Technol.* **2018**, *11*, 311–320. [CrossRef]
3. Masterson, D.M. State of the art of ice bearing capacity and ice construction. *Cold Reg. Sci. Technol.* **2009**, *58*, 99–112. [CrossRef]
4. Blaisdell, G.L.; Lang, R.M.; Crist, G.; Kurtti, K.; Harbin, R.J.; Flora, D. *Construction, Maintenance, and Operation of a Glacial Runway, McMurdo Station, Antarctica*; ERDC/CRREL M-98-1; US Army Corps of Engineers, Engineer Research and Development Center: Washington, DC, USA, 1998.
5. Sharp, R.P. Suitability of ice for aircraft landings. *Trans. Am. Geophys. Union* **1947**, *28*, 111–119. [CrossRef]
6. Flight Safety Foundation. Landing Distances. In *FSF ALAR Briefing Note*; Flight Safety Foundation: Alexandria, VA, USA, 2000.
7. Swithinbank, C. *Ice Runways Near the South Pole*; ERDC/CRREL R-89-19; US Army Corps of Engineers, Engineer Research and Development Center: Washington, DC, USA, 1989.

8. Squire, V.A.; Robinson, W.H.; Langhorne, P.J.; Haskell, T.G. Vehicles and aircraft on floating ice. *Nature* **1988**, *333*, 159–161. [CrossRef]
9. McCallum, A.B. Movement and expected lifetime of the Casey ice runway. In Proceedings of the 13th International Conference on Cold Regions Engineering, Orono, ME, USA, 23–26 July 2006; American Society of Civil Engineers (ASCE): Reston, VA, USA, 2006; pp. 1–19.
10. Editorial Committee of Encyclopedia of Rivers and Lakes in China. *Encyclopedia of Rivers and Lakes in China: Section of Heilong River and Liaohe River Basins*; China Water & Power Press: Beijing, China, 2014; p. 17. (In Chinese)
11. Lu, J.; Xia, C. Investigation for emphasis cold water domain of inland in China Investigation for the cold water domain and its resources in Neimonggu autonomous region. *Chin. J. Fish.* **2005**, *18*, 47–52. (In Chinese)
12. Graf, R. Estimation of the Dependence of Ice Phenomena Trends on Air and Water Temperature in River. *Water* **2020**, *12*, 3494. [CrossRef]
13. Mousa, R.M.; Mumayiz, S.A. Optimization of runway orientation. *J. Transp. Eng.* **2000**, *126*, 228–236. [CrossRef]
14. Ashford, N.; Wright, P.H. *Airport Engineering*, 3rd ed.; John Wiley & Sons, Inc: New York, USA, 1992.
15. Sarsam, S.I. Runway orientation problems–A case study of Middle Euphrates International Airport MEIA. In Proceedings of the 2nd International Conference of Buildings, Construction and Environmental Engineering (BCEE2-2015), Beirut, Lebanon, 17–18 October 2015.
16. Haehnel, R.B.; Blaisdell, G.L.; Melendy, T.D.; Shoop, S.; Courville, Z. *A Snow Runway for Supporting Wheeled Aircraft: Phoenix Airfield, McMurdo, Antarctica*; ERDC/CRREL TR-19-4; US Army Corps of Engineers, Engineer Research and Development Center: Washington, DC, USA, 2019.
17. Lotsari, E.; Lind, L.; Kämäri, M. Impacts of Hydro-Climatically Varying Years on Ice Growth and Decay in a Subarctic River. *Water* **2019**, *11*, 2058. [CrossRef]
18. Brooks, R.N. Quantifying Peak Freshwater Ice Across the Northern Hemisphere Using a Regionally Defined Degree-Day Ice-Growth Model. Master's Thesis, University of Victoria, Victoria, BC, Canada, 2012.
19. Korhonen, J. Long-term changes in lake ice cover in Finland. *Hydrol. Res.* **2006**, *37*, 347–363. [CrossRef]
20. Solarski, M.; Rzetala, M. A Comparison of Model Calculations of Ice Thickness with the Observations on Small Water Bodies in Katowice Upland (Southern Poland). *Water* **2022**, *14*, 3886. [CrossRef]
21. Stefan, J. Ueber die Theorie der Eisbildung, insbesondere über die Eisbildung im Polarmeere. *Ann. Phys.* **1891**, *278*, 269–286. [CrossRef] (In German)
22. Leppäranta, M. A review of analytical models of sea-ice growth. *Atmos. Ocean* **1993**, *31*, 123–138. [CrossRef]
23. Husselmann, M.L.; Van Zyl, J.E. Effect of stand size and income on residential water demand. *J. S. Afr. Inst. Civ. Eng.* **2006**, *48*, 12–16. Available online: https://hdl.handle.net/10520/EJC26987 (accessed on 9 December 2024).
24. Lin, L. Study on Effect of Ice Deformation on Reservoir Piers. Master's Thesis, Dalian University of Technology, Dalian, China, 2023. (In Chinese)
25. Schiff, B. Antonov AN-2 The Flying Bear. Available online: https://www.aopa.org/news-and-media/all-news/2001/october/pilot/antonov-an-2 (accessed on 9 December 2024).
26. Aero Corner. Bombardier Q400. Available online: https://aerocorner.com/aircraft/bombardier-q400/ (accessed on 9 December 2024).
27. Austrian Airlines. Bombardier Q400. Available online: https://www.austrian.com/nl/en/bombardier-q400 (accessed on 9 December 2024).
28. GlobalAir.com. Gulfstream G650ER. Available online: https://www.globalair.com/aircraft-for-sale/specifications?specid=1520 (accessed on 9 December 2024).
29. Rocco. Lockheed C-5 Galaxy vs. Lockheed C-130 Hercules. Available online: https://aerocorner.com/comparison/lockheed-c-5-galaxy-vs-lockheed-c-130-hercules/ (accessed on 9 December 2024).
30. Ultimate Specs. C-17 Globemaster III Technical Data. 1991. Available online: https://www.ultimatespecs.com/aircraft-specs/boeing/boeing-c-17-globemaster-iii-1991 (accessed on 9 December 2024).
31. GlobalAir.com. CZAW SportCruiser LSA. Available online: https://www.globalair.com/aircraft-for-sale/specifications?specid=1378 (accessed on 9 December 2024).
32. GlobalAir.com. Cessna 172N. Available online: https://www.globalair.com/aircraft-for-sale/specifications?specid=1237 (accessed on 9 December 2024).
33. Textron Aviation Inc. Citation CJ4 Gen2. Available online: https://cessna.txtav.com/citation/cj4-gen2#_model-specs (accessed on 9 December 2024).
34. English Wikipedia. List of Airliners by Maximum Takeoff Weight. Available online: https://en.wikipedia.org/wiki/List_of_airliners_by_maximum_takeoff_weight#cite_note-27 (accessed on 9 December 2024).
35. Li, Z.; Fu, X.; Shi, L.; Huang, W.; Li, C. Recent Advances and Challenges in the Inverse Identification of Thermal Diffusivity of Natural Ice in China. *Water* **2023**, *15*, 1041. [CrossRef]

36. Shi, L.; Li, Z.; Lu, P.; Feng, E. Influence of ice temperature on thermal diffusivity of natural freshwater ice. *J. Hydraul. Eng.* **2013**, *44*, 1112–1117. [CrossRef] (In Chinese)
37. Wang, T. Exploring of the Relationship between Cold Ice Thickness Reservoir of Growth and Negative Accumulated Temperature. Bohai University: Jinzhou, China, 2011. (In Chinese)
38. Shen, H.T.; Yapa, P.D. A unified degree-day method for river ice cover thickness simulation. *Can. J. Civ. Eng.* **1985**, *12*, 54–62. [CrossRef]
39. Gu, H.; Yu, Z.; Li, G.; Ju, Q. Nonstationary Multivariate Hydrological Frequency Analysis in the Upper Zhanghe River Basin, China. *Water* **2018**, *10*, 772. [CrossRef]
40. Lu, Y. Impacts of Climate Change on Hydrological Hazards: Mechanisms, Predictions and Coping Strategies. *Appl. Math. Nonlinear Sci.* **2024**, *9*, 1–17. [CrossRef]
41. Mo, C.; Jiang, C.; Lei, X.; Cen, W.; Yan, Z.; Tang, G.; Li, L.; Sun, G.; Xing, Z. Optimal Scheduling of Reservoir Flood Control under Non-Stationary Conditions. *Sustainability* **2023**, *15*, 11530. [CrossRef]
42. Xu, K.; Ma, C.; Lian, J.; Bin, L. Joint Probability Analysis of Extreme Precipitation and Storm Tide in a Coastal City under Changing Environment. *PLoS ONE* **2014**, *9*, e109341. [CrossRef]
43. Cui, Y.; Tang, H.; Jin, J.; Zhou, Y.; Jiang, S.; Chen, M. System Structure–Based Drought Disaster Risk Assessment Using Remote Sensing and Field Experiment Data. *Remote Sens.* **2022**, *14*, 5700. [CrossRef]
44. Liu, Z.; Yang, H.; Wang, T.; Yang, D. Estimating the annual runoff frequency distribution based on climatic conditions and catchment characteristics: A case study across China. *Int. Soil Water Conserv. Res.* **2023**, *11*, 470–481. [CrossRef]
45. Ke, Q.; Jonkman, S.N.; Van Gelder, P.H.A.J.M.; Bricker, J.D. Frequency Analysis of Storm-Surge-Induced Flooding for the Huangpu River in Shanghai, China. *J. Mar. Sci. Eng.* **2018**, *6*, 70. [CrossRef]
46. Minh, H.V.T.; Lavane, K.; Lanh, L.T.; Thinh, L.V.; Cong, N.P.; Ty, T.V.; Downes, N.K.; Kumar, P. Developing Intensity-Duration-Frequency (IDF) Curves Based on Rainfall Cumulative Distribution Frequency (CDF) for Can Tho City, Vietnam. *Earth* **2022**, *3*, 866–880. [CrossRef]
47. Ashton, G.D. Thin ice growth. *Water Resour. Res.* **1989**, *25*, 564–566. [CrossRef]
48. Ashton, G.D. River and lake ice thickening, thinning, and snow ice formation. *Cold Reg. Sci. Technol.* **2011**, *68*, 3–19. [CrossRef]

water

MDPI

Article

Optimization of a Snow and Ice Surface Albedo Scheme for Lake Ulansu in the Central Asian Arid Climate Zone

Xiaowei Cao [1,*], Miao Yu [2], Puzhen Huo [2], Peng Lu [2], Bin Cheng [3], Wei Gao [1], Xingyu Shi [1] and Lijun Wang [4,*]

1 School of Architectural Engineering, Qingdao Huanghai University, Qingdao 266555, China; gaow@qdhhc.edu.cn (W.G.); shixy11@qdhhc.edu.cn (X.S.)
2 State Key Laboratory of Coastal and Offshore Engineering, Dalian University of Technology, Dalian 116023, China; yumiao2411@dlut.edu.cn (M.Y.); hpz_ice@mail.dlut.edu.cn (P.H.); lupeng@dlut.edu.cn (P.L.)
3 Finnish Meteorological Institute, 00101 Helsinki, Finland; bin.cheng@fmi.fi
4 Huacheng Boyuan Engineering Technology Group Co., Ltd., Beijing 100043, China
* Correspondence: caoxw11@qdhhc.edu.cn (X.C.); 13901212966@sina.cn (L.W.)

Abstract: Surface albedo measurements of snow and ice on Lake Ulansu in the Central Asian arid climate zone were conducted during the winter of 2016–2017. Observations were categorized into three stages based on the ice growth and surface condition: bare ice, snow cover, and melting. During the bare ice stage, the mean surface albedo was 0.35 with a decreasing trend due to the accumulation of wind-blown sediment on the ice surface (range: 0.99–1.87 g m^{-2}). Two snowfall events occurred during the snow cover stage, significantly increasing the surface albedo to 0.91. During the melting stage, the albedo decreased at a decay rate of 0.20–0.30/day. Four existing albedo schemes were evaluated but found unsuitable for Lake Ulansu. A new surface albedo scheme was proposed by incorporating the existing albedo schemes with the measured data. This scheme incorporated the effect of sediment content on bare ice albedo for the first time. It demonstrated a modelling efficiency of 0.933 over the entire 3-month period, which was used to evaluate the fit between the predicted and observed values. When validated with albedo observations from other winters, it achieved a modelling efficiency of 0.940. The closer the value is to 1, the better the model's predictive accuracy, indicating a higher level of reliability in the model's performance. This scheme has potential applicability to other lakes in the Central Asian arid climate zone, which is characterized by low precipitation, frequent sandstorms, and intense solar radiation.

Keywords: albedo scheme; lake ice; snow; sediment content; optimization

Academic Editor: Hung Tao Shen

Received: 25 December 2024
Revised: 22 January 2025
Accepted: 26 January 2025
Published: 12 February 2025

Citation: Cao, X.; Yu, M.; Huo, P.; Lu, P.; Cheng, B.; Gao, W.; Shi, X.; Wang, L. Optimization of a Snow and Ice Surface Albedo Scheme for Lake Ulansu in the Central Asian Arid Climate Zone. *Water* **2025**, *17*, 523. https://doi.org/10.3390/w17040523

1. Introduction

Lake ice and snow albedo, which refers to the fraction of solar energy reflected by these surfaces, have a substantial impact on the energy balance of lakes and their surrounding environments [1–6]. Seasonal ice-covered lakes are a key component of the terrestrial landscape in the Northern Hemisphere [7,8]. Due to rising global air temperatures, numerous studies have reported a significant reduction in the annual freezing period of lakes [7,9–11]. Predictive studies on Northern Hemisphere temperatures indicate that the duration of seasonal ice cover will continue to decrease in the future [12–14]. This reduction in lake ice during winter can significantly impact water resources, ecological stability, transportation, fisheries, and human activities [15]. The Central Asian arid climate zone plays a critical role in the global climate system [16], with warming rates in this region occurring at a pace faster than the global average [17]. This has further exacerbated the reduction of seasonal lake ice in the zone [18]. When studying lake ice and snow in this area, it is essential to consider its unique climatic characteristics, such as low precipitation (approximately 8 mm during the winter), frequent sandstorms (approximately 15 days per year), and intense solar radiation (approximately 200 W/m^2 during the freezing period) [3,19–21]. For winter lake ice surfaces, two main conditions exist: with snow cover and without. Key factors affecting snow cover albedo include snow water content, impurity level [22], grain size [23], snow density, and surface roughness [24,25]. Dry snow has an albedo as high as 0.9 that decreases with increasing liquid water, while for wet snow, it is typically around 0.5 [26–28]. Factors affecting the albedo of lake ice include the presence of air bubbles within the ice, ice crystal structure, ice thickness, and surface roughness [29]. Dry ice has an albedo of approximately 0.5, decreasing to 0.2–0.3 during the melting season [3,8]. Many existing snow and ice albedo schemes are empirical and rely on easily obtainable climatic data [30–33]. These schemes typically estimate snow albedo using parameters such as snow depth and the number of days since snowfall [30,34] and estimate lake ice albedo using ice thickness and surface temperature [35–37]. While these parameterizations offer reasonable approximations of snow and ice albedo, their applicability is limited. They rely on the statistical fitting of albedo measurements that reflect specific snow, ice, and atmospheric conditions at particular times and locations [38]. These parameterization schemes are primarily applied to high-latitude lakes and do not take into account the climatic characteristics of the Central Asian arid climate zone. Additionally, these schemes have not been validated for lake ice and snow in this particular region.

In summary, although lake ice albedo in the Central Asian arid climate zone plays an important role in climate system studies, a snow and ice albedo scheme for its entire ice period is lacking. In this study, a field investigation was conducted to measure the snow and ice cover albedo of a typical lake (Lake Ulansu [3,18,39]) in the Central Asian arid climate zone and to validate existing snow and ice albedo schemes. Additionally, a new snow and ice albedo scheme was optimized by refining the existing schemes and incorporating climatic characteristics and the measured data. The accuracy of the new scheme was further validated using 3 years of winter surface albedo measurements.

2. Materials and Methods

2.1. Study Area

Lake Ulansu is located in Inner Mongolia, China at coordinates 40°36′–41°30′ N, 108°43′–108°70′ E. It covers an area of 306 km^2, with maximum length and width of 35.4 km and 12.7 km, respectively. The water volume of the lake can reach 2.5–3 × 10 m^3 [40], and the maximum and mean water depths are 2.5–3.0 and 1.0–1.5 m [18], respectively. Existing survey results show that the maximum ice thickness in winter ranges from 0.3 to 0.6 m [17,40,41]. This field study was conducted from 1 January to 9 March in 2017 in the eastern part of the lake (40.9° N, 108.9° E), approximately 200 m from the shore (Figure 1). The water depth at the observation point was 1.7 m, and the maximum ice thickness was 0.57 m, both of which fall within the average range for the Lake Ulansu during the winter and are representative.

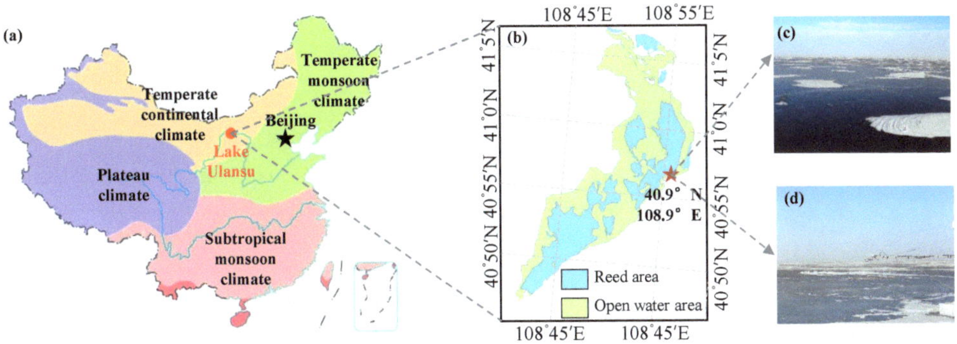

Figure 1. Location of Lake Ulansu (**a**) [18], measurement site (Universal Transverse Mercato projection and plotted using QGIS (Quantum GIS 3.34)) (**b**), ice cover on the lake (**c**), and solar radiation measurement (**d**).

2.2. Data Collection

In situ measurements of solar radiation, atmospheric surface layer conditions (temperature, humidity, and wind speed and direction), and snow and ice thickness were conducted. Figure 2 illustrates the equipment used to measure solar radiation and atmospheric conditions, with further details provided in a previous study [39]. Solar radiation was measured using two pyranometers (TBQ-2, Jinzhou Sunshine Meteorological Technology Co., Ltd., Jinzhou, China, http://pvsb.china-nengyuan.com/member_product/78658.html, accessed on 1 January 2017.): one for incoming radiation and one for reflected radiation. The TBQ-2 measurement range was 300–3000 nm, with a sensor accuracy of 5%, and its manufacturing and calibration standards followed the guidelines specified in QX/T 55-2007.

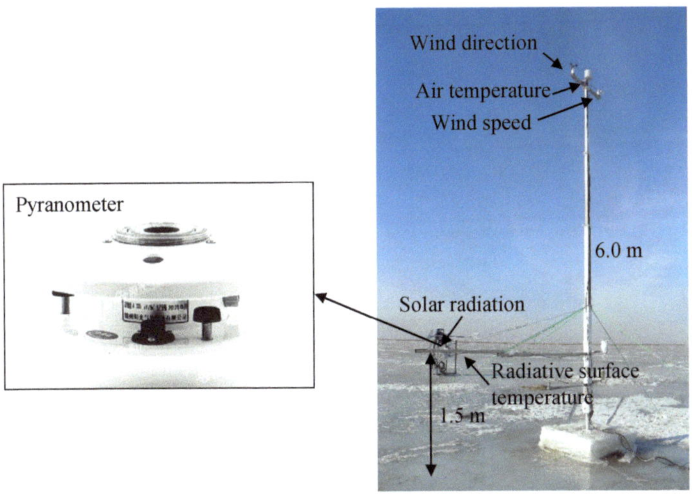

Figure 2. Meteorological tower measurements at the site, and the pyranometers used, in winter 2016–2017.

In arid and semi-arid regions, environmental conditions often result in the adherent of a layer of sediment to the lake ice surface. To quantify the sediment content, the sediment per unit area was measured. A 0.5 × 0.5 m area was delineated on the ice surface (Figure 3), and surface sediment was collected from this area. The sediment was then dried and weighed to calculate its content per unit area (g m^{-2}), following the procedures outlined in the GB11901–1989 standard [42]. The detailed steps are as follows:

1. Place the filter membrane in a drying oven and dry it at a temperature of 103–105 °C for 2 h. Remove it and place it in a desiccator to cool to room temperature, then weigh it to obtain the net weight of the filter membrane.
2. Press the filter membrane tightly against the inner wall of the funnel. Using a vacuum filtration method, filter the water sample containing surface sediment.
3. After filtration, the sediment will adhere to the filter membrane. Place the filter membrane back into the drying oven and dry it again at the same temperature for 2 h. Remove it and place it in a desiccator to cool to room temperature, then weigh it to obtain the total weight of the filter membrane and the sediment.
4. Subtract the net weight of the filter membrane from the total weight to obtain the weight of the sediment.

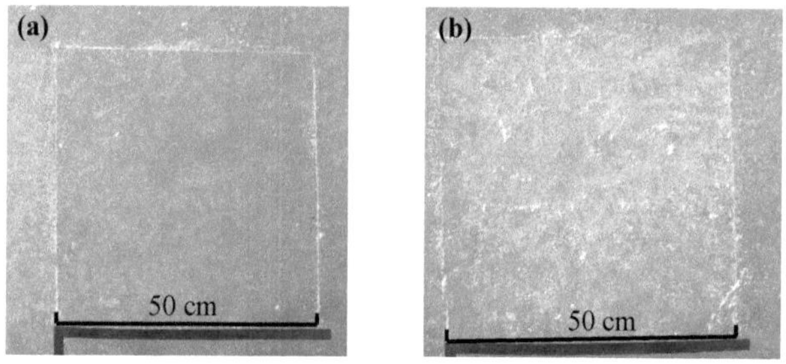

Figure 3. Sediment collection area: pre- (**a**) and post-collection (**b**).

Sediment content measurements were taken daily during the snow-free observation stage, with each day's sampling location adjacent to the previous day's location.

2.3. Existing Surface Albedo Schemes

Albedo schemes previously applied to predict lake ice and snow albedo were selected. These included those proposed by Gabison [36], Shine and Henderson-Sellers (SH), Flato and Brown (FB) [35], and Henneman and Stefan (HS) [30].

2.3.1. Gabison Scheme

The Gabison scheme, developed based on the method suggested by Maykut [35] and combined with field measurement data, requires parameters such as ice and snow thickness and surface temperature. Albedo calculations are divided into two stages based on ice surface temperature: freezing ($T_s \leq -3$ °C) and melting ($T_s \geq -3$ °C) stages. During the freezing stage, the albedo of the ice surface (α_i) is derived using quadratic regression with respect to ice thickness (h_i), and is within the range 0.05 m $\leq h_i \leq$ 1 m.

$$\alpha_i = 0.21 + 1.026h_i - 0.561h_i^2, \tag{1}$$

The albedo of snow (α_s) is calculated based on the ice albedo and snow thickness:

$$\alpha_s = \alpha_i + (0.8 - \alpha_i)\frac{h_s}{0.05}, \tag{2}$$

where h_s is the depth of the snow cover.

During the melting stage, the albedo of ice with a thickness of 1 m or more decreases linearly from 0.72 to 0.47. A similar linear decrease is applied to thinner ice (0.05 m $\leq h_i \leq$ 1 m) during the melting stage (Figure 4).

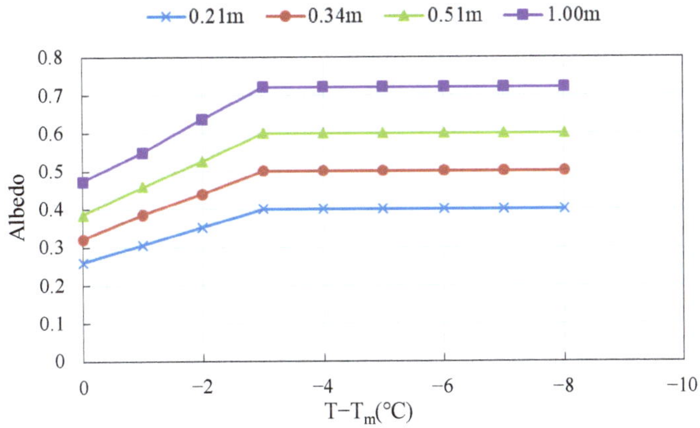

Figure 4. Dependence of ice albedo on ice thickness and melting temperature deficit [36].

2.3.2. Shine and Henderson-Sellers (SH) Scheme

Shine and Henderson-Sellers summarized previous albedo schemes and proposed new schemes for different ice surface types. These schemes require parameters such as snow depth and ice thickness. The specific classifications and calculation formulas are presented in Table 1.

Table 1. Shine and Henderson-Sellers scheme.

Albedo Class	Value
Dry snow	$\alpha_s = 0.8$
Melting snow	$\alpha_s = 0.65$
Thin melting snow on bare ice ($h_s \leq 0.1$ m)	$\alpha_s = 0.53 + 1.2h_s$
Bare ice ($h_i \geq 1.5$ m)	$\alpha_i = 0.53$
Thin melting ice ($1.5 \geq h_i \geq 0$ m)	$\alpha_i = 0.47 + 0.21(h_i - 1.0),\ 1.5 \geq h_i \geq 1.0$ m $\alpha_i = 0.25 + 0.70h_i - 0.86h_i^2 + 0.38h_i^3,\ 1.0 \geq h_i \geq 0.05$ m $\alpha_i = 0.1 + 3.6h_i,\ \ h_i \leq 0.05$ m
Thin forming ice ($1.5 \geq h_i \geq 1.0$ m)	$\alpha_i = 0.47 + 0.50(h_i - 1.0),\ 1.0 \geq h_i \geq 0.05$ m $\alpha_i = 0.25 + 0.70h_i - 0.86h_i^2 + 0.38h_i^3,\ 1.0 \geq h_i \geq 0.05$ m $\alpha_i = 0.1 + 3.6h_i,\ \ h_i \leq 0.05$ m
Bare frozen ice ($h_i \geq 1.5$ m)	$\alpha_i = 0.72$
Snow on frozen ice	$\alpha_s = 0.8,\ h_s \geq 0.5$ m $\alpha_s = \alpha_i + \frac{h_s(0.8-\alpha_i)}{0.05},\ h_s \leq 0.5$ m; $h_i \leq 1.5$ m

2.3.3. Flato and Brown (FB) Scheme

In the FB scheme, the freezing ice albedo is derived from Maykut [43], while the melting ice albedo is based on Heron and Woo [44]. Surface temperature (T_s) is incorporated, and ice and snow are classified into two types (freezing and melting) based on the surface temperature.

$$\text{Albedo} = \begin{cases} \min[\alpha_s,\ \alpha_i + h_s(\alpha_s - \alpha_i)/0.1] & h_i > 0.001\text{ m}\ \ h_s < 0.1\text{ m} \\ \alpha_s & h_i > 0.001\text{ m}\ \ h_s > 0.1\text{ m} \end{cases}, \tag{3}$$

$$\begin{cases} \alpha_i = \max\left(0.15,\ 0.44h_i^{0.28} + 0.08\right) \\ \alpha_s = 0.75 \end{cases},\ T_s < 0\,^{\circ}\mathrm{C},$$
$$\begin{cases} \alpha_i = \min\left(0.55,\ 0.075h_i^{2} + 0.15\right) \\ \alpha_s = 0.65 \end{cases},\ T_s = 0\,^{\circ}\mathrm{C}, \quad (4)$$

2.3.4. Henneman and Stefan (HS) Scheme

The HS scheme improves upon the FB snow albedo scheme by calculating snow albedo based on accumulated solar radiation and air temperature. It introduces parameters such as accumulated solar radiation, air temperature, and the number of days since the last snowfall. The accumulation of solar radiation and air temperature is reset with each new snowfall day, which is defined as a day on which snowfall exceeds 2.5 mm (i.e., a snowfall greater than a trace amount).

Accumulated solar radiation is the sum of daily radiation received from the time of a new snowfall until the next snowfall day. Accumulated air temperature is calculated using a degree–day method with a calibrated base temperature, T_{base}. A temperature index, T_{index}, for each day is computed as:

$$T_{index} = T_a - T_{base}, \quad (5)$$

where T_a is the average daily air temperature (°C) and $T_{base} = -18$°C.

Scheme I:

This requires inputs such as average daily air temperature, accumulated solar radiation, accumulated air temperature, and daily snowfall data.

For the freezing stage, the α_s is given by:

$$\alpha_s = -0.0015R + 0.83, \quad (6)$$

For the melting stage, the α_s is given by:

$$\alpha_s = \begin{cases} 0.0029R - 0.009T + 0.95, & T_a \geq 0\,^{\circ}\mathrm{C} \\ \alpha_{s_{-1}} - 0.00036T + 0.95, & T_a < 0\,^{\circ}\mathrm{C} \end{cases}, \quad (7)$$

where R is accumulated incoming daily solar radiation since the last snowfall (MJ m^{-2}), $\alpha_{s_{-1}}$ is the albedo on the previous day, and T is the accumulated daily air temperature index since the last snowfall (°C).

Scheme II:

This requires only snowfall and average air temperature data to predict the albedo.

For the freezing stage, the α_s is calculated as:

$$\alpha_s = -0.011d + 0.83, \quad (8)$$

For the melting stage, the α_s is given by:

$$\text{Melting stage: } \alpha_s = \begin{cases} \alpha_{s_{-1}} - 0.17, & T_a \geq 0\,^{\circ}\mathrm{C} \\ \alpha_{s_{-1}} - 0.013, & T_a < 0\,^{\circ}\mathrm{C} \end{cases}, \quad (9)$$

where d is the number of days since the last snowfall.

The input parameters, output albedo, and a brief analysis of each scheme are presented in Table 2. From the table, it is evident that none of these schemes consider the effect of sediment content on the albedo of the ice surface. Therefore, the accuracy of their albedo calculations for the ice and snow on Lake Ulansu remains to be validated.

Table 2. Summary of existing albedo schemes.

Albedo Scheme	Input Parameters	Output Albedo	Analysis
Gabison	Ice surface temperature; Ice thickness; Snow thickness	Ice and snow	Does not provide a calculation formula for snow albedo during the melting period.
SH	Ice thickness; Snow thickness	Ice and snow	For snow albedo greater than 0.5 m, only reference values are provided, without considering variations in snow properties.
FB	Ice surface temperature; Ice thickness; Snow thickness	Ice and snow	Similar to the SH scheme, the snow albedo is provided only as a reference, without accounting for variations in snow properties.
HS I	Accumulated solar radiation; Accumulated air temperature	Snow	The acquisition of cumulative temperature and radiation is challenging.
HS II	Days since last snowfall; Average air temperature	Snow	The calculation is relatively simple, but it can only calculate the snow albedo.

3. Results of Existing Schemes

3.1. General Ice Conditions

Figure 5 presents the observed data on incident and reflected solar radiation, average albedo, ice and snow thickness, and daily average air temperature from 1 January to 6 March 2017 on Lake Ulansu. The average daily albedo was calculated by dividing the total daily reflected solar radiation by the total daily incoming solar radiation [30]. The total daily solar radiation was obtained by integrating 10-minute radiation measurements over daylight hours.

Based on the ice and snow thickness data shown in Figure 5a, the observation stage was divided into three distinct phases:

1. Bare ice stage (1 January to 6 February):

During this phase, the ice thickness increased from 33.2 to 50.9 cm, with an average growth rate of 0.49 cm/day. Sediment content on the lake ice surface was measured throughout this stage.

2. Snow cover stage (7 February to 3 March):

During this stage, the ice thickness increased from 50.9 cm to its maximum thickness of 56.9 cm. Two snowfall events occurred in this stage: on 7 February and 21 February. Following the first snowfall, snow thickness decreased from 13 cm to <1 cm within a week. The second snowfall was limited to 2.3 cm and had melted completely by 3 March.

3. Melting stage (4 March to 9 March):

In this stage, the ice thickness decreased from 56.9 to 54.2 cm, with a melting rate of 0.4 cm/day.

Throughout the observation stage, the average sediment content was 1.39 ± 0.21 g m^{-2}, exhibiting a clear increasing trend, with a regression coefficient of 0.02. The average incident and reflected solar radiation were 91.75 and 49.89 W m^{-2}, respectively. During the snow-free freezing stage, the average reflected solar radiation was approximately 58.3 W m^{-2}. Following the snowfalls on 7 February and 21 February, the average reflected irradiance exceeded 200 W m^{-2}, subsequently decreasing as the snow melted. The albedo changes followed the same trend as the reflected solar radiation.

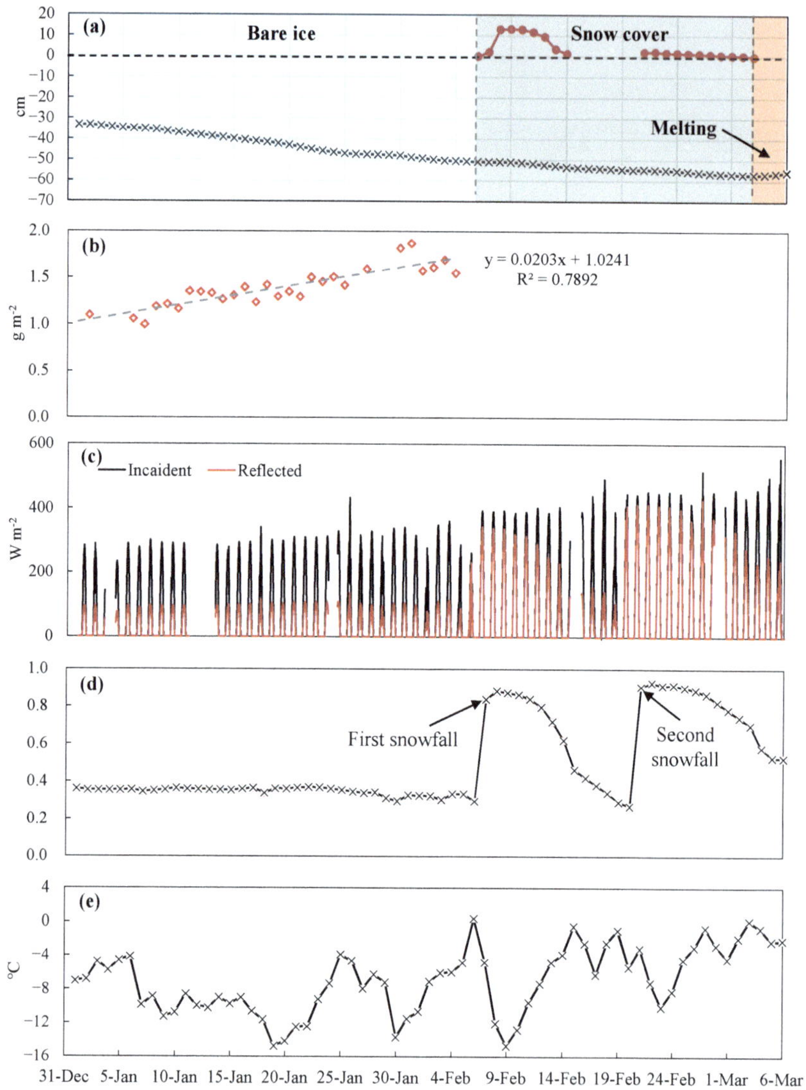

Figure 5. (**a**) Snow and ice thickness, (**b**) sediment content before snowfall, (**c**) incident and reflected solar radiation, (**d**) average albedo, and (**e**) daily average air temperature between 1 January and 6 March 2017.

In the freezing stage, the average albedo was 0.35 ± 0.02. After each snowfall, the albedo increased to approximately 0.91, then gradually decreased as the snow melted. During the melting stage, the average albedo was 0.55 ± 0.07. The average daily air temperature during the observation stage was −6.74 ± 3.95 °C, with the daily average temperature ranging from a maximum of 0.44 °C to a minimum of −14.73 °C.

3.2. Prediction Results for Different Stages

Using the albedo schemes described in Section 2.3, the observational data from Figure 5 were applied to predict the albedo. To assess the accuracy of the predictions, several error analysis metrics were used, including the mean absolute error (MAE), root mean square error (RMSE), mean absolute percentage error (MAPE), and modeling efficiency (EF). These coefficients provide a comprehensive evaluation of scheme performance. The EF serves as an overall indicator of the goodness of fit, with values closer to 1 indicating a better fit between the predicted and observed results. The specific calculation formulas for these metrics are provided in Appendix A.

3.2.1. Bare Ice Stage

The predicted albedo values for the bare ice stage are shown in Figure 6. The SH and FB schemes provided predictions closer to the observed values, while the Gabison scheme predictions exhibited a larger deviation, mainly due to a higher initial albedo value. All three schemes predicted a gradual increase in albedo, in contrast to the observed trend. Table 3 presents the error analysis of the scheme predictions relative to the observed data. According to this analysis, the SH scheme yielded the most accurate results, with MAE, RMSE, MAPE, and EF values of 0.073, 0.078, 21.471%, and −14.574, respectively. Although this scheme provided the closest fit, it still did not meet the required error analysis standards. Specifically, an EF value less than zero indicates that the scheme is not suitable for simulation. All EF values were negative because all predicted values were higher than the observed values. This discrepancy is attributed to the presence of sediment on the ice surface of Lake Ulansu, which reduces albedo compared to that on typical high-latitude lakes [3]. Therefore, none of these three schemes were appropriate for predicting albedo during the bare ice stage at Lake Ulansu.

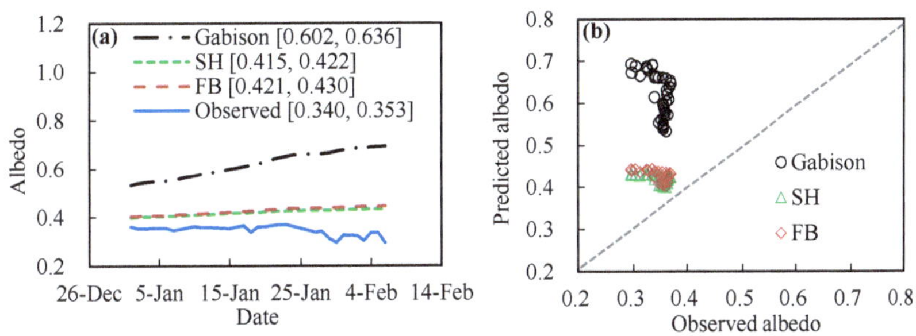

Figure 6. Albedo values observed and predicted by the Gabison, Shine and Henderson-Sellers (SH), and Flato and Brown (FB) schemes during the bare ice stage (**a**), with 95% confidence intervals in brackets, and the comparison between predicted and observed albedo (**b**).

Table 3. Mean absolute error (MAE), root mean square error (RMSE), mean absolute percent error (MAPE), and modeling efficiency (EF) for the Gabison, SH, and FB schemes during the bare ice stage.

Statistical Parameter	Gabison	SH	FB
MAE	0.273	0.073	0.080
RMSE	0.281	0.078	0.085
MAPE	79.866%	21.471%	23.506%
EF	−203.179	−14.574	−17.662

3.2.2. Snow Cover Stage

During the snow cover stage, the HS scheme (Equations (6) and (7)) was also incorporated. As shown in Figure 7, all schemes captured the general increasing and decreasing albedo trends following snowfall. However, the albedo predictions from the FB and Gabison schemes exhibited larger deviations, primarily due to unrealistic initial value settings. The SH scheme predicted a decline in albedo that was too rapid, with both the minimum and maximum values deviating to some extent from the observed data. Among the two HS scheme variants, scheme II provided predictions that aligned more closely with the observed data than those of scheme I. However, the minimum albedo value predicted by scheme II approached 0, which is significantly lower than the observed minimum value of 0.26.

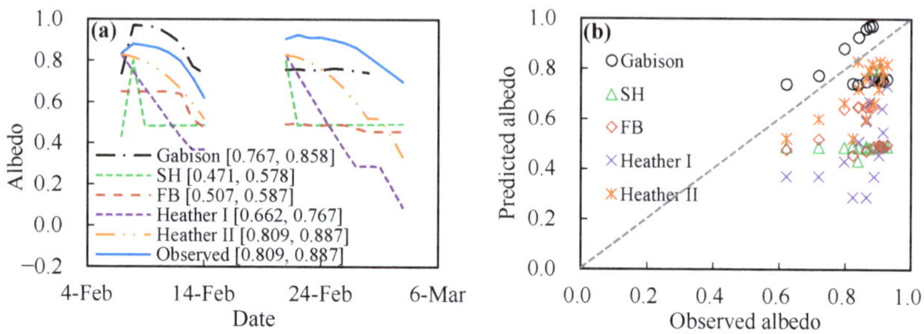

Figure 7. Albedo values observed and predicted by the Gabison, SH, FB, and Hennemen and Stefan (HS) I and II schemes during the snow cover stage (**a**), with 95% confidence intervals in brackets, and the comparison between predicted and observed albedo (**b**).

The error analysis results in Table 4 show that SH scheme II produced the smallest deviation from the observed values. The EF values for SH, FB, and HS I were all less than 0, indicating that they were unsuitable for predicting snow albedo on Lake Ulansu during winter. Although the EF value for the Gabison scheme was greater than 0, it remained lower than that of HS II. From the observation results, it can be seen that the albedo increased to 0.91 after snowfall and then gradually decreased as the snow melted. Notably, the second snowfall was about 8 cm less than the first, which indicates a low correlation between snow depth variation and albedo reduction in Lake Ulansu. Consequently, the prediction results from the Gabison, SH, and FB schemes, which rely on snow depth to calculate albedo, are less accurate. On the other hand, the HS I scheme relies on cumulative temperature and radiation for its calculations. However, the temperature and cumulative radiation in Lake Ulansu differ significantly from those in high-latitude lakes, making the HS I scheme less accurate than the HS II scheme.

Table 4. MAE, RMSE, MAPE, and EF for the Gabison, SH, FB, and HS schemes during the snow cover stage.

Statistical Parameter	Gabison	SH	FB	HS I	HS II
MAE	0.180	0.271	0.244	0.281	0.137
RMSE	0.220	0.303	0.279	0.314	0.155
MAPE	39.040%	38.600%	33.682%	43.583%	23.620%
EF	0.108	−0.740	−0.478	−0.874	0.546

3.2.3. Melting Stage

For the melting stage, the albedo results (Figure 8) indicated that the FB and SH schemes were numerically closer to the measured values than the Gabison scheme. However, all three schemes significantly diverged from the observed trend. This was confirmed by the error analysis in Table 5, which shows that the EF values for all schemes were below 0, indicating that none of these were suitable for simulating ice albedo on Lake Ulansu during the melting stage.

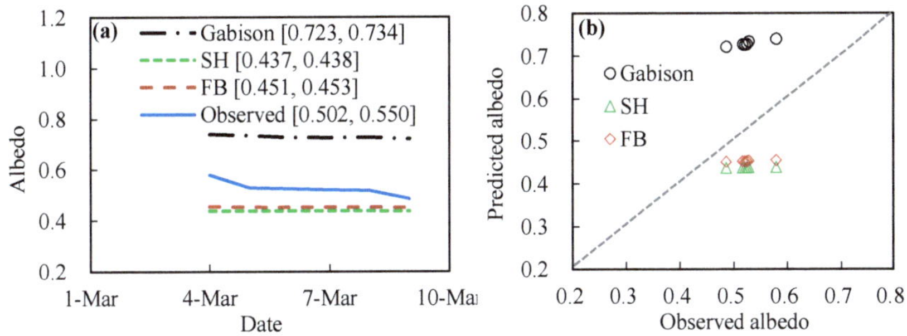

Figure 8. Albedo values observed and predicted by the Gabison, SH, and FB schemes during the snow melting stage (**a**), with 95% confidence intervals in brackets, and the comparison between predicted and observed albedo (**b**).

Table 5. MAE, RMSE, MAPE, and EF for the Gabison, SH, and FB schemes during the melting stage.

Statistical Parameter	Gabison	SH	FB
MAE	0.144	0.160	0.144
RMSE	0.167	0.191	0.178
MAPE	27.111%	24.635%	22.001%
EF	−1.515	−2.276	−1.839

4. A New Albedo Scheme Based on Observations

4.1. New Albedo Scheme Development

Based on the results from Section 3.2, a new albedo scheme for lake ice and snow was optimized for the bare ice, snow cover, and melting stages.

4.1.1. Bare Ice Stage

As shown in Figure 5, while the scheme predictions were relatively close to the measured values, they exhibited an opposite trend compared to the observed data. This discrepancy was primarily due to the sediment layer on the ice surface, which showed a clear increasing trend (Figure 5). To address this, the effect of surface sediment content on albedo was incorporated into the parameterization for the bare ice stage. Based on the error analysis in Table 2, the SH scheme was selected for optimization during this stage. In the SH scheme, albedo is calculated using ice thickness as the primary parameter. Given that changes in ice thickness closely align with variations in surface sediment content, a linear fit of ice thickness (h_i) and surface sediment content (G_s) was performed using the least squares method. The resulting formula is shown in Equation (10), with the fitting results presented in Figure 9.

$$G_s = -0.13 + 3.56h_i, \quad (10)$$

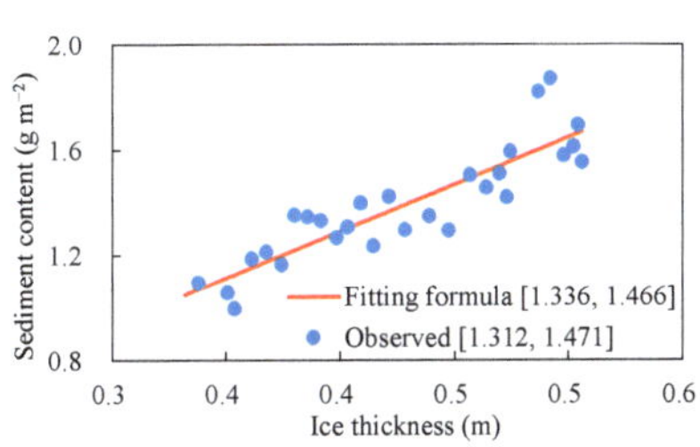

Figure 9. Fitting result of the ice thickness and sediment content before snowfall, with 95% confidence intervals in brackets.

Figure 9 shows the linear fit between ice thickness and sediment content prior to snowfall. The fit was highly accurate, with an EF value of 0.778. Using the sediment content and ice thickness data, a least squares method was applied to develop the albedo scheme, resulting in a best-fitting equation of (11):

$$\alpha_i = -0.10 - 0.18 \times h_i + 0.49 \times G_s - 0.18 \times G_s^2, \quad (11)$$

By combining Equations (10) and (11), the albedo scheme for the bare ice stage of Lake Ulansu was derived as:

$$\alpha_i = \max\left(0.2, 0.03 + 1.74 \times h_i - 2.29 \times h_i^2\right), \ 0 < h_i \leq 0.60 \text{ m}, \quad (12)$$

The predicted albedo results, shown in Figure 10, demonstrated that after adjusting the formula coefficients to incorporate surface sediment content, the calculated albedo trend aligned well with the observed data, and the values also closely matched. The error analysis further revealed that the adjusted scheme provided satisfactory results, with an MAE of 0.009, RMSE of 0.012, MAPE of 2.748%, and EF of 0.695.

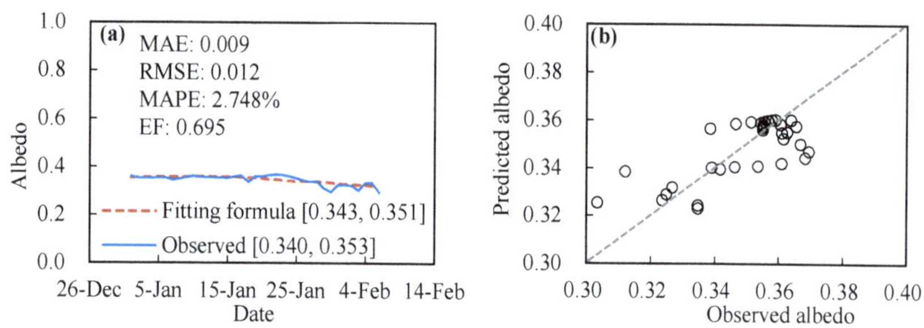

Figure 10. Observation and fitting results of the albedo during the bare ice stage (**a**), with 95% confidence intervals in brackets, and the predicted vs. observed albedo (**b**).

4.1.2. Snow Cover Stage

The main factors influencing snow albedo are snow thickness and its physical properties. As shown in Figure 5a,d, although the snow thickness varied by 8 cm between the two snowfall events, the albedo consistently increased to 0.91 after each snowfall. This suggests that the impact of snow thickness on snow cover albedo at Lake Ulansu is relatively small. Based on the error analysis in Figure 7 and Table 3, the HS II scheme, using the number of days since snowfall as a key parameter, showed the best performance in terms of both trend accuracy and error metrics. Therefore, the HS II scheme was selected for optimization, and the number of days since snowfall was used for parabolic fitting, as shown in Equation (13):

$$\alpha_s = \max\left(0.4, 0.884 + 0.025 \times d - 0.004 \times d^2\right), \ h_s > 0 \text{ m}, \quad (13)$$

The albedo prediction results are shown in Figure 11. The optimized albedo scheme for the snow cover stage accurately captured the trend of albedo changes following snowfall and was numerically close to the observed values, with an EF of 0.796.

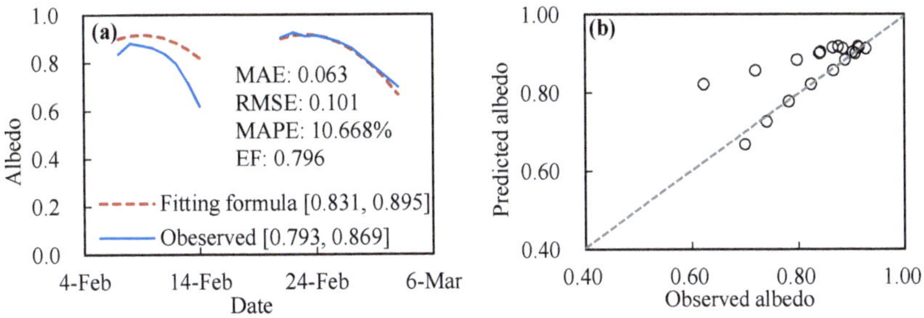

Figure 11. Observation and fitting results of the albedo during the snow cover stage (**a**), with 95% confidence intervals in brackets, and the predicted vs. observed albedo (**b**).

4.1.3. Melting Stage

As discussed in Section 3.2.3, the existing albedo schemes produced unsatisfactory results for calculating albedo during the ice melting stage on Lake Ulansu. Based on the bare ice stage, the albedo during the melting stage was re-fitted as a function of ice thickness, as shown in Equation (14):

$$\alpha_i = \max\left(0.1, -0.25 + 1.40h_i\right),\ 0\ \text{m} < h_i \leq 0.60\ \text{m}, \tag{14}$$

The results, shown in Figure 12, demonstrated that the re-fitted scheme yielded satisfactory calculation results in both trend and numerical values. The error analysis also showed positive results, with a MAE of 0.012, RMSE of 0.018, MAPE of 2.402%, and EF of 0.531.

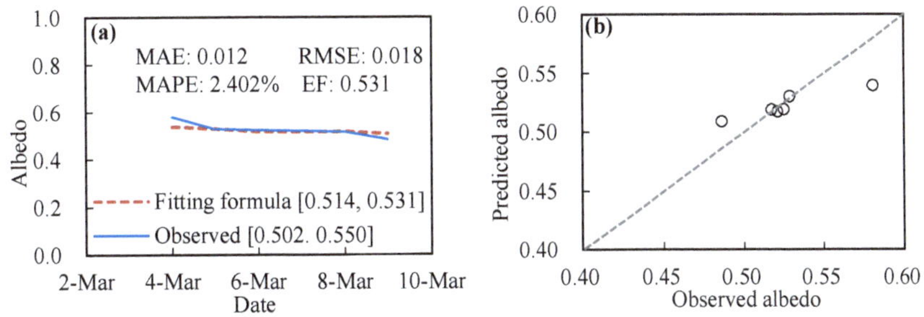

Figure 12. Observation and fitting results of the albedo during the melting stage (**a**), with 95% confidence intervals in brackets, and the predicted vs. observed albedo (**b**).

4.1.4. Whole Observation Stage

By integrating the scheme optimization results from the three stages with the field measurement data, the lake ice albedo scheme for Lake Ulansu was summarized, as shown in Equation (14).

Bare ice stage:

$$\alpha_i = \max\left(0.2, 0.03 + 1.74 \times h_i - 2.29 \times h_i^2\right),\ 0 < h_i \leq 0.60\ \text{m},$$

Snow cover stage:

$$\alpha_s = \max\left(0.4, 0.884 + 0.025 \times d - 0.004 \times d^2\right),\ h_s > 0\ \text{m},$$

Melting stage:

$$\alpha_i = \max(0.1, -0.25 + 1.40h_i),\ 0 < h_i \leq 0.60\ \text{m},$$

Figure 13 shows the results of the albedo scheme for the winter of 2016–2017 compared to the observed data. The simulated values closely matched the observed values, showing a generally consistent trend. Error analysis indicated an improvement over the separate calculations for each phase, with a MAE of 0.028, RMSE of 0.059, MAPE of 5.561%, and a notable increase in EF to 0.933.

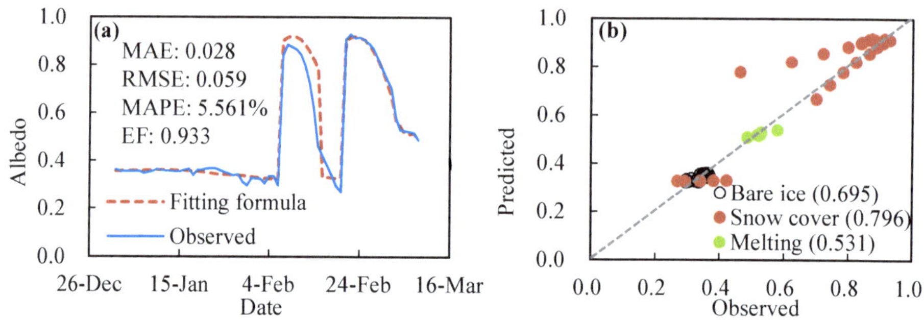

Figure 13. Observation and fitting results of the albedo in winter 2016–2017 (**a**) and the predicted vs. observed albedo (**b**), with the EF values calculated individually for each stage in parentheses.

4.2. Albedo Scheme Validation

To further validate the applicability of the albedo scheme (Equation (14)) for Lake Ulansu, portions of the albedo observation data from the winters of 2015–2016, 2017–2018, and 2022–2023 were used. For the bare ice and snow cover stage data, in contrast to the winter of 2016–2017, the snow on the ice surface was unevenly distributed due to wind effects, as shown in Figure 14. Among the three winters, only the winter of 2017–2018 had snow distribution data for beneath the sensors, recorded using drones. Therefore, for the winters of 2015–2016 and 2022–2023, only the observation data from the bare ice stages were used for validation, while the bare ice and snow cover stages were uniformly classified as the freezing stage.

Figure 14. Uneven snow distribution on the lake ice surface during the winter 2015–2016 (**a**) and 2017–2018 (**b**).

When the snow within the effective observation radius of the pyranometer is unevenly distributed, both the snow and the bare ice contribute to the albedo [3]. To assess the impact of this composite surface morphology on albedo, the snow cover proportion beneath the pyranometer was extracted from drone-recorded images of snow cover, as illustrated in

Figure 15. Based on the proportions of snow and bare ice, the albedo was calculated using Equation (15):

$$\text{Albedo} = S_s \times \alpha_s + S_i \times \alpha_i, \tag{15}$$

where S_s is the snow cover proportion, and S_i is the bare ice proportion.

Figure 15. Snow cover area within the effective radius (3 m) of the pyranometer on 16 January, 26 January, and 6 February 2018.

The albedo predicted by Equations (12)–(14), compared to the observed values, is shown in Figure 16, and the error analysis is presented in Table 6. For the predicted data from the freezing, melting, and entire stages, the MAE value was less than 0.05, with an average RMSE of 0.048 and MAPE of 10.793%. The EF for the freezing stage was 0.890, higher than the 0.826 for the melting stage. This difference was primarily due to observational limitations, because the data for the melting stage are mostly concentrated in the early stages of melting, which limits its ability to validate predictions for the later melting stages. After analyzing the data from both the freezing and melting stages, the overall EF reached 0.940, consistent with the results from the winter of 2016–2017.

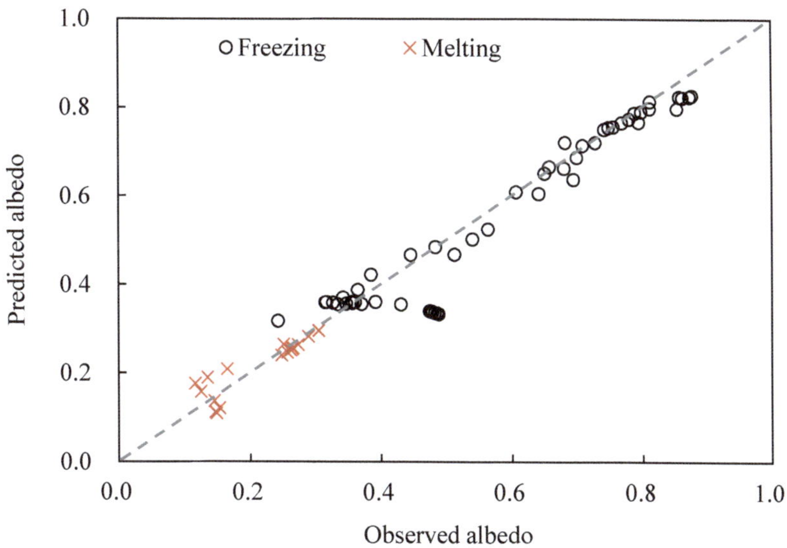

Figure 16. Observation and predicted results of the lake ice albedo.

Table 6. MAE, RMSE, MAPE, and EF for the scheme validation.

	MAE	RMSE	MAPE	EF
Freezing	0.041	0.062	8.607%	0.890
Melting	0.022	0.028	13.812%	0.826
All data	0.036	0.055	9.962%	0.940

5. Conclusions

Snow and ice albedos were measured on Lake Ulansu over a 68-day period during the winter of 2016–2017, alongside simultaneous atmospheric data observations. The observations were categorized into three distinct stages: bare ice, snow cover, and melting. The average albedo during the bare ice stage was 0.35, increasing to 0.91 following snowfall. The maximum ice thickness reached 46.4 cm, with two snowfall events occurring during the observation stage. The daily average temperature was −6.74 °C, with maximum and minimum temperatures of 0.44 and −14.73 °C, respectively. Due to the environmental conditions in arid and semi-arid regions, a sediment layer adhered to the ice surface, with an average sediment content of 1.39 g m^{-2}, showing a clear increasing trend.

Four existing albedo schemes (Gabison, SH, FB, and HS) were selected to predict the ice and snow albedo on Lake Ulansu during the winter of 2016–2017. During the bare ice stage, the predicted albedo trend showed the opposite to the observed trend, primarily due to the increasing sediment content on the ice surface. The HS II scheme performed the best for predicting snow cover albedo, achieving an EF of 0.55. For the melting stage, none of the existing schemes were suitable for accurately predicting albedo.

A new albedo scheme for ice and snow was developed based on the observed data and existing schemes. For the bare ice stage, the parameter of sediment content on the lake ice surface, which had not been considered in the previous surface albedo schemes, was incorporated into the new scheme. To enhance its applicability, the sediment content was integrated into the new scheme in the form of a function of ice thickness. This albedo scheme was derived from the SH scheme and had an EF of 0.695. The HS II scheme was adapted to develop the snow cover albedo scheme, with an EF of 0.796. The albedo scheme for the melting stage was based on the ice thickness and observed albedo data, with an EF of 0.531. Integration of the schemes for all three stages produced a comprehensive albedo scheme for Lake Ulansu, with an overall EF of 0.933. The albedo scheme was validated using albedo observation data from the winters of 2015–2016, 2017–2018, and 2022–2023, achieving an EF of 0.940 when comparing predicted and observed values. Validation analysis showed that the new albedo scheme closely matched the observed lake ice and snow albedo on Lake Ulansu. During the validation process, it was found that when snow is unevenly distributed, the contribution of bare ice to the albedo should also be considered alongside the snow. Therefore, more wind and snow cover data are needed to establish general patterns of blowing snow cover on Lake Ulansu, which would further improve the applicability of the new scheme.

The new albedo scheme has potential applicability for snow and ice that encounter surface sediment during winter in the Central Asian arid climate zone. However, additional field observation data are required to develop a more generalized parameterization formula for sediment content on the ice surface, such as examining its potential relationship with wind, and the contribution of sediment content in snow to albedo changes. Moreover, further field data are needed to validate the applicability of this albedo scheme for other lakes in the Central Asian arid climate zone. Additionally, we will include a comparative study of the measured and predicted albedo with satellite-derived data to enhance the accuracy and robustness of the scheme. These aspects should be key focuses of future research.

Author Contributions: Conceptualization, X.C.; investigation, X.C. and P.H.; writing—original draft preparation, X.C. and P.L.; writing—review and editing, M.Y., B.C., W.G., X.S., and L.W. All authors have read and agreed to the published version of the manuscript.

Funding: This research was supported by the Qingdao Huanghai University Doctoral Research Fund Project (2024boshi02, 2024boshi03, 2022boshi03), the National Natural Science Foundation of China (42406244, U24A20582), the Postdoctoral Fellowship Program of CPSF (Grant number GZB20230102), the Joint Funds of the National Natural Science Foundation of Liaoning Province (2023-BSBA-021), the Shandong Provincial Natural Science Foundation (ZR2024QD256), the Qingdao City Natural Science Foundation (24-4-4-zrjj-48-jch), and the 2024 Shandong Province College Students Innovation and Entrepreneurship Training Program Project (S202413320023).

Data Availability Statement: The data presented in this study are available on request from the corresponding authors.

Acknowledgments: The authors would like to thank the research group from Water Conservancy and Civil Engineering College (Inner Mongolia Agricultural University), for their contributions to the in-situ measurement.

Conflicts of Interest: Author Lijun Wang was employed by the company Huacheng Boyuan Engineering Technology Group Co., Ltd. The remaining authors declare that the research was conducted in the absence of any commercial or financial relationships that could be construed as a potential conflict of interest.

Abbreviations

The following abbreviations are used in this manuscript:

SH	Shine and Henderson-Sellers
FB	Flato and Brown
HS	Henneman and Stefan
MAE	Mean absolute error
RMSE	Root mean square error
MAPE	Mean absolute percent error
EF	Modeling efficiency

Appendix A

Error calculation formulas:

Mean absolute error (MAE):

$$\text{MAE} = \frac{(\sum|y_i - \hat{y}_i|)}{n},$$

Root mean square error (RMSE):

$$\text{RMSE} = \sqrt{\frac{\sum(y_i - \hat{y}_i)^2}{n}},$$

Mean absolute percent error (MAPE):

$$\text{MAPE} = \frac{1}{n}\sum\left|\frac{y_i - \hat{y}_i}{y_i}\right|,$$

Modeling efficiency (EF):

$$\text{EF} = 1 - \frac{\sum(y_i - \hat{y}_i)^2}{\sum(y_i - \overline{y})^2},$$

where y_i is the observed value, $\hat{y}_i$ is the simulated value, and n is the number of data points.

References

1. Austin, J.A.; Colman, S.M. Lake Superior summer water temperatures are increasing more rapidly than regional air temperatures: A positive ice-albedo feedback. *Geophys. Res. Lett.* **2007**, *34*, L06604. [CrossRef]
2. Robinson, A.L.; Ariano, S.S.; Brown, L.C. The influence of snow and ice albedo towards improved lake ice simulations. *Hydrology* **2021**, *8*, 11. [CrossRef]
3. Cao, X.; Lu, P.; Leppäranta, M.; Arvola, L.; Huotari, J.; Shi, X.; Li, G.; Li, Z. Solar radiation transfer for an ice-covered lake in the central Asian arid climate zone. *Inland Waters* **2021**, *11*, 89–103. [CrossRef]
4. Lang, J.; Lyu, S.; Li, Z.; Ma, Y.; Su, D. An investigation of ice surface albedo and its influence on the high-altitude lakes of the Tibetan Plateau. *Remote Sens.* **2018**, *10*, 218. [CrossRef]
5. Raoult, N.; Charbit, S.; Dumas, C.; Maignan, F.; Ottlé, C.; Bastrikov, V. Improving modelled albedo over the Greenland ice sheet through parameter optimisation and MODIS snow albedo retrievals. *Cryosphere* **2023**, *17*, 2705–2724. [CrossRef]
6. Shao, D.; Jin, Y.; Ji, W.; Li, H.; Feng, T.; Li, H.; Wang, J.; Hao, X. Estimating broadband snow albedo by an asymptotic radiative transfer model-constrained deep learning models. *IEEE Trans. Geosci. Remote Sens.* **2024**, *62*, 4302016. [CrossRef]
7. Brown, L.C.; Duguay, C.R. The response and role of ice cover in lake-climate interactions. *Prog. Phys. Geogr.* **2010**, *34*, 671–704. [CrossRef]
8. Leppäranta, M. *Freezing of Lakes and the Evolution of Their Ice Cover*; Springer: Berlin/Heidelberg, Germany, 2015.
9. Benson, B.J.; Magnuson, J.J.; Jensen, O.P.; Card, V.M.; Hodgkins, G.; Korhonen, J.; Livingstone, D.M.; Stewart, K.M.; Weyhenmeyer, G.A.; Granin, N.G. Extreme events, trends, and variability in Northern Hemisphere lake-ice phenology (1855–2005). *Clim. Change* **2012**, *112*, 299–323. [CrossRef]
10. Sharma, S.; Richardson, D.C.; Woolway, R.I.; Imrit, M.A.; Bouffard, D.; Blagrave, K.; Daly, J.; Filazzola, A.; Granin, N.; Korhonen, J. Loss of ice cover, shifting phenology, and more extreme events in Northern Hemisphere lakes. *J. Geophys. Res. Biogeosci.* **2021**, *126*, e2021JG006348. [CrossRef]
11. Sharma, S.; Blagrave, K.; Filazzola, A.; Imrit, M.A.; Hendricks Franssen, H.J. Forecasting the permanent loss of lake ice in the Northern Hemisphere within the 21st century. *Geophys. Res. Lett.* **2021**, *48*, e2020GL091108. [CrossRef]
12. Shuter, B.; Minns, C.K.; Fung, S.R. Empirical models for forecasting changes in the phenology of ice cover for Canadian lakes. *Can. J. Fish. Aquat. Sci.* **2013**, *70*, 982–991. [CrossRef]
13. Leppäranta, M. Modelling the formation and decay of lake ice. In *The Impact of Climate Change on European Lakes*; Springer: Berlin/Heidelberg, Germany, 2010; pp. 63–83.
14. Knoll, L.B.; Sharma, S.; Denfeld, B.A.; Flaim, G.; Hori, Y.; Magnuson, J.J.; Straile, D.; Weyhenmeyer, G.A. Consequences of lake and river ice loss on cultural ecosystem services. *Limnol. Oceanogr. Lett.* **2019**, *4*, 119–131. [CrossRef]
15. Hampton, S.E.; Galloway, A.W.; Powers, S.M.; Ozersky, T.; Woo, K.H.; Batt, R.D.; Labou, S.G.; O'Reilly, C.M.; Sharma, S.; Lottig, N.R. Ecology under lake ice. *Ecol. Lett.* **2017**, *20*, 98–111. [CrossRef] [PubMed]
16. Lioubimtseva, E.; Cole, R. Uncertainties of climate change in arid environments of Central Asia. *Rev. Fish. Sci.* **2006**, *14*, 29–49. [CrossRef]
17. Yu, Y.; Pi, Y.; Yu, X.; Ta, Z.; Sun, L.; Disse, M.; Zeng, F.; Li, Y.; Chen, X.; Yu, R. Climate change, water resources and sustainable development in the arid and semi-arid lands of Central Asia in the past 30 years. *J. Arid Land* **2019**, *11*, 1–14. [CrossRef]
18. Huang, W.; Zhao, W.; Zhang, C.; Leppäranta, M.; Li, Z.; Li, R.; Lin, Z. Sunlight penetration dominates the thermal regime and energetics of a shallow ice-covered lake in arid climate. *Cryosphere* **2022**, *16*, 1793–1806. [CrossRef]
19. Yang, Y.; Leppäranta, M.; Cheng, B.; Li, Z. Numerical modelling of snow and ice thicknesses in Lake Vanajavesi, Finland. *Tellus A Dyn. Meteorol. Oceanogr.* **2012**, *64*, 17202. [CrossRef]
20. You, Y.; An, L.; Li, S.; Zhang, B.; Zhang, J. Analysis and Research on the Differences in Observed Data of Sand–Dust Weather between China and Mongolia. *Atmosphere* **2023**, *14*, 1401. [CrossRef]
21. Fang, Y.; Changyou, L.; Leppäranta, M.; Xiaonghong, S.; Shengnan, Z.; Chengfu, Z. Notable increases in nutrient concentrations in a shallow lake during seasonal ice growth. *Water Sci. Technol.* **2016**, *74*, 2773–2783. [CrossRef]
22. Conway, H.; Gades, A.; Raymond, C. Albedo of dirty snow during conditions of melt. *Water Resour. Res.* **1996**, *32*, 1713–1718. [CrossRef]
23. Grenfell, T.C.; Warren, S.G.; Mullen, P.C. Reflection of solar radiation by the Antarctic snow surface at ultraviolet, visible, and near-infrared wavelengths. *J. Geophys. Res. Atmos.* **1994**, *99*, 18669–18684. [CrossRef]
24. Dirmhirn, I.; Eaton, F.D. Some characteristics of the albedo of snow. *J. Appl. Meteorol. (1962–1982)* **1975**, *14*, 375–379. [CrossRef]
25. Bergstrom, A.; Gooseff, M.N.; Myers, M.; Doran, P.T.; Cross, J.M. The seasonal evolution of albedo across glaciers and the surrounding landscape of Taylor Valley, Antarctica. *Cryosphere* **2020**, *14*, 769–788. [CrossRef]
26. Jakkila, J.; Leppäranta, M.; Kawamura, T.; Shirasawa, K.; Salonen, K. Radiation transfer and heat budget during the ice season in Lake Pääjärvi, Finland. *Aquat. Ecol.* **2009**, *43*, 681–692. [CrossRef]
27. Leppäranta, M.; Terzhevik, A.; Shirasawa, K. Solar radiation and ice melting in Lake Vendyurskoe, Russian Karelia. *Hydrol. Res.* **2010**, *41*, 50–62. [CrossRef]

28. Chen, Q.-X.; Gao, Z.-Y.; Huang, C.-L.; Dong, S.-K.; Lin, K.-F. Simulation of Spectral Albedo and Bidirectional Reflectance over Snow-Covered Urban Canyon: Model Development and Factor Analysis. *Remote Sens.* **2024**, *16*, 2340. [CrossRef]
29. Li, Z.; Wang, Q.; Tang, M.; Lu, P.; Li, G.; Leppäranta, M.; Huotari, J.; Arvola, L.; Shi, L. Diurnal cycle model of lake ice surface albedo: A case study of Wuliangsuhai Lake. *Remote Sens.* **2021**, *13*, 3334. [CrossRef]
30. Henneman, H.E.; Stefan, H.G. Albedo models for snow and ice on a freshwater lake. *Cold Reg. Sci. Technol.* **1999**, *29*, 31–48. [CrossRef]
31. Rogers, C.K.; Lawrence, G.A.; Hamblin, P.F. Observations and numerical simulation of a shallow ice-covered midlatitude lake. *Limnol. Oceanogr.* **1995**, *40*, 374–385. [CrossRef]
32. Brock, B.W.; Willis, I.C.; Sharp, M.J. Measurement and parameterization of albedo variations at Haut Glacier d'Arolla, Switzerland. *J. Glaciol.* **2000**, *46*, 675–688. [CrossRef]
33. Pedersen, C.A.; Winther, J.-G. Intercomparison and validation of snow albedo parameterization schemes in climate models. *Clim. Dyn.* **2005**, *25*, 351–362. [CrossRef]
34. Gray, D.; Landine, P. Albedo model for shallow prairie snow covers. *Can. J. Earth Sci.* **1987**, *24*, 1760–1768. [CrossRef]
35. Flato, G.M.; Brown, R.D. Variability and climate sensitivity of landfast Arctic sea ice. *J. Geophys. Res. Ocean.* **1996**, *101*, 25767–25777. [CrossRef]
36. Gabison, R. A thermodynamic model of the formation, growth, and decay of first-year sea ice. *J. Glaciol.* **1987**, *33*, 105–119. [CrossRef]
37. Shine, K.; Henderson, S. The sensitivity of a thermodynamic sea ice model to changes in surface albedo parameterization. *J. Geophys. Res. Atmos.* **1985**, *90*, 2243–2250. [CrossRef]
38. Gardner, A.S.; Sharp, M.J. A review of snow and ice albedo and the development of a new physically based broadband albedo parameterization. *J. Geophys. Res. Earth Surf.* **2010**, *115*, F01009. [CrossRef]
39. Lu, P.; Cao, X.; Li, G.; Huang, W.; Leppäranta, M.; Arvola, L.; Huotari, J.; Li, Z. Mass and heat balance of a lake ice cover in the central Asian arid climate zone. *Water* **2020**, *12*, 2888. [CrossRef]
40. Shi, X.; Yu, H.; Zhao, S.; Sun, B.; Liu, Y.; Huo, J.; Wang, S.; Wang, J.; Wu, Y.; Wang, Y. Impacts of environmental factors on Chlorophyll-a in lakes in cold and arid regions: A 10-year study of Wuliangsuhai Lake, China. *Ecol. Indic.* **2023**, *148*, 110133. [CrossRef]
41. Huang, W.; Zhang, Z.; Li, Z.; Leppäranta, M.; Arvola, L.; Song, S.; Huotari, J.; Lin, Z. Under-ice dissolved oxygen and metabolism dynamics in a shallow lake: The critical role of ice and snow. *Water Resour. Res.* **2021**, *57*, e2020WR027990. [CrossRef]
42. Zhang, X.; Zhang, C.; Meng, F.; Wang, C.; Ren, P.; Zou, Q.; Luan, J. Near-zero liquid discharge of desulfurization wastewater by electrodialysis-reverse osmosis hybrid system. *J. Water Process Eng.* **2021**, *40*, 101962. [CrossRef]
43. Maykut, G.A. Large-scale heat exchange and ice production in the central Arctic. *J. Geophys. Res. Ocean.* **1982**, *87*, 7971–7984. [CrossRef]
44. Heron, R.; Woo, M.-K. Decay of a High Arctic lake-ice cover: Observations and modelling. *J. Glaciol.* **1994**, *40*, 283–292. [CrossRef]

MDPI AG
Grosspeteranlage 5
4052 Basel
Switzerland
Tel.: +41 61 683 77 34

Water Editorial Office
E-mail: water@mdpi.com
www.mdpi.com/journal/water

www.ingramcontent.com/pod-product-compliance
Lightning Source LLC
Chambersburg PA
CBHW041410210926
51850CB00013B/3234
* 9 7 8 3 7 2 5 8 3 8 1 4 1 *